Advances in Intelligent Systems and Computing

Volume 1432

T0175434

The series "Advances in Intelligent Systems and Computing" contains publications on theory, applications, and design methods of Intelligent Systems and Intelligent Computing. Virtually all disciplines such as engineering, natural sciences, computer and information science, ICT, economics, business, e-commerce, environment, healthcare, life science are covered. The list of topics spans all the areas of modern intelligent systems and computing such as: computational intelligence, soft computing including neural networks, fuzzy systems, evolutionary computing and the fusion of these paradigms, social intelligence, ambient intelligence, computational neuroscience, artificial life, virtual worlds and society, cognitive science and systems, Perception and Vision, DNA and immune based systems, self-organizing and adaptive systems, e-Learning and teaching, human-centered and human-centric computing, recommender systems, intelligent control, robotics and mechatronics including human-machine teaming, knowledge-based paradigms, learning paradigms, machine ethics, intelligent data analysis, knowledge management, intelligent agents, intelligent decision making and support, intelligent network security, trust management, interactive entertainment, Web intelligence and multimedia.

The publications within "Advances in Intelligent Systems and Computing" are primarily proceedings of important conferences, symposia and congresses. They cover significant recent developments in the field, both of a foundational and applicable character. An important characteristic feature of the series is the short publication time and world-wide distribution. This permits a rapid and broad dissemination of research results.

Indexed by DBLP, INSPEC, WTI Frankfurt eG, zbMATH, Japanese Science and Technology Agency (JST).

All books published in the series are submitted for consideration in Web of Science.

For proposals from Asia please contact Aninda Bose (aninda.bose@springer.com).

Subarna Shakya · Ke-Lin Du · Klimis Ntalianis
Editors

Sentiment Analysis and Deep Learning

Proceedings of ICSADL 2022

Volume 2

 Springer

Editors
Subarna Shakya
Department of Electronics and
Computer Engineering
Pulchowk Campus, Institute
of Engineering, Tribhuvan University
Lalitpur, Nepal

Ke-Lin Du
Department of Electrical and
Computer Engineering
Concordia University
Montreal, QC, Canada

Klimis Ntalianis
University of West Attica
Aigaleo, Greece

ISSN 2194-5357   ISSN 2194-5365 (electronic)
Advances in Intelligent Systems and Computing
ISBN 978-981-19-5442-9   ISBN 978-981-19-5443-6 (eBook)
https://doi.org/10.1007/978-981-19-5443-6

This Springer imprint is published by the registered company Springer Nature Singapore Pte Ltd.
The registered company address is: 152 Beach Road, #21-01/04 Gateway East, Singapore 189721,
Singapore

We, the organizers, dedicate this ICSADL 2022 proceeding to the worldwide community of artificial intelligence, information, and communication technologies [ICT] researchers. We also dedicate this proceeding to the authors and editorial team of the conference, who have highly discussed the solutions to problems that exist in the international research community.

Preface

The 2nd International Conference on Sentimental Analysis and Deep Learning [ICSADL 2022] aims to bring together the researchers from the relevant fields of artificial intelligence and other computing technologies to provide a broad forum to discuss the recent innovations in deep learning and behavioral analysis domain. A specific goal of ICSADL 2022 is to study how the proposed research idea of context may be understood and used to create fully interactive information and computing systems.

ICSADL 2022 has achieved this goal by inviting research contributions that approached intelligent information system contexts from many perspectives such as deep learning, social media mining, behavioral analysis, sentiment analysis, and its instances in social media analysis. We received 332 papers from various countries in total and accepted 75 of them. Further, the oral presentations were delivered over two days by organizing different parallel sessions along with a keynote session delivered by plenary speakers.

Collectively, the research works discussed at the conference contributed to our knowledge and basic understanding on the state-of-the-art mechanisms involved in the design, development, and analysis of the intelligent information systems. This knowledge will lead to progress in computing and communication research, as well as improved behavioral analysis and online data mining technologies. All that is evident when looking at the papers presented in the conference under four different categories: sentiment analysis, deep learning, big data analytics, and real-time applications.

We acknowledge the distinguished speaker of the conference: Dr. Manu Malek for the acceptance to deliver a keynote talk on the respective fields of expertise. We also thank all organizing and program committee members and national and international referees for their excellent work. Our high appreciation is to all the principal speakers at the conference event and also the participants for making this conference successful and enjoyable.

Prof. Dr. Subarna Shakya
Professor
Department of Electronics
and Computer Engineering
Pulchowk Campus, Institute
of Engineering, Tribhuvan University
Lalitpur, Nepal

Dr. Ke-Lin Du
Affiliate Associate Professor
Department of Electrical and
Computer Engineering
Concordia University
Montreal, QC, Canada

Dr. Klimis Ntalianis
Professor
University of West Attica
Aigaleo, Greece

Contents

Contents

About the Editors

Prof. Dr. Subarna Shakya is currently a Professor of Computer Engineering, Department of Electronics and Computer Engineering, Central Campus, Institute of Engineering, Pulchowk, Tribhuvan University, Coordinator (IOE), LEADER Project (Links in Europe and Asia for engineering, eDucation, Enterprise and Research exchanges), Erasmus Mundus. He received M.Sc. and Ph.D. degrees in Computer Engineering from the Lviv Polytechnic National University, Ukraine, 1996 and 2000 respectively. His research area includes E-Government system, Computer Systems and Simulation, Distributed and Cloud Computing, Software Engineering and Information System, Computer Architecture, Information Security for E-Government, and Multimedia system.

Dr. Ke-Lin Du has been a research scientist at Center for Signal Processing and Communications, Department of Electrical and Computer Engineering, Concordia University since 2001, where he became an Affiliate Associate Professor in 2011. He received a Ph.D. degree in electrical engineering from Huazhong University of Science and Technology in 1998. He has published five textbooks, 50 papers, and holds six granted U.S. Patents. Presently, he is on the editorial boards of IEEE Spectrum Chinese Edition, IET Signal Processing, Circuits Systems and Signal Processing, Mathematics, Scientific Reports, and AI. He served as program chair of dozens of conferences. His research area includes wireless communications, signal processing, and machine learning. He has been a Senior Member of the IEEE since 2009.

Dr. Klimis Ntalianis is a full professor at the University of West Attica, Athens, Greece. He has worked as Senior Researcher in multi-million research and development projects, funded by the General Secretariat of Research and Technology of Greece (GSRT), the Research Promotion Foundation of Cyprus (RPF), the Information Society S.A. of Greece and the European Union. He is also serving as a Senior Project Proposal Evaluator for GSRT, RPF, the European Union, the Natural Sciences and Engineering Research Council of Canada and the National Science Center of Poland. In parallel he is a member of several master theses and Ph.D. evaluation committees in Greece, Cyprus, Germany and India. He is also serving as promotion evaluator for Saudi Arabia and academic staff. He has served as general chair in several conferences (IEEE etc.). Dr. Ntalianis has published more than 160 scientific papers in international journals, books and conferences. His main research interests include social computing, multimedia analysis and information security.

Design of a Real-Time Obstacle Avoiding and Trajectory Generation Algorithm for an Unmanned Aerial Vehicle

Hernando González, Alhim Vera, Juan Monsalve, and Diego Valle

Abstract The paper presents an object tracking algorithm implemented through an unmanned aerial vehicle (UAV). The system generates the trajectory taking into account the image obtained by the front camera of the drone, the object to be detected is a moving red box and using color segmentation techniques the object is detected. The red box is continuously tracked by centering the image frame. Open-source computer vision libraries (OpenCV) are used to process the images obtained from the drone. The software was verified by simulations with Gazebo and Rviz on the robot operating system (ROS) and compared with the real drone.

Keywords Unmanned aerial vehicle (UAV) · Computer vision · Trajectory tracking · Robot operating system (ROS)

1 Introduction

In recent decades, the process of recognition of areas difficult to access for humans is increasingly becoming a need for quick solution and efficiency, so the unmanned aerial vehicles (UAV) or quadcopters, facilitate the exploration of areas by means of a camera and have great mobility in different terrains, these aerial vehicles are controlled directly by a user. UAVs are mobile robots that are able to perform certain tasks that depend on their capabilities. Before performing a task, the robot must plan it, usually breaking it down into several sequential or simultaneous subtasks.

H. González (✉) · A. Vera · J. Monsalve · D. Valle
Universidad Autónoma de Bucaramanga, Bucaramanga, Colombia
e-mail: hgonzalez7@unab.edu.co

A. Vera
e-mail: avera488@unab.edu.co

J. Monsalve
e-mail: jmonsalve858@unab.edu.co

D. Valle
e-mail: dvalle@unab.edu.co

© The Author(s), under exclusive license to Springer Nature Singapore Pte Ltd. 2023   503
S. Shakya et al. (eds.), *Sentiment Analysis and Deep Learning*,
Advances in Intelligent Systems and Computing 1432,
https://doi.org/10.1007/978-981-19-5443-6_38

These subtasks must be consistent with the capabilities of the drone, that is, they must be possible considering the dynamic behavior of the mobile and its conditions (sampling times, the physical limits generated by the actuators, etc.). A subtask can be to plan the movement to follow of the drone, thus mediating a trajectory to find collision-free paths, this process of tracking trajectories is called trajectory planning. Robotic autonomy is becoming more and more important and is becoming a key point for its study and development, and these systems are accompanied by artificial vision algorithms [1, 2].

Trajectory planning emphasizes the limitations of the static environment and the drone, so it is important to know the hardware fully and have a wide enough field to obtain possible trajectories and subsequently track it, as shown in [3]. Drone 2.0 is used to generate trajectories avoiding possible collisions thanks to its sensors and cameras. An additional distinction can be made, the drone could use the information given to it by its cameras to make a total recognition of the scenario that surrounds it, and with this information make the decision on the trajectory to follow just as shown in [4] or it could also make the decision about its movement every certain interval of time, that is, it can make a global planning of the trajectory or only local planning [5]. This paper is structured as follows: in Sect. 2, it presents the hardware and software used in the project, in Sect. 3, the development of the trajectory generation algorithms, Sect. 4 shows the results of the experiment, and Sect. 5 presents the conclusions of all the research work.

2   UAV Hardware

Due to the demands of the drone in processing and operating time, it was decided to create a customized drone, divided into its main parts: chassis, motors, propellers, sensor, battery, flight control, microprocessor, depth sensor, camera, radio, and GPS. Figure 1 shows the drone hardware implemented in this research. The structure of the drone is characterized by its carbon fiber design, with a diameter of 250 mm, and weight of 122 g, its bushels engines have an advantage in their high efficiency, high torque, and good control dynamics, its propellers or propellers have a low wind resistance, which reduces the vibrations of this. One of the main characteristics is the flight duration, the weight was considered and looking for an autonomy of 12 min, a 4.000 mAh, 4-cell, 14.8 V Lipo battery was used, with a 30-40C discharge cell, where C is an indicator of the continuous discharge rate of the Lipo. Finally, landing gear was added to give it more stability, it was designed with a height of 8 cm in order to leave a space between the Lipo battery and the ground.

Holybro GPS
Neo – M8N

Jetson
Nano

Frame
QAV-250

Genius
Facecamx
1000

PixHawk
4 Mini

fr-sky d4r
receiver

Fig. 1 UAV hardware

2.1 Flight Control

To optimize the quadrotor's response times, the PIXHAWK4 mini flight control and the NVIDIA Jetson Nano embedded system have been integrated into the drone. The PIXHAWK4 card is an embedded system that allows the pilot to interact with the drone controls, electrically or mechanically, its main objectives are to balance the drone and adopt a certain flight attitude. In turn, it contains four sensors: an accelerometer (ICM 20,689), a gyroscope (BMI055), a magnetometer (IST8310), and a barometer (MS5611). The PIXHAWK4 [6] operates under the open-source PX4 framework, all configuration and calibration are performed under the Qground-Control framework [7]. On the other hand, the main computer is an Nvidia Jetson Nano, its operation is focused on the development of artificial intelligence software, its main feature is that it provides 472 GFLOP, which allows it to have an optimal time of floating point operations and in turn allows to run deep learning algorithms, on the other hand, it runs several neural networks in parallel and processes several high resolution sensors in a matter of seconds. One way to optimize the processing time is to execute everything using a hardware-in-the-loop technique, which means the flight controller and the central computer are in the drone. This technique will reduce the communication time between them.

2.2 Depth Sensor, Camera, and GPS

In the area of obstacle avoidance, it used the real sense r200 sensor, this sensor was designed by Intel and allows us to generate a map of points that will be used for obstacle avoidance, and it is compatible with operating systems such as Linux and Python programming language which are used throughout the development. The integrated camera is Genius facecamx1000, the main advantage of this is its size and ease of integration, and its main features are the HD 720p video resolution of 1Mpx compatible with Windows, Linux, and Mac operating systems. The GPS module used was Ublox neo m8n dual. The GPS is a fundamental piece for the development of the project because it allows to know the current location of the drone and the relative location of the object with respect to the drone, this sensor has an error rate of ± 5 cm.

2.3 Communication System

To optimize the information processing time, *in-the-loop* hardware was used, which allows flight control and processing to be performed directly on the drone, reducing data sending and receiving times. When integrating the complete drone system, it is necessary to define the internal (Pixhawk 4 mini and Jetson Nano) and external (Jetson Nano and external computer) communication protocols. Figure 2 shows the schematic and connection system of the drone. The Jetson Nano card is connected via USB protocol to the PIXHAWK, with a frequency of 57,600 baud. The Jetson Nano receives three main connections: real sense R200 (depth sensor), USB camera, and the Wi-Fi module.

For the communication between the central computer and the flight control, the MavLink communication protocol was used, which allows the transfer of messages in a light way because only the headers are transferred by XML files, through this protocol the vehicle's orientation, GPS location, speed, among others, are transferred. For the activation of the codes in the Jetson Nano, it was necessary to create a SCREEN section which allows to keep the code running even if the window is closed. In addition to this, it connected to the Jetson Nano through the secure shell communication protocol or SSH, which allows remote access to a central computer or server through a secure channel that contains all the encrypted information. The main task of the central computer (Jetson Nano) is articulate all the modules and execute them in parallel.

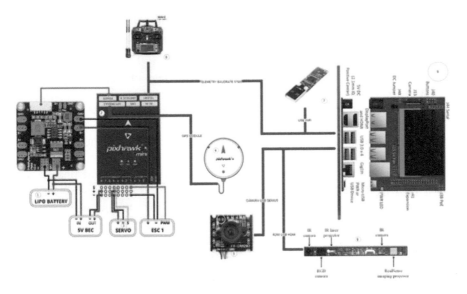

Fig. 2 Electrical schematic

3 Trajectory Planning

For the development of algorithm, the different flight modes of the PX4 mini were taken into account, being beneficial for this project the flight mode OFFBOARD, its operation is based on entering consignment points to the drone, that is, the place where it takes off (point 0), is taken as the beginning, and from data given in meters, in its coordinate axes that in this case are latitude, longitude, and altitude, the drone moves until it reaches that setpoint at a constant speed previously stipulated. We send a command to change the states of the drone and assign it the OFFBOARD flight mode with Python from the central computer (Jetson Nano) to the flight controller (PX4).

3.1 Nodes of Algorithms in ROS

ROS is an open-source software that provides libraries for the design and visualization of virtual environments with UAVs [8, 9]. The library that we integrate with ROS is Gazebo. Gazebo is an open-source 3D robotics simulator. It integrated the ODE physics engine, OpenGL rendering, and support code for sensor simulation and actuator control. With Gazebo, we can design and visualize a virtual environment with the UAV. For the internal communication between the Jetson nano central computer and the Px4 mini flight control, the MAVROS communication protocol was used, which allows the transfer of messages in a lightweight way because only the headers are transferred by XML files, by this protocol is transferred mainly the orientation of the vehicle, its GPS location, its speed and the following nodes developed for this project:

- Cameras Node. It is responsible for obtaining the map of points and publishing them so that other nodes can subscribe to them.
- Evasion Node. It connects to the camera node to obtain the point map and then perform obstacle avoidance.
- Trajectory Planning Node. Configures the drone's internal parameterization and OFFBOARD flight mode.
- Gazebo Node [10]. It simulated the real environment in which the final tests of the algorithms were carried out.
- Rviz Node [11]. It provides the 3D data obtained by the cameras, such as images and point maps.

For the validation of the algorithms, basic geometric figures were made (square, triangle, and 3D rhombus) and the graphs obtained from their position at each instant of time were analyzed to later generate a table of errors which makes the comparison of the real vs the experimental trajectory and conclude its operation. Figure 3 shows 3D behavior of drone during the trajectory of a 2D rhombus; the real and experimental data were superimposed to calculate the mean squared error.

When analyzing the track records of the generated trajectories, it is observed that the drone takes off smoothly until it reaches a height of 3 m, makes three turns describing the same geometric figure, then the tracking algorithm generates a

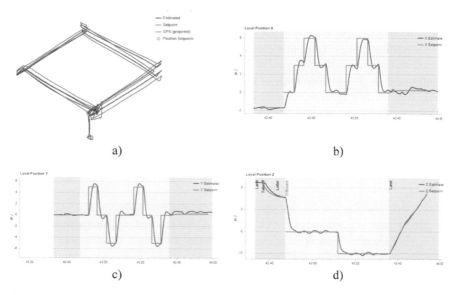

Fig. 3 **a** Trajectories generated in the logs of the PX4, **b** local position X, **c** local position Y, and **d** local position Z [12]

straight line that decreases until it reaches the ground again. Once the experimental data generated during the tracking of the trajectory is known, it is compared with the setpoint values, finding that there is a 20% error in the position due to changes in the wind speed, also an oscillation in the x and y axis is observed.

4 Obstacle Evasion

This section presents a methodology for detecting a specific color object and how the algorithm can be validated by means of the ROS software.

4.1 Red Box Recognition

Due to the requirement of recognizing an object, whose dimensions and colors are constant, a color segmentation technique was implemented in order to detect the red object. The first step is to obtain the frame of the camera integrated to the drone, next it applied a nonlinear transformation from RGB (red, green, and blue) to the color space called HSV (hue, saturation, and value) [13]. To segment the box, select the range of colors in which the element is located in the hsv color space, *low_reds* = (120, 110, and 255), and *high_reds* = (179, 255, and 255) and eliminate from the image everything outside this range. The segmentation can result in discontinuous segmentation and small false positives somewhere in the image, for this reason a morphological transformation called opening and closing was applied, which is useful to eliminate the noise present in the segmentation and to close the small black holes inside the foreground object. Figure 4 shows the results of each step of the described algorithm. Finally, the centroid of the object is calculated and the depth with respect to the centroid. If there is more than one object in the image, it prioritized the object with the smallest distance from the drone.

After obtaining the depth and the location of the object centroid, the distance of the object centroid from the frame centroid was calculated. If the centroid of the object is to the right of the image centroid, the drone will make a movement to the left side, until the object disappears from the visual area of the drone and can continue its trajectory, in the same way, it works on the opposite side. For safety reasons, the algorithm indicates as null the depths greater than 10 m, thus ensuring that the drone executes short trajectories and obtains greater control in the test.

4.2 Robot Operating System (ROS)

The algorithm was validated on the ROS platform. The first step to create the simulation environment was to create the computational graph, which is presented in

Fig. 4 **a** Image captured by the drone's 1280 × 720 on-board camera, **b** image in HSV heat space, **c** image after color preprocessing system, and **d** image with perimeter joining

Fig. 5. The graph is generated with the ROS function *rosrun rqt_graph,* and it allows to visualize all the nodes that are executed in parallel and in turn each of the active topics.

Simulation tests are performed using the 3D simulation environment Gazebo. Gazebo allows the creation of sensors for the simulation of data input. Gazebo is compatible with ROS, which allows the artificial sensor data to be extracted and published in a ROS theme, as explained in the previous section. This means that, from the point of view of the local planner node, there is no difference in simulated sensors or with data coming from a real camera which allows us to validate the performance before taking it to a real environment. To simulate the UAV, we used a model of the 3DR Iris Quadcopter to which it was necessary to add a node with the monocular camera, which is inside the ROS simulation environment. And for the environment, we took some photos of the real environment to create the simulated environment. Then, we create the trees, mountains, and other objects in Gazebo; we also measure the number of wind knots to apply the same measure to the software.

- Node Real Sense. It is responsible for obtaining the map of points and publishing them so that other nodes can access the information.
- Gazebo Node. Generates a simulation environment where physical spaces can be recreated.
- Rviz Node. Rviz is a 3D visualization tool for ROS applications. It provides a view of the robot model, captures information from the robot sensors, and plays back

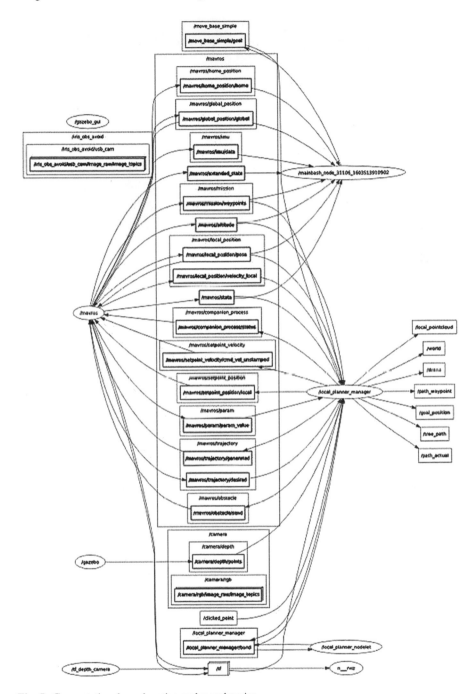

Fig. 5 Computational graph active nodes and topics

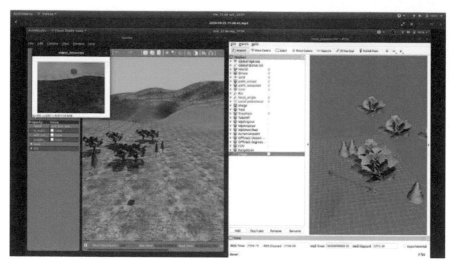

Fig. 6 Parallel algorithm testing: left: Gazebo simulation—right: RViz simulation

the captured data. It can display data from camera, lasers, 3D and 2D devices, such as images and point clouds.

- Vision Node. Its main function is to subscribe to the /tf_depth_camera node which supplies it with the image from the main camera and is processed by the computer vision algorithms.
- Local Planner Node. Its main function is to connect to the avoidance node and receive the point map to evaluate the cost function and determine the most optimal route.

Figure 6 shows the working environment created in Gazebo, including a red object, the drone detects the object, moves toward the object and evades it until it reaches a point defined by the user.

5 Results

The objective of this section is to integrate the trajectory algorithm and the artificial vision algorithm. The drone climbs 3 m high and starts to rotate on its own axis until it finds the object, once located a trajectory is generated between the initial point to the object, as can be seen in Fig. 7. The system generates set points while approaching the object, the set points are represented by the orange dots in the illustration, in this way, we can see how the system automatically generates a point in a direction in front of the drone and then performs the trajectory, that is the blue line. When the object disappears, it performs the search by turning on its own axis until it finds

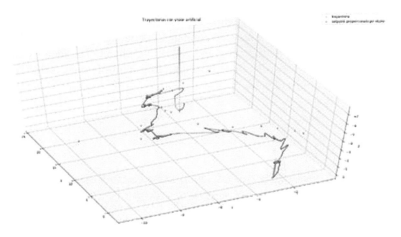

Fig. 7 Logs PX4 tracking algorithm red box, orange line setpoints, and blue line trajectories [12]

it again and continues its tracking process. At the end of its tracking, it performs a return home.

According to the flights performed outdoors, it was demonstrated that the system can generate trajectories autonomously, detecting possible obstacles present in the trajectory, avoiding these obstacles and returning to the initial trajectory autonomously, in a continuous and controlled flight. Even though the drone has the corresponding positioning sensors, it is observed that there is still an error in the position estimated by the GPS with respect to the experimental one. It is proposed as future work, to integrate a more robust structure as the TAROT, adding a power supply system of a 10,000 mAh battery and 24 V, which allows flights of up to 40 min and reduce the vibrations that occur in the system, a depth sensor D400 which is a sensor with algorithmic developments easily accessible to the entire community and with better hardware compared to the R200 worked initially. It is important to distribute the electronic elements in the drone in a balanced way, to achieve symmetry of weights and thus avoid various disturbances in the system that were reflected during the take off and flight.

6   Conclusions

According to the flight tests performed outdoors, it was demonstrated that the system is able to detect the object, in this case the red box, generate the target and perform the trajectory avoiding obstacles, executing a smooth and controlled flight. A minimal difference was obtained with respect to the simulation and the real environment due to the integration of the ROS + Gazebo + Rviz simulation system, which allowed a correct operation of the drone before sending it to a real environment. In the future works, it will seek to add objects that move around the environment that allow the drone to modify its trajectory while approaching the target point. Besides, it is intended to introduce a more advanced computer vision algorithm to detect specific classes of objects, create a trajectory, and follow the object. The idea would be to use pre-trained neural networks that allow detecting multiple objects and fulfilling tasks such as inspections of cattle areas.

References

1. Yu, Y., Wang, X., Zhong, Z., & Zhang, Y. (2017). ROS-based UAV control using hand gesture recognition. In *29th Chinese control and decision conference (CCDC)* (pp. 6795–6799).
2. Meier, L., Honegger, D., & Pollefeys, M. (2015). PX4: A node-based multithreaded open source robotics framework for deeply embedded platforms. In *IEEE international conference on robotics and automation (ICRA)* (pp. 6235–6240).
3. Rivera, D. (2012). Desarrollo de un algoritmo de evasión de obstáculos para Quadrotors en ambientes dinámicos utilizando una cámara de profundidad. Last accessed March 7, 2022, Homepage https://repositorio.unal.edu.co/handle/unal/21096?show=full
4. Gao, F., & Shen, S. (2016). Online quadrotor trajectory generation and autonomous navigation on point clouds. In *IEEE international symposium on safety, security, and rescue robotics (SSRR)*, ISBN:978–1–5090–4349–1. https://doi.org/10.1109/SSRR.2016.7784290
5. Vanneste, S., Bellekens, B., & Weyn, M. (2014). 3DVFH+: Real-time three-dimensional obstacle avoidance using an octomap. In *CEUR workshop proceedings* (p. 1319).
6. PX4 Autopilot User Guide. Last accessed March 7, 2022, Homepage https://docs.px4.io/master/en/
7. QGROUNDCONTROL. Last accessed March 7, 2022, Homepage http://qgroundcontrol.com/
8. Wei, Y., & Lin, Z. (2017). Vision-based tracking by a quadrotor on ROS**this work was supported in part by the U.S. Army Research Office under grant W911NF1510275. *IFAC—PapersOnLine, 50*(1), 11447–11452, ISSN 2405–8963. https://doi.org/10.1016/j.ifacol.2017.08.1814
9. Olalekan, A. F., Sagor, J. A., Hasan, M. H., & Oluwatobi, A. S. (2021). Comparison of two SLAM algorithms provided by ROS (Robot Operating System). In *2021 2nd International conference for emerging technology (INCET)* (pp. 1–5). https://doi.org/10.1109/INCET51464.2021.9456164
10. Gazebo: Robot simulation made easy. Last accessed March 8, 2022, Homepage http://gazebosim.org/
11. AWS RoboMaker Guía para desarrolladores. Last accessed March 8, 2022, Homepage https://docs.aws.amazon.com/es_es/robomaker/latest/dg/aws-robomaker-dg.pdf#simulation-tools-rviz
12. Vera, A. A., & Valle, D. F. (2020). Diseño de un sistema de visión artificial para la detección y control de presencia de zopilote negro en aeródromos. Recuperado de: http://hdl.handle.net/20.500.12749/12743
13. Feng, L., Xiaoyu, L., & Yi, C. (2014). An efficient detection method for rare colored capsule based on RGB and HSV color space. In *2014 IEEE international conference on granular computing (GrC)* (pp. 175–178). https://doi.org/10.1109/GRC.2014.6982830

Taguchi Based Design of Sequential Convolution Neural Network for Classification of Defective Fasteners

Manjeet Kaur, Krishan Kumar Chauhan, Tanya Aggarwal, Pushkar Bharadwaj, Renu Vig, Isibor Kennedy Ihianle, Garima Joshi, and Kay Owa

Abstract Fasteners play a critical role in securing various parts of machinery. Deformations such as dents, cracks, and scratches on the surface of fasteners are caused by material properties and incorrect handling of equipment during production processes. As a result, quality control is required to ensure safe and reliable operations. The existing defect inspection method relies on manual examination, which consumes a significant amount of time, money, and other resources; also, accuracy cannot be guaranteed due to human error. Automatic defect detection systems have proven impactful over the manual inspection technique for defect analysis. However, computational techniques such as convolutional neural networks (CNN) and deep learning-based approaches are evolutionary methods. By carefully selecting the design parameter values, the full potential of CNN can be realised. Using Taguchi based design of experiments and analysis, an attempt has been made to develop a robust automatic system in this study. The dataset used to train the system has been created manually for M14 size nuts having two labeled classes: Defective and Non-defective. There are a total of 264 images in the dataset. The proposed sequential CNN comes up with a 96.3% validation accuracy, 0.277 validation loss at 0.001 learning rate.

Keywords Fasteners · Sequential convolution neural network · Defects · Taguchi analysis

NTU-PU Collaborative Research Grant.

M. Kaur (✉) · K. K. Chauhan · T. Aggarwal · P. Bharadwaj · R. Vig · G. Joshi
UIET, South Campus Panjab University, Sector 25, Chandigarh 160014, India
e-mail: mnkur96385@gmail.com

I. K. Ihianle · K. Owa
School of Science and Technology, Nottingham Trent University, 50 Shakespear Street, Nottingham NG1 4FQ, UK

1 Introduction

Bolts and nuts are common fasteners in the mechanical and automotive industries. Cold forming, hot forming, thread production, machining, hardening and tempering are the procedures that fasteners go through throughout production [3]. Changes in the material's intrinsic characteristics, effect of vibrations, tool damage, and improper process management can result in flaws in the end products as fasteners because of the processes they go through [5]. Fastener raw materials might often develop cracks. Wrinkles are a form of imperfection on fasteners caused by material displacement during the forging process of nuts in particular. Deformation, dents, wrinkles, scratches, fractures, rough surface, missing and misaligned threads on the fastener surface are all faults generated by processing [20]. Small and medium-sized industries are the largest manufacturers of these fasteners but the process of inspection is done manually, which requires a lot of people, money, and time. Even with all of that work, the potential for errors still exists, and defective products may reach clients. As a result of the vibration, these defective fasteners may lose up, or even break in long-term [21]. The development of automation technologies to detect bad steel fasteners could help in overcoming this challenge. The machine learning defect detection system reduced human error and effort, it makes the procedure more accurate by detecting defects that might otherwise go unnoticed by humans [4]. The process involved calculating a collection of hand-crafted textural characteristics, which were afterward trained on a classifier [14]. Although the automated detection process based on image processing and machine learning techniques offered great benefits it has some major drawbacks. One of the most significant is that the appearance of abnormalities changes in terms of form, size, colour, geometry, and other factors even within the same inspection work [15]. Due to these reasons computer vision applications and deep learning based convolution neural network (CNN) can be used for better efficiency and precise detection rate.

2 Related Work

Computer vision-based applications for object detection and classification using transfer learning and traditional CNNs are emerging and have become very popular in software, mechanical and electrical industrial sectors. The amount of speed, efficiency, and robustness provided by these applications has made day-to-day work simple [11]. There are a lot of examples where the use of such deep learning models for the detection and classification of industrial fasteners has proved to be successful in achieving the goal of high accuracy in less time. Some of the examples where computer vision has exceptional contributions are taken into consideration before proposing the model. Several algorithms have been used for metal crack detection and texture feature extraction. Different edge detection operators have been used to obtain thin edge features on defective samples [19]. Similarly several machine learning based techniques, have been explored for classification of surface defects on rolled steel [3]. Sharifzadeh et al. applied Gaussian functions and histogram approaches to detect various flaws on steel components, with a detection rate of 88.4% for holes, 78% for scratches, 90.4% for coil breaks, and 90.3% for rust problems [16]. Ashour et al. applied support vector machine to inspect the visual texture feature from the sample data, also different kernel functions has been investigated for best performance classifier [1]. Park et al. proposed machine learning based imaging based system for defect detection for dirt, scratches, and wears on surface. CNN has been applied to check the existence of defect in the target region on an input sample image. The proposed method has been proven to be advantageous in terms of time, cost and performance as compared to manual inspection techniques [13]. Bhandari and Deshpande proposed a modified scheme based on heuristics algorithm used by human inspectors for identifying surface imperfections to compute the features, then used SVM to classify surface images into two classes: defective and defect-free. The system was tested on a surface texture database and achieved a classification accuracy of 94.19% [2]. Zhao et al. investigated computer vision to detect pin missing problems in transmission lines that employed bolts to link different components. Defects in these bolts can cause major problems, such as grid breakdown. To solve this problem, a CNN model was proposed, which included three key improvements for extracting small-scale bolt features and achieved a 71.4% accuracy [22]. A version of CNN architecture was used by Liu et al. to detect catenary support components (CSCs). It involved the integration of a detection network for CSCs utilising large scale optimized and improved Faster R-CNN with a cascade network for the detection of CSCs with small scales. The model has a good accuracy of 92.8% [9]. Taheritanjani et al. created a system that automatically recorded, preprocessed data and compared it to a variety of supervised and unsupervised machine learning models for detecting damage in 12 different fasteners. The method also helped in determining the type of fastener used. The supervised model achieved 99%, whereas the unsupervised model achieves 84% [18]. Giben et al. proposed a visual inspection method for semantic segmentation and classification of material using deep convolution neural network (DCNN). In this approach they identified ten kinds of materials in total using this method, and the proposed model had a classification accuracy of 93.35%. The detection rate for chipped and crumbling ties was 92.11% and 86.06%, respectively [7]. To extract rich feature information, Kou et al. used specifically built dense convolution blocks in their defect detection model, which significantly increases feature reuse, feature propagation, and the network's characterization ability. The suggested model generated 71.3% mean Average Precision (mAP) on the GC10-DET dataset, which is publicly available, and 72.2% mAP on the NEUDET dataset [8]. Song et al. presented a CNN based technique in which they consider damage screw, surface dirt and stripped screw as defective classes, images has been taken up with industrial camera. Proposed model achieve 98% accuracy with 1.2s of average time taken to process per image [17]. Gai et al. proposed a CNN technique for detecting defects on steel surfaces in industrial parts. An industrial camera was used to collect and pre-process the data, and then the VGG architecture was used to improve network features, classification, and defect recognition capability. The proposed model was then compared to other traditional strategies and found to be more effective [6]. Computer vision-based applications such as object identification and classification using transfer learning or traditional CNNs have grown increasingly popular in all industries. The speed, efficiency, and robustness provided by these approaches have simplified day-to-day tasks. Researchers have employed the transfer learning approach in defect categorization but these models were trained on datasets for object detection and have no resemblance to the defect detection domain. Due to the fundamental difference between the object detection and defect detection domains, CNN models must be built from the scratch and trained on good quality dataset. Furthermore, the lack of standard datasets leads to erroneous and inconclusive findings, and there is currently no model that can be right away used for transfer learning. This work proposes a dataset of defects on the nuts and a design of sequential CNN model to handle these two key difficulties. This study also demonstrates the efficacy of the Taguchi strategy, which is a simple yet effective method for handling parameter optimization problems. The proposed method aids in determining the sequential CNN parameter value for defect classification in fasteners, according to the results.

3 Proposed Framework for Taguchi Based Design of Sequential Convolution Neural Network (CNN)

Figure 1 shows the proposed framework for designing an optimized sequential CNN model. The details of each step are as presented below.

3.1 Dataset Preparation

In order prepare a practical dataset, samples of M14 size nuts with six side faces have been collected from the fastener industries. The images of size 1844 * 4000 pixels were captured by mobile phone camera in variable light conditions by varying the level of illumination. Image of each face of nut is captured. The real world captured dataset had the problem of class imbalance, as the number of non-defective sides of nut were more than the defective samples. To enhance the number of defective images in the dataset, augmentation in terms of scaling and brightness variation was applied. This resulted in 132 images each in 'Defective' and 'Non-Defective' class. 80% of images were used for the training set and 20% for the test set. Figure 2 shows the sample images from dataset. The most common types of defects are crack, dent, patches, scratches and wrinkles [6]. Here, all the defects are considered as a single class as 'Defective'.

STEP 1: Dataset Preparation

- Number of Classes: 2 ('Defective' and 'Non-Defective')
- Data Augmentation: Brightness Variation and Rotate
- Images per Class: 132
- Total number of images in dataset: 264

STEP 2: Identify Sequential CNN Design Parameters

- Number of Convolution Layer
- Input Image size
- Filter Size
- Loss function
- Optimizer
- Activation Function

STEP 3: Taguchi's based Grid search for CNN Design Parameters

- Identify the number of levels and desired variations of CNN Design parameters.
- Select an appropriate Taguchi's Orthogonal Array
- Perform experimental Runs.
- Determine Response parameters
- Taguchi Analysis
- Select optimum set of parameters

STEP 4: Analysis of Proposed CNN Model

- Perform Confirmation Experimental Run
- Results and Discussion

Fig. 1 Proposed workflow for Taguchi based design of sequential convolution neural network (CNN)

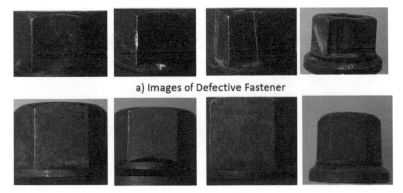

a) Images of Defective Fastener

b) Images of Non-Defective Fastener

Fig. 2 Sample images M14 size nuts in dataset

3.2 Sequential CNN Design Parameters

The architecture of a basic sequential CNN is as depicted in Fig. 3. These architectures can be created from scratch. To create a CNN model for a given application, early layers learn low-level characteristics, while the end-layer conducts classification based on the feature map.

Number of CNN Layers The CNN structures are motivated by the fact that increase in number of layers help in better approximation of target function and enhances the capability to learn from more detailed feature set. This makes, the number of layer as one of the essential dimension in regulating learning ability of the sequential architecture.

Input Image Size Image size is chosen such that the fine defects such as cracks and scratches are clearly visible but larger input image size leads to more number of input is visible. However, having larger size input image leads to increase in complexity of CNN. Therefore, $[100 \times 100]$ and $[200 \times 200]$ are selected for the study. If a coloured image of 100 pixels in height and width is supplied, each pixel has one single channel, and the input layer has the form (100, 100, 3).

Optimizer An optimizer updates the model with respect to the loss function and help in minimizing it by calculating partial derivative of the loss corresponding to weights and updating the weights in the direction opposite of the obtained gradient. This process is repeated until the loss function's minima is reached. Gradient descent takes into account the learning rate and takes larger steps if the loss is large and a smaller steps if loss is decreasing. The purpose of small learning rate is to come close to minimum value. The objective function is optimised using Stochastic Gradient Descent (SGD). It replaces the gradient with a value computed from a randomly picked data

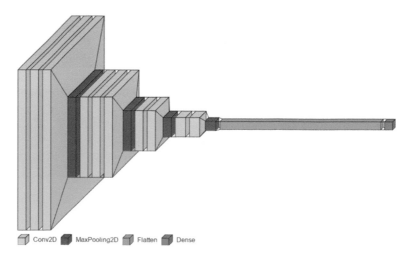

Fig. 3 Structure of sequential CNN

subset, this can be considered a probabilistic estimate to steepest descent. As a result, specific samples are chosen rather than the entire dataset. The global minimum can be readily attained, but when dealing with a large dataset becomes more difficult, hence SGD takes approximation of gradient descent by selecting a subset of samples. It can be particularly helpful in the case of large dataset. Adaptive Moment Estimation (ADAM) scales the learning rate using an exponentially weighted gradient. It is one of the most popular gradient descent optimizer algorithms. It keeps the record of the average of earlier gradients that are decreasing. It is a very effective optimizer that requires very little memory and has a low training cost.

Loss Function Hinge rank loss aims to reduce the spread between the output of the model and the target vector while isolating it from all the other vectors, thus penalizing equally all the errors. Squared hinge loss computes the square of hinge loss, resulting in flat error function's surface and also making it numerically simple to handle. If a hinge loss improves performance on a binary classification issue, most probably a squared hinge loss will improve performance even more.

Activation Function Rectified Linear Unit (ReLU) is a non-linear activation function to calculate activations of convolved feature map. ReLU response maximizes beyond a certain point away from zero and has a V-shape, whereas the response is capped to the maximum value of 6 giving it a Z-shape.

Filter Size The purpose of implementing CNN is to extract key features from the input images that can characterize a class. Convolutional operation considers the local subset of pixels, therefore different levels of subsets can be explored in the image by using different filter sizes and variation in filter size capture different details. A small size filters find fine details by exploring the small regions in the image. While, large size filter finds coarse information and the model tries to find features in a large area of the input at each computation. Hence, spatial filters can be explored to improve performance with regard to learning aspect of the network. By carefully adjusting the filters, CNN can perform well both on coarse and fine details. In this research, an attempt is made to experiment with these basic building elements of sequential CNN in order to identify the best combination that will allow the model to perform well. The influence of each parameter adjustment and its combinations can be better understood by carefully designing the experiments.

3.3   Taguchi Based Grid Search for CNN Design Parameters

Taguchi based grid search approach is relevant to all the real-world design problems which depends on number of control factors or hyperparameters. Taguchi's orthogonal array has been used in the design of experiments [10]. It is found out to be proficient when paralleled to many other statistical designs [13]. Value of control factors must be determined in order to achieve the optimal results. Taguchi analysis aims to enhance quality as a way to deliver resilient configurations and design. The

Table 1 Parameters and variations

Factors (6)	Name of parameter	Variations/levels
A	No. of CNN layers	6, 8, 10, 12
B	Image size	[100 × 100], [200 × 200]
C	Optimizer	Adam, sgd
D	Loss function	Hinge, squared hinge
E	Activation function	ReLU, ReLU6
F	Filter size	[2 × 2], [3 × 3]

term "signal" refers to improved performance with little variance, which is referred to as "noise". The consistency of performance is a metric of robustness, which is achieved by making the design immune to the effects of uncontrollable characteristics. A two-step optimization procedure is used in Taguchi grid search. Selection of hyper parameters and desired modifications is the initial stage. Next, the orthogonal array is chosen to ensure that these control components participate equally and in a planned manner [12]. The best potential performance is produced in the second stage by identifying the best arrangement of hyperparameters.

Selection of Hyperparameters Variations in object size, and distance from the camera are sources of variation (noise) in the defect categorization domain. The aim is to find the optimum hyperparameter combination that delivers high performance when subjected to fluctuations. Table 1 lists the description, value of parameters variation while performing trials. Since there is one component with four levels and five factors with two levels, L16 is the best choice among the Taguchi mixed level design possibilities for six parameters. A total of 36 trials will be required if the complete factorial design with all possible combinations is used. However, utilising Taguchi's orthogonal array to properly design trials, the total number of trials that must be examined for testing is reduced to 16 as indicated in Table 2.

Selection of Response Parameters In this work, the response parameters or performance metrics that are used to assess performance are test accuracy, validation accuracy, test loss and validation loss. The purpose is to maximize accuracy and minimize loss.

4 Analysis of Results

The model is implemented with Python software, and the DOE analysis is done with Minitab. The recommended set of parameters has been fine-tuned for the fastener dataset. The outcome of experiments for the possible set of combinations are recorded, each training is conducted for 500 epochs and the best result are recorded. Table 2 is analysed using Taguchi's analysis in a Minitab tool. The highlighted text indicates the best results in Table 2.

Table 2 Experimental outcome as per Taguchi's L16 orthogonal array

Exp. Run	Parameter variation						Response values			
	A	B	C	D	E	F	Train loss	Train accuracy	Val. Loss	Val. Accuracy
1.	6	[100 × 100]	adam	Hinge	ReLU	[2 × 2]	0.3885	0.9251	0.9303	0.9145
2.	6	[100 × 100]	adam	Hinge	ReLU	[3 × 3]	0.6127	0.8508	0.7954	0.8629
3.	6	[200 × 200]	sgd	Sqd. Hinge	ReLU6	[2 × 2]	0.3539	0.7552	0.6096	0.6721
4.	6	[200 × 200]	sgd	Sqd. Hinge	ReLU6	[3 × 3]	0.5114	0.9091	0.8036	0.8966
5.	8	[100 × 100]	adam	Sqd. Hinge	ReLU6	[2 × 2]	0.1441	0.8449	1.6285	0.8497
6.	8	[100 × 100]	adam	Sqd. Hinge	ReLU6	[3 × 3]	0.5757	0.8545	0.7945	0.8397
7.	8	[200 × 200]	sgd	Hinge	ReLU	[2 × 2]	0.275	0.1983	0.579	0.1883
8.	8	[200 × 200]	sgd	Hinge	ReLU	[3 × 3]	0.2206	0.9027	0.4843	0.9151
9.	10	[100 × 100]	sgd	Hinge	ReLU6	[2 × 2]	0.3332	0.9455	0.6477	0.9433
10.	10	[100 × 100]	sgd	Hinge	ReLU6	[3 × 3]	0.2258	0.9273	0.5374	0.9201
11.	10	[200 × 200]	adam	Sqd. Hinge	ReLU	[2 × 2]	0.0762	0.8949	0.3033	0.9021
12.	10	[200 × 200]	adam	Sqd. Hinge	ReLU	[3 × 3]	0.3889	0.9536	0.7223	0.9433
13.	12	[100 × 100]	sgd	Sqd. Hinge	ReLU	[2 × 2]	0.3889	0.9455	0.6691	0.9433
14.	**12**	**[100 × 100]**	**sgd**	**Sqd. Hinge**	**ReLU**	**[3 × 3]**	**0.3626**	**0.9455**	**0.4624**	**0.9433**
15.	12	[200 × 200]	adam	Hinge	ReLU6	[2 × 2]	0.1161	0.8545	0.5709	0.8528
16	12	[200 × 200]	adam	Hinge	ReLU6	[3 × 3]	0.1911	0.9159	0.3713	0.9216

Interval Plot for variation in values in train loss, train accuracy, validation loss and validation accuracy is shown in Fig. 4. It is used to evaluate and compare confidence intervals for result means. An interval plot depicts a 95% confidence interval for each group's mean. The plot indicates that the mean value of train and validation accuracy is almost equal with similar variability, while the loss values show clear displacement in terms of mean average value. The validation loss has a lot of variability and is considerably larger. Further, by analysis of outcome, efforts are undertaken to reduce loss and enhance accuracy.

Main Effect Plot: Figure 5 shows the Main Effect Plot for the adjustment of six hyper-parameters. The main effect graphs show how each factor affects the system's overall

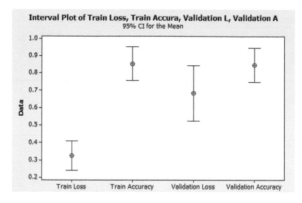

Fig. 4 Interval plot

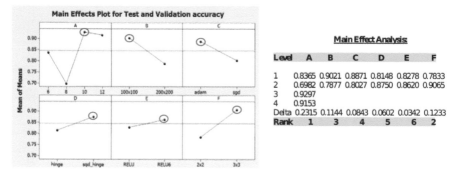

Fig. 5 Main effect plot and main effect analysis from minitab tool

performance. 'Main effect' is defined as when different degrees of a parameter have varying effects on performance. When the line is horizontal, it signifies that the characteristic average is the same across all variants of that hyperparameter. If the plot is not horizontal, and changes in the factor's values affect the characteristic differently, there is a significant influence [12]. The major effect plot of test and validation accuracy in Fig. 5 is used to determine the best set of values for each factor. The element with the greatest fluctuation has the greatest effect on the system response, and hence has Rank 1 and so on, according to the ranks of main effect analysis. The optimizer and loss function have the largest impact on system performance, while the type of activation function has the least.

4.1 Optimized Sequential CNN Model

Taking into account the hyperparameters listed in Table 1, the proposed values of parameters on the basis of Taguchi analysis are selected as shown in main effect plot to maximize accuracy in Fig. 5. The number of CNN layers is 10, image size is [100 × 100], Activation function is Relu6, loss function is squared hinge, optimizer is adam, filter size [3 × 3], filters in each layer are [32, 32, 32, 64, 64, 64, 128, 128, 256, 256]. Proposed sequential CNN architecture is designed and confirmation run is performed taking into consideration these parameter and summarized in Table 3 (Fig. 6).

Table 3 Proposed sequential CNN architecture

Layer	Type	Output shape	No. of parameters
conv_1	2D, Conv	(100, 100, 32)	416
conv_2	2D, Conv	(100, 100, 32)	4128
conv_3	2D, Conv	(100, 100, 32)	4128
max_pooling_1	Pooling, max	(50, 50, 32)	0
conv_4	2D, Conv	(50, 50, 64)	18496
conv_5	2D, Conv	(50, 50, 64)	36928
conv_6	2D,Conv	(50, 50, 64)	36928
max_pooling_1	Pooling, max	(25, 25, 64)	0
conv2d_7	2D, Conv	(25, 25, 128)	73856
conv2d_8	2D, Conv	(25, 25, 128)	147584
max_pooling_3	Pooling	(12, 12, 128)	0
conv2d_9	2D, Conv	(12, 12, 256)	295168
conv2d_10	2D, Conv	(12, 12, 256)	590080
max_pooling_4	Pooling, max	(6, 6, 256)	0
flatten_1	Flatten	(9216)	0
dense_1	Dense	(1)	9217
	Total parameters: 1,216,929		
	Trainable parameters: 1,216,929		
	Non-trainable parameters: 0		
	No. of CNN Layers: 10		
	No. of Filters in each Layer: [32, 32, 32, 64, 64, 64, 128, 128, 256, 256]		
	Activation Function: ReLU6		
	Loss Function: Squared Hinge		
	Optimizer: Adaptive Moment Optimizer (adam)		
	Batch Size: 32		
	Filter size in each layer: [3 × 3]		
	Classifier: Binary Support Vector Machine		

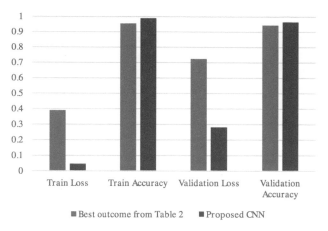

Fig. 6 Comparison of results for proposed design

5 Conclusion

Automatic defect detection systems have proven to be more effective than manual inspection procedures in quality control. The goal of this work is to use Taguchi-based design of trials and analysis to create a robust autonomous system for visual inspection. A dataset of pictures of nut surface defects is developed and used. A sequential CNN architecture is created using Taguchi design of experiments and analysis. The findings suggest that choosing the right hyperparameters is crucial for attaining low loss values and higher accuracy in a specific application. Training accuracy is 98.69%, training loss is 0.0449, validation loss is 0.2794, and validation accuracy is 96.3% in this study. System resilience is seen with consistent training and validation outcomes, that is low loss and high accuracy are achieved.

In future work, more comprehensive study will be carried out taking into various categories of defects.

References

1. Ashour, M. W., Halin, A. A., Khalid, F., Abdullah, L. N., & Darwish, S. H. (2015). Texture-based classification of workpiece surface images using the support vector machine. *International Journal of Software Engineering and Its Applications, 9*(10), 147–160.
2. Bhandari, S. H., Deshpande, S., & Deshpande, S. (2008). A simple approach to surface defect detection. *2008 IEEE Region, 10*, 8–10.
3. Caleb, P., & Steuer, M. (2000). Classification of surface defects on hot rolled steel using adaptive learning methods. In *KES'2000. Fourth International Conference on Knowledge-Based Intelligent Engineering Systems and Allied Technologies. Proceedings (Cat. No. 00TH8516)* (Vol. 1, pp. 103–108). IEEE.

4. Chen, J., Liu, Z., Wang, H., Núñez, A., & Han, Z. (2017). Automatic defect detection of fasteners on the catenary support device using deep convolutional neural network. *IEEE Transactions on Instrumentation and Measurement, 67*(2), 257–269.
5. Elangovan, S. B. T. K. R. (2019). Fabrication and analysis of polymer bolt and nut assembly by additive manufacturing system.
6. Gai, X., Ye, P., Wang, J., & Wang, B. (2020). Research on defect detection method for steel metal surface based on deep learning. In *2020 IEEE 5th Information Technology and Mechatronics Engineering Conference (ITOEC)* (pp. 637–641). IEEE.
7. Giben, X., Patel, V. M., & Chellappa, R. (2015). Material classification and semantic segmentation of railway track images with deep convolutional neural networks. In *2015 IEEE International Conference on Image Processing (ICIP)* (pp. 621–625). IEEE.
8. Kou, X., Liu, S., Cheng, K., & Qian, Y. (2021). Development of a yolo-v3-based model for detecting defects on steel strip surface. *Measurement, 182*, 109454.
9. Liu, W., Liu, Z., Nunez, A., & Han, Z. (2020). Unified deep learning architecture for the detection of all catenary support components. *IEEE Access, 8*, 17049–17059.
10. Montgomery, D. C. (2017). *Design and analysis of experiments*. Wiley.
11. Neogi, N., Mohanta, D. K., & Dutta, P. K. (2014). Review of vision-based steel surface inspection systems. *EURASIP Journal on Image and Video Processing, 2014*(1), 1–19.
12. Oehlert, G. W. (2010). A first course in design and analysis of experiments.
13. Park, J. K., Kwon, B. K., Park, J. H., & Kang, D. J. (2016). Machine learning-based imaging system for surface defect inspection. *International Journal of Precision Engineering and Manufacturing-Green Technology, 3*(3), 303–310.
14. Saiz, F. A., Serrano, I., Barandiarán, I., & Sánchez, J. R. (2018). A robust and fast deep learning-based method for defect classification in steel surfaces. In *2018 International Conference on Intelligent Systems (IS)*, (pp. 455–460). IEEE.
15. Shaheen, M. A., Foster, A. S., Cunningham, L. S., & Afshan, S. (2020). Behaviour of stainless and high strength steel bolt assemblies at elevated temperatures-a review. *Fire Safety Journal, 113*, 102975.
16. Sharifzadeh, M., Amirfattahi, R., Sadri, S., Alirezaee, S., & Ahmadi, M. (2008). Detection of steel defect using the image processing algorithms. In *The International Conference on Electrical Engineering* (pp. 1–7). Military Technical College.
17. Song, L., Li, X., Yang, Y., Zhu, X., Guo, Q., & Yang, H. (2018). Detection of micro-defects on metal screw surfaces based on deep convolutional neural networks. *Sensors, 18*(11), 3709.
18. Taheritanjani, S., Schoenfeld, R., & Bruegge, B. (2019). Automatic damage detection of fasteners in overhaul processes. In *2019 IEEE 15th International Conference on Automation Science and Engineering (CASE)* (pp. 1289–1295), IEEE.
19. Xue, B., & Wu, Z. (2021). Key technologies of steel plate surface defect detection system based on artificial intelligence machine vision. *Wireless Communications and Mobile Computing, 2021*.
20. Xue, P., Jiang, C., & Pang, H. (2021). Detection of various types of metal surface defects based on image processing. *Traitement du Signal, 38*(4).
21. Yun, J. P., Choi, S., Kim, J. W., & Kim, S. W. (2009). Automatic detection of cracks in raw steel block using gabor filter optimized by univariate dynamic encoding algorithm for searches (UDEAS). *NDT and E International, 42*(5), 389–397. https://doi.org/10.1016/j.ndteint.2009.01.007
22. Zhao, Z., Qi, H., Qi, Y., Zhang, K., Zhai, Y., & Zhao, W. (2020). Detection method based on automatic visual shape clustering for pin-missing defect in transmission lines. *IEEE Transactions on Instrumentation and Measurement, 69*(9), 6080–6091.

The Impact of Online Game Addiction on Students Taking Online Classes

Richard Septianus Frendi, Renaldi, Kevin Sandjaya, Richard Sanjaya, Ford Lumban Gaol, Tokuro Matsuo, and Chew Fong Peng

Abstract Online games are very popular among teenagers, and such e-sports have an attraction that makes the players prefer to play games instead of taking lessons. Students who do not attend class and those who attend class but do not pay attention instead play online games, disturb lecture activities. The purpose of this study is to describe the factors that makes students addicted to playing virtual games and to explain its impacts. The work uses quantitative methods for the selection of informants, i.e., people are considered based on the criteria that have been determined by the researchers according to the background and objectives of the study, and therefore, the informants in this study are addicted online game players. The effect of games on teenagers' studies is the decline in adolescent learning achievement. Ways to overcome online game addiction is to look for other more useful activities such as exercising or socializing with other people.

Keywords Impact of online games · College student · Game online addiction

R. S. Frendi (✉) · Renaldi · K. Sandjaya · R. Sanjaya
School of Information System, Bina Nusantara University, Alam Sutera, Indonesia
e-mail: richard.frendi@binus.ac.id

K. Sandjaya
e-mail: kevin.sandjaya@binus.ac.id

R. Sanjaya
e-mail: richard.sanjaya@binus.ac.id

F. L. Gaol
Binus Graduate Program—Doctor of Computer Science, Jakarta, Indonesia
e-mail: fgaol@binus.edu

T. Matsuo
Advanced Institute of Industrial Technology, Tokyo, Japan
e-mail: matsuo@aiit.ac.jp

C. F. Peng
University of Malaya, Kuala Lumpur, Malaysia
e-mail: fpchew@um.edu.my

1   Introduction

The term "online game" is a game that can only be played if the device being used to play is connected to the Internet [1]. Online gaming, according to Andrew Rollings and Ernest Adams, is a technology rather than a game genre; it is a mechanism that connects players rather than a specific pattern in the game. An online game is defined as a computer game that may be played by many players over a computer network (LAN or Internet). Games that may be accessed directly through systems offered by Internet service providers and are supplied as supplementary services from online service providers are referred to as direct access games. There are many different sorts of online games, ranging from simple text-based games to complicated graphics-based games that create virtual worlds that can be played by multiple people at the same time [2].

Nexia online, an RPG game released by MayGames with rudimentary 2D-based visuals, was the first online game to arrive in Indonesia in 2001. Nexia simply requires a modest computer configuration; it can even be run on a Pentium 2 with minimal 3D graphics. In Indonesia, this Korean game was the first to introduce gameplay and conversation. Unfortunately, due to the non-renewal of the license, the game was shut down in 2004. Since then, several other game providers have entered the Indonesian online gaming market, including Redmoon (2002), Laghaim in early 2003, Ragnarok Online (RO) in mid-2003, and Gunbound in 2004 [3].

According to Pande and Marheni (2015), playing online games causes players to become interested in lingering in front of the device, causing them to forget about studying, eating, sleeping, and engaging with the outside world. Someone who spends more time playing online games will spend less time communicating and interacting with others in their environment. Playing games is usually done alone and for a long period of time, thus it is possible to get addicted [4].

According to Wan and Chiou (2006)'s findings, adolescent addiction to online games is driven by a number of other reasons, one of which is psychological requirements and motivation. Entertainment and recreation, emotional coping (diversion from loneliness, isolation, boredom, releasing stress, relaxation, venting anger, and frustration), escaping from reality, fulfilling interpersonal and social needs (making friends, strengthening friendships, creating a sense of belonging, and recognition from others), the need for achievement, providing pleasure and challenge (superiority, desire for control, and to increase self-confidence) are the seven themes that psychological and motivational needs are divided into [5].

Online lectures are currently being held in Indonesia since the pandemic hit the country. The COVID pandemic began in Indonesia on March 2, 2020, and since then, the virus' spread has been accelerated, prompting the government to establish online lectures through various online meeting applications. These virtual classes prevent students from directly socializing with other students [1].

Bad habits of students during online lectures:

- Fun alone when lecture is in progress

The habit of watching YouTube, Netflix, playing online games, and sleeping, while studying online should be avoided since it has the possibility to overlook material. In addition, important information from the lecturer will also be left behind [6].

- Cannot set time

  Usually, when carrying out activities from home such as playing online games, it makes us underestimate some things that can in turn cause time delay. Managing time needs to be done so that the activities you do, such as lectures, doing assignments, organization, and others can run in a balanced manner [6].

- Underestimating the responsibilities given

  Another bad thing to avoid is underestimating the responsibilities given. E-learning students often postpone assignments and do them on days near the dead-line. In addition, when participating in other organizations or events, be it intern-ships or committees, it's a good idea to contribute and not just ride in discussions [6].

- Often stay up late

  Due to the assumption that online lectures provoke students have flexible time, it becomes an opportunity to stay up late by playing online games. Sleeping on time is required so that the next day can be productive, and the tasks given do not pile up [6].

- In and out of online class without permission

  Going in and out of online classes without permission should be desisted. Even though learning is carried out virtually, good manners have to be maintained. If signal problems occurs, ask the class leader or lecturer for excuse [6].

  Research Question:

- How much time do you spend playing online games?
- What days do you usually play online games? On weekdays or weekends?
- What is the age range of students who like to play online games?

The following are some of the consequences faced by individuals who are addicted to online games.

1.1 Health Issues

In terms of physical health, online games cause sleep issues that damage the metabolic system, fatigue syndrome, stiff neck and muscles, and carpal turner syndrome. Furthermore, the inclination to live a sedentary lifestyle and to prioritize playing games over other important tasks (such as eating) leaves online gaming addicts thirsty, skinny, or even fat, and in danger of developing non-communicable diseases (e.g., heart disease) [7–11].

1.2 Inducing Laziness to Study

Students will take time from their study schedule to play online games. Helping parents after school hours will be lost, money will be lost, snacks or school fees will be diverted, lose track of time, disrupting eating routines, and disrupting their emotions due to playing games [10].

1.3 Lack of Concentration

People who are hooked to online videos/games will lose focus when doing other things, which will affect their performance and productivity because they keep thinking about the game they are about to play [4–9].

1.4 Lack of Discipline

The desire to complete tasks in games and the enjoyment of playing, often leads to the abandonment of other pursuits. Because of playing games or thinking about them, time for worship, school tasks, college assignments, or work is disregarded. Furthermore, many games continue to function even when not connected to the web [12, 13].

1.5 Difficult to Think Clearly

According to these studies, video games alter not just the way the brain functioning but also its structure. The use of video games, for example, has been shown to impact the brain's ability to focus and think. Furthermore, most research findings suggest that youngsters who do not play online games are more concentrated than those who do. [16-18]

This paper is being presented in order to determine how many pupils are addicted to Internet games [20-24]. The goal in presenting a paper titled "The Effect of Online Game Addiction on Students Taking Online Lectures" is to lessen the amount of online game addiction that students are experiencing [7].

2 Research Technique

The technique is to use online surveys using online media such as google form, which is a medium that helps to give survey questions to the targeted respondents. This survey is distributed to fellow students who are undergoing online lectures as in Fig. 1. The survey is targeted for 50 people. The respondents are from Bina Nusantara University who play online games every day. The questionnaire 1 contains query as "Do you like playing online games?" Various sources on the Internet have been searched to determine how to overcome the addiction of playing online games, and it is found that by limiting the playing time, looking for new hobbies, refraining from gaming devices, and undergoing psychotherapy, addiction can be reduced. After finding the research methods that can be used, results of "How much time students spend for entertainment after using their time for more useful things?" is determined based on form 2. Results from respondents are compared between form 1 and form 2, to get the progress from respondents.

3 Research Method

The research has been conducted in a quantitative manner. From Fig. 2, quantitative research can be defined as a research method that examines a specific population or sample, collects data using research equipment, and analyzes quantitative/statistical data with the goal of testing a hypothesis. Data is gathered based on the amount of time kids spend playing online games on a daily basis. Most of the respondents believe they play online games frequently, thus they need a means to avoid being addicted to them. This can be accomplished by limiting the amount of time spent playing games, finding new hobbies, placing gadgets away from sight, and seeking counsel.

Researchers used Bina Nusantara University's student population since they are now attending online courses, and to determine how severe online game addiction is at the University. This study's data have been gathered twice. At first, it is the basics where the students are given form 1 to fill. After discovering the duration, the respondent spends playing online games, he may read the solutions given in the form to overcome the addiction, and it is hoped that the respondent would follow the directions. After a week, the second form is distributed to check if the time period of playing online games has changed.

4 Research Result

Among 50 students, 31 students are aged 18–20 years while 19 others are aged 21–22 years. Based on the results of the survey conducted from December 20, 2021,

Fig. 1 Flowchart of data
collection for the research

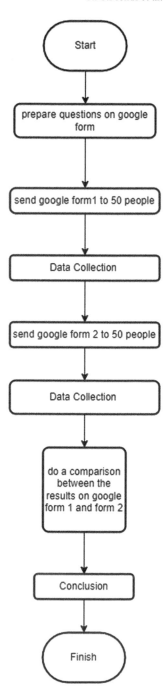

Fig. 2 Flowchart of the
steps involved in the research

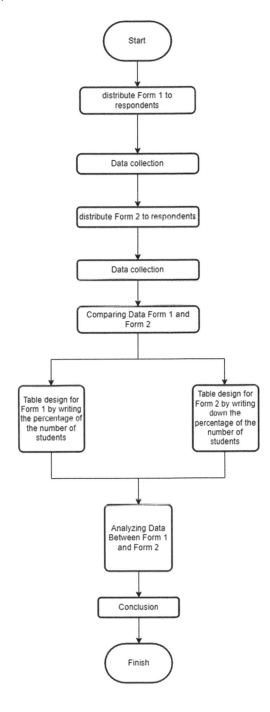

Table 1 Survey result of students who play online games as per the first form

Day	Number of students playing for <3 h (%)	Number of students playing for >3 h (%)
Monday	31 (62%)	19 (38%)
Tuesday	28 (56%)	22 (44%)
Wednesday	28 (56%)	22 (44%)
Thursday	25 (50%)	25 (50%)
Friday	23 (46%)	27 (54%)
Saturday	11 (22%)	39 (78%)
Sunday	5 (10%)	45 (90%)

until December 27, 2021, it is found that students spend less time playing online games on weekdays compared to the weekends. Some students spend less than 3 h a day in playing, whereas some of them spend more than 3 h playing online games each day.

Table 1 and Figs. 3 and 4 show the survey results obtained from students who responded to the form 1 which questioned them regarding the time duration they spend each day playing online games. These respondents are given suggestions to learn how to overcome this addiction. In Table 2, the survey conducted from January 1, 2022, until January 8, 2022, shows the results obtained from students, who responded to the form 2 which questioned them regarding the time duration they spend each day playing online games after reading how to overcome the online game addiction, have been tabulated to determine whether the respondents have experienced a decrease in time playing online games.

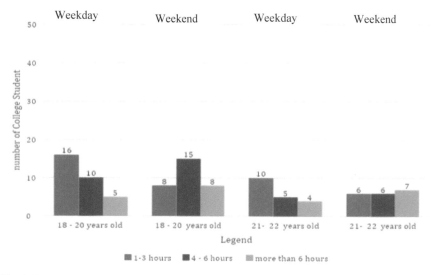

Fig. 3 Representation of the results of questionnaire 1 in graphical form

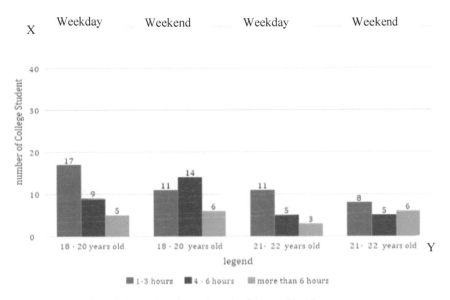

Fig. 4 Representation of the results of questionnaire 2 in graphical form

	Day	Number of students playing for <3 h (%)	Number of students playing for >3 h (%)
Table 2 Survey result of students who play online games as per the second form	Monday	33 (66%)	17 (34%)
	Tuesday	30 (60%)	20 (40%)
	Wednesday	29 (58%)	21 (42%)
	Thursday	28 (56%)	22 (44%)
	Friday	25 (50%)	25 (50%)
	Saturday	15 (30%)	35 (70%)
	Sunday	10 (20%)	40 (80%)

From the comparative results between form 1 and form 2, not much difference could be observed. However, there is a significant decrease in the respondent's time of playing online games. The number of students who play online games for less than 3 h, on Monday has been increased by 4%, on Tuesday has been increased by 4%, on Wednesday has been increased by 2%, on Thursday has been increased by 6%, on Friday has been increased by 4%, on Saturday has been increased by 8%, and on Sunday has been increased by 10%. The increase in the number of students who play for less than 3 h show that the students who spend more than 3 h in playing has been decreased simultaneously. Therefore, this portrays the success in overcoming the problem of online game addiction to a considerable extent.

5   Discussion

The results of the survey and research are discussed in accordance with the decrease in virtual game time after the students are provided with suggestions to overcome the addiction. Although there is no significant decrease in online play time based on the survey taken ten days apart, the students who are addicted must be educated about the harmful effects of playing games frequently, which may result in a considerable decrease in the number of students who misuse a lot of their valuable hours. From this, it is determined that respondents feel lonely due to online lectures, and therefore, they are more prone to experience addiction to online games.

This study discovered that playing online games is linked to addiction because they play without thinking about the other things they need to do. Moreover, they are willing to sacrifice their time every single day to play the game although they are loaded with educational assignments. Furthermore, the more addicted a person is, the harder it will be for them to socialize with others.

Previous research [13] obtained the same results as this current work, i.e., many students play online games to fill their spare time and some are addicted to the point of forgetting their obligations as students. Most students spend about 3 h or even more per day playing online games [13].

6   Conclusion

Online game addiction is a mental illness characterized by the need to play games for hours or even days without tiring. There are various methods for overcoming this online game addiction, such as restricting game time, discovering other interests, abstaining from game equipment, and engaging in psychotherapy, etc. While this strategy may not work for everyone, it is something to be tried at first to guarantee that they are no longer addicted to online games. People between the ages of 18 and 22 are more likely to spend their time on the weekends playing online games after finishing their work, in order to forget about their hard work and to relax. Therefore, the weekend is the ideal time for them to spend 1–4 h in playing. Relaxing and getting entertained by the game to forget about their worries can be replaced with healthy alternatives. [22–24]

Acknowledgements This research was supported by Mr. Ford Lumban Gaol, a lecturer in Research Method in Information Systems at the University of Bina Nusantara and the agreement term: February 4, 2022.

References

1. Redondo, R. (2021, september). pengaruh game online bagi mahasiswa.
2. Orami. (2016). kecanduan game. https://www.orami.co.id/magazine/kecanduan-game/
3. Abdi, H. (2019, january). 6 Jenis Game Online Seru dan Terpopuler yang Harus Dike-
   tahui. https://hot.liputan6.com/read/4053559/6-jenis-game-online-seru-dan-terpopuler-yang-
   harus-diketahui
4. Sandra, M. (2020, maret). Fenomena Kecanduan Game Online pada mahasiswa. http://reposi
   tory.unej.ac.id/handle/123456789/62440
5. Putry, B. (2019, oktober). Game online bisa membuat malas anak muda. https://jamberita.com/
   read/2019/10/02/5953524/game-online-bisa-membuat-malas-anak-muda/
6. Riani, A. (2021, oktober). 5 Kebiasaan Buruk Selama Kuliah Online yang Harus Dihindari,
   Bisa Bikin Auto Ngulang Matkul.
7. Handayani, V. V. (2021, febuari). Ini Efek Kecanduan Bermain Game Online bagi Otak
   melalui. https://www.halodoc.com/artikel/harus-tahu-ini-efek-kecanduan-bermain-game-onl
   ine-bagi-otak
8. Data, D. (2020, febuary). Inilah Dampak Positif dan Negatif Bermain Game. URL: https://
   blog.dimensidata.com/inilah-dampak-positif-dan-negatif-bermain-game/
9. Rokom. (2018, october). Inilah Dampak Kecanduan Dari Game Online. https://sehatnegeriku.
   kemkes.go.id/baca/umum/20180706/4226605/inilah-dampak-kecanduan-game-online/
10. Adrian. (2016, juli). Mental kecanduan main game online fungsi otak. https://hellosehat.com/
    mental/kecanduan/main-game-online-fungsi-otak/
11. Vebrina, R. (2019, juni). kecanduan game online penyakit gangguan otak. https://indonesia
    baik.id/infografis/kecanduan-game-online-penyakit-gangguan-otak
12. helsel. (2018, oktober). Pengaruh Game terhadap Kedisiplinan. https://helmiselo.blogspot.
    com/2018/11/pengaruh-game-terhadap-kedisiplinan.html
13. Gervase, A. (2019). pengaruh terhadap game online kepada mahasiswa.
14. Lebho, M. A. dkk. (2020, juli). Perilaku Kecanduan Game Online dari Kesepian dan Kebutuhan
    Berafiliasi pada Remaja hlm 202
15. Kautsar. (2019, oktober). pengaruh game online terhadap prestasi akademik peserta didik di
    man3 aceh besar.
16. Ondang, G. L., dkk. (2020). dampak game online terhadap motivasi belajar mahasiswa jurusan
    sosiologi fispol unsrat.
17. Meidy, D. (2021). dampak kecanduan game online mahasiswa bimbingan dan konseling (studi
    tentang kebiasaan mahasiswa bermain game online di universitas hein namotemo.
18. Natalie. (2021). kecanduan game online pada remaja.
19. Soraya. (2019). dampak kecanduan game online di bawah umur.
20. Jacklind. (2009). Kecanduan game online di usia 15–16 tahun.
21. Syazira, S. N., dkk. (2021). Pengaruh Intensitas Bermain Game Online Terhadap Prokrastinasi
    Akademik Mahasiswa.
22. Anggraini, P. A. (2020, june). 4 Bahaya Game Online untuk Kesehatan, dari Kecan-
    duan hingga Obesitas. 2021 https://health.kompas.com/read/2020/01/25/073300368/4-bah
    aya-game-online-untuk-kesehatan-dari-kecanduan-hingga-obesitas?page=all
23. Adrian, K. (2020, mei). Ini Ciri-Ciri Kecanduan Game Online dan Cara Mengatasinya. https://
    www.alodokter.com/ini-ciri-ciri-kecanduan-game-online-dan-cara-mengatasinya
24. Sandra, D. (2018). dampak game online pada remaja sejak dini.

Application's Impact on the SepuisLife Company's Insurance Sales Process During the Pandemic

M. Rifki Ananda, Gybrnd Tasawuf, Steven, Revie Muhammad Hanif, Michael Jonathan, Ford Lumban Gaol, Tokuro Matsuo, and Natalia Filimonova

Abstract Insurance companies are generally considered as one of the largest types of corporations in the world; prior to the pandemic, their status was already high in the world of business and economics. Insurance sales are already at an all-time high, particularly for health insurance. One of the most crucial components of our lives is our health. People are always trying to find new ways to stay healthy. In late 2019, the SARS-CoV-2 virus was discovered in Wuhan, China. This virus, known as SARS-CoV-2, has spread over the world and is now known as Covid-19 illness. This disease has now become a pandemic with countless victims. Many companies such as SepuisLife, who perform on the matters of Health Insurance, would obviously get affected by Covid-19 pandemic. This causes an increase in the insurance sales and must have a new approach to sell the insurance policy during this pandemic. This paper reviews different related research works proposed by different teams but

M. Rifki Ananda (✉) · G. Tasawuf · Steven · R. M. Hanif · M. Jonathan
School of Information System, Bina Nusantara University, Alam Sutera, Jakarta, Indonesia
e-mail: m.ananda@binus.ac.id

G. Tasawuf
e-mail: gybrnd.tasawuf@binus.ac.id

R. M. Hanif
e-mail: revie.hanif@binus.ac.id

M. Jonathan
e-mail: michael.jonathan002@binus.ac.id

F. L. Gaol
Binus Graduate Program—Doctor of Computer Science, Jakarta, Indonesia
e-mail: fgaol@binus.edu

T. Matsuo
Advanced Institute of Industrial Technology, Tokyo, Japan
e-mail: matsuo@aiit.ac.jp

N. Filimonova
Vladimir Branch of Russian Academy of National Economy and Public Administration, Vladimir, Russia
e-mail: natal_f@mali.ru

541

S. Shakya et al. (eds.), *Sentiment Analysis and Deep Learning*,
Advances in Intelligent Systems and Computing 1432,
https://doi.org/10.1007/978-981-19-5443-6_41

with similar topics by taking the advantage of their perspectives. Several surveys are provided to help us understand the needs from people. For proceeding with future research in this topical area, this paper will provide some information regarding the insurance and any of its different types like health insurances and vehicle insurance.

Keywords Pandemic · Insurance company strategy · Insurance

1   Introduction

SepuisLife is a company engaged in selling insurance for the social community; the insurance categories offered are quite diverse, ranging from physical and mental health, safety when traveling, to inpatient compensation. The existing pandemic condition makes people to take health insurance. This generates big data and leads to opening an enlarged insurance branch office. Since SepuisLife still operates in one main branch, it is difficult to enlarge the branch because many system changes have to be made, and SepuisLife doesn't have the required resources to do so and with a limiting manpower and skill set of its employee [1, 2].

The case faced by the SepuisLife Company is that the workflow within the company is unstable, and the productivity level of SepuisLife itself is very low. The main problem that is being experienced by SepuisLife is that with the ongoing Covid-19 pandemic, the insurance sales process carried out by SepuisLife has decreased because during the pandemic the company was unable to carry out their business activities as usual (before the pandemic started), for example, they had to follow the rules of WFH; the insurance sales process can only be done through the Superme application (SepuisLife application) and requires an expansion of the department to manage the Superme application due to being the main place of purchase during the pandemic.

Even though SepuisLife has an application where customers can buy insurance online, customers still buy insurance and go directly to the SepuisLife company; in this condition, SequisLife could enlarge their branch to make it easier to do a lot of transactions, but SequisLife has problems expanding the branch due to the difficulty of system reengineering by SepuisLife. As a result, this density is difficult to overcome during a pandemic that could result in the spread of Covid-19 [1].

The last obstacle for SepuisLife is now they need to pump the marketing for their application which is Superme as the company's main insurance sales process because due to the requirement to follow the rules and health procedures during the pandemic which is customer isn't allowed to attend directly to buy the insurance. Therefore, the recognition of the application throughout the usual customer or potential customer is a must if the company want to survive and is still running even in the pandemic situation [3].

Research Question:

In this research, for the SepuisLife case, there will be several questions to find out the conclusions, namely:

1. Does the person know, SepuisLife Company has an application to access the insurance?
2. Does the current pandemic situation affect the overall sales on insurance?
3. Does the customer prefer using applications more than offline services after the pandemic occurred?
4. What should the SepuisLife company do during this pandemic?

1.1 Insurance

Insurance is a contract made by two or more people in whom the responsible party pays the previously agreed fee, which is due to a loss due to the use of insurance, one of which causes a loss, damage, and loss of something caused by an unexpected event. Now, insurance can be obtained through insurance companies that provide various types of insurance according to the customer's needs for future events [1].

The main purpose from an insurance is acting as protection toward any unexpected accidents; insurance could also maintain business longevity if you choose to use business insurance; besides that insurance could also help with the lower class citizen that still need assistance on their economy. With the presence of insurance, any damages or unexpected accidents will get minimize that is why having insurance could be a saving grace, especially on the lower class of citizen that need assistance on their economy [4].

The policy revolves around insurance works for certain period usually monthly, where the terms end the insurance holder with an option to renew their policy or maybe terminate and end their insurance policy; when owning an insurance, the holders must knew what does the insurance cover [4].

1.2 Role of Application Inside the Company Sales Process

The role of application in business sales is very important because it affects the effectiveness and efficiency of the process cycle. Imagine your business didn't have an application to support the process for the sale. It means your employee must do the process manually. They need to calculate the price manually, check the stockpile or services manually, create the invoice manually, etc. This will only make the process more time-consuming; therefore, reducing the company's effectiveness and the customer will surely be unhappy about it [4].

Surely, with the presence of an application, it will also help in continuing the business process of the company while obeying the Covid-19 regulation with the *Work from Home*. Regarding the pandemic situation, SepuisLife will heavily rely on their application to continue the transaction process of the insurance. If a company has the application to support the process, there will be a lot of benefit they will get from it, like the problems that are already mentioned above, and there will be no possible human error that can happen since all the processes are automated by a computer and system.

1.3 *The New Insurance Marketing and Sales Process in this Pandemic Situation*

Regarding the marketing for insurance's sales in the current pandemic, the company needs to develop and adapt to a new strategy if they want to strive even in a pandemic situation. The strategy that needs to be develop must have several purposes like first it needs to be efficient and easy for customer to obtain the insurance; second, it needs to be an easy adaptable method for the company's employee to serve the customer without developing unnecessary time-consuming process, and last but not least, the third is the company needs to develop a new application that is easy to use by customer and vice versa for the customer can use the company's every services that are provided from whenever they are [5, 6].

The perspective of the customer needs to know about this application's development existence therefore the company needs to make an advertisement that is easy to be spotted by the customer and potential customer on everyday like necessities such as television, smartphone, or in a mall advertisement board where people could easily see and access the advertisement [5].

2 Research Technique

The research will be performed based on the flowchart as shown in Fig. 1, which started with observation of the SepuisLife Company, literature study on previous study on similar topics which are relevant with insurance companies such as Sepuis-Life to be used as basic knowledge of the research and reference. Following that, a requirement gathering strategy based on a literature review and sending questionnaires to research volunteers, the results of the surveys, and using the major literature as the fundamental knowledge to conduct the research will be implemented. With the final part of the research on performing a discussion about the results of our research with previous research on similar topics, and finally on creating an action plan [5].

The research conducted on the project during this pandemic will use a case study type research method, where the problem is present in the form of a case in which there is a single unit, the unit is the SepuisLife Company which is a company engaged in insurance in Indonesia. In-depth research from a variety of perspectives, including the context of the SepuisLife Company itself, as well as the company profile, and then the risk factors and problems encountered by the SepuisLife Company. Then, the data analysis technique will be carried out using the quantitative method where the calculation of the data obtained from the analysis of the data that has been obtained

Fig. 1 Flowchart research
technique

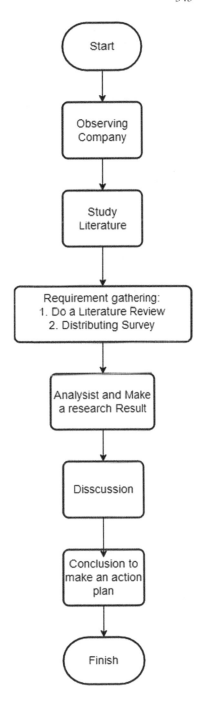

during SepuisLife's operation while the pandemic took place with the pandemic before it took place [6, 7].

2.1 Literature Review

A literature review is a search and evaluation of the existing literature in our selected topic area or subject. It provides the most up-to-date information about the subject or topic we are writing about. It compiles a summary of the information found in the literature. Knowing this, we need literature that can make our research more viable and effective, and before we get criticized by somebody else, it is more assuring if we criticize ourselves by reviewing the previous paper and preventing any similar mistakes.

The existing research article describes the financial services industry of life insurance companies, where the finance is growing so rapidly, marked by so many institutions established both private and government parties that offer services in the financial sector, not only from domestic but also those established by foreign parties. Especially, in today's era of digital technology, information is so easy to obtain. The presence of financial services companies is expected to provide benefits to the community. These various service industries can be in the form of banking services or other financial institutions.

Insurance service companies in practice offer risk transfer services to the insured by paying a premium. Insurance business present in Indonesia has long been pioneered by insurance companies Bumiputera 1912 and Jiwasraya, both of which were born since Indonesia was colonized by the Dutch. According to (Siswanto & Hasanah, 2019), the presence of insurance service companies is motivated by the increasing awareness of the Indonesian people regarding the uncertainty and misfortune that can occur in the future. The funds obtained from the deposit of the insured premium must be able to be managed properly by the company, of course in investment assets that are able to provide profits so that it does not cause problems. The government as a regulator pays attention to insurance companies in carrying out their business activities as stated in Law no. 2 1992. These rules must be used as guidelines so that the business can run smoothly.

The agreement made by the insured and the insurer related to the rights and obligations with the emergence of a premium burden for the insured calculated with the amount of risk, the greater the risk, the greater the premium obligation according to the level of risk insured. The lesser the risk, the lower the premium the insured must pay. Financial risk can arise in the insurance business resulting from speculative actions in investing; this speculative risk has two possibilities of profit or loss. Most of the funds collected were allocated for investment activities; this activity is commonplace because apart from financing its operations, it is equally important to take into account the obligation to pay the insured's claims in the future.

The another existing research study aims to identify and analyze the investment decisions of the beauty clinic business in the form of opening a new branch which will be established by Dr. Laurensia Leoni. Data collection methods used in this study were interviews, observation, and documentation. Analysis of investment decisions in beauty clinics is carried out using financial aspects. Analysis of branch expansion investment decisions in the financial aspect includes the following steps: calculating the initial investment of the business, analyzing cash inflows, analyzing cash outflows, assessing net cash disbursement, calculating discount rates, and assessing investment decisions from the financial aspect using the net present value (NPV) method, as well as an estimate of the strategy for the location of branch development.

Establishment of a beauty clinic on Jl. Sutan Syahrir in Solo was used as the object of this research. Based on data on the number of beauty clinics in Solo (source: www. google.com), Bver Salon is the only beauty clinic in the Jebres area. From the results of field surveys and interviews with prospective business actors, Bver Salon has advantages in the field of home service, namely treatment methods that can be done at home and private tutoring services (beauty class private) for beginners who want to increase their knowledge and skills in the field of beauty. Bver Salon provides air-conditioned room facilities, mattresses. And for this article's analysis technique, quantitative methods are use.

There are 2 characteristics in making investment decisions, namely involving a fairly long period of time because it involves assets that can be depreciated and requires substantial funds. According to Garrison et al. (2013), there are several types of capital budgeting decisions, including.

3   Standard Work from a Lean Theory Perspective

This thesis describes and illustrates numerous concepts of standardized working conditions processes that can be adopted by a concrete construction company. Work practices and knowledge transfer employing industrial and lean concepts in conjunction with program training will reduce variability in the operation of constructions.

Work process variability raises the risk of breakdowns or any deviation from expected results, errors, and negative iteration, all of which lead to schedule and cost overruns. Standardizing work processes reduce the likelihood of breakdowns while enhancing throughput and laying the groundwork for learning from any that do occur. It also serves as a starting point for experimenting with various work approach designs.

The findings of an exploratory study into job standardization in a concrete construction division are presented, as well as the cultural and organizational hurdles that must be overcome in order to shift the current paradigm. The study produced two outcomes: first, a better understanding of standard work processes and how they differ from preconceived notions, and second, the creation of standard work procedures as a baseline for the improvement can be continuous. This study can assist practitioners in learning how to assess processes, improve them, and convey critical information.

4 Research Method

The research design that we determined for this project is a case study with a single unit in the form of the SepuisLife Company itself; this is done to measure the scope of the condition of SepuisLife when there are many customers during the pandemic with other companies, there are measurements to measure comparisons from companies After the literature review company discussed, the data analysis technique chosen was quantitative data analysis based on data during the pandemic and before the pandemic took place.

First, count the score most chosen by respondent, and there are 2 types of respondent; it is using insurance and not using, so the total of type should be counted together by numbering the score most chosen by response, so some formula needed to make the research result.

Modus Formula

HI   High respondent for using insurance
Hin   High respondent for not using insurance

Median Formula

$$ME = \frac{X(n+1)}{2}$$

Mean Formula for Population

The term population mean, which is the average score of the population, we survey and the target is around 20 respondents on a given variable, and we, is represented by:

$$\mu = \frac{\left(\sum xi\right)}{N}$$

μ represents the population mean
ΣXi represents the sum of all present score in the population/$X1 + X2 + X3 \cdots + Xn$
N represents the total number of individuals or cases in the population

Research Sample Size Formula by Slovin

Based on the notation of the minimum research sample size formula by Slovin above, people of population if we have around 20, so we can determine the minimum sample we get to study, set margin of error is around 5% or 0.05.

The calculation is as follows:

$$n = \frac{N}{\left(1 + \left(N \times e^2\right)\right)}$$

n   *The minimum sample to be studied*
N   *Total people in a population*
e   *margin error*

5   Research Result

From the results of the performed survey and literature review from several papers, a research result will be formulated to answer the research questions.

The questions 1–3 will be explained by Figs. 2, 3, 4, 5, 6 and 7 to obtain a result.

From Fig. 2, total responses and percentage who use insurance has been obtained from 23 responses, 52.2% says they are not using insurance. For this pandemic in Indonesia, many responses are not using insurance due to one condition.

Figure 3 depicts the respondents' opinions, who utilize insurance but discuss whether insurance is required or not, in the form of a score. Responses prefer a score of 4, which shows that insurance is needed during a pandemic. (Y/↑ = Responses number, X/→ = the score 1 mean not agree to 5 is mean very agree).

From Fig. 4, the respondents' opinion is not using insurance but talking about whether insurance is needed or not in the form of a score.

Fig. 2 Use insurance

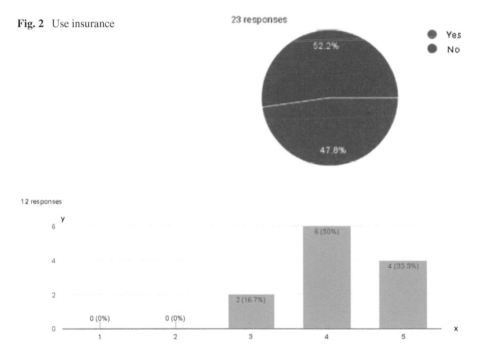

Fig. 3 Customer opinion about insurance needed (using insurance)

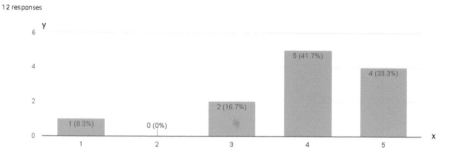

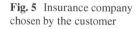

Fig. 4 Customer opinions about insurance needed (not using insurance)

Fig. 5 Insurance company chosen by the customer

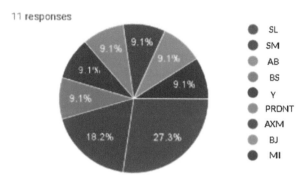

Fig. 6 Option to buy insurance

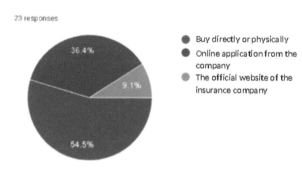

Fig. 7 People's recognition regarding the existence of SepuisLife's application

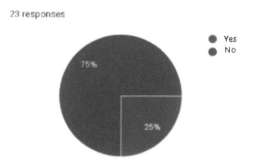

Responses prefer a score of 4, and one of them chooses 1/not agree which shows that insurance is needed during a pandemic.

(Y/↑ = Responses number, X/→ = the score 1 mean not agree to 5 is mean very agree).

From Fig. 5, respondent modus is using SL as insurance; it is 27, 3%. So, insurance SL is most trusted or good with marketing for customers.

From Fig. 6, most respondents choose to buy/subscribe the insurance directly to the company office rather than using application and website.

From Fig. 7, most of the respondents don't know about SepuisLife has insurance application to sell some type of insurance it is 75%.

For the question number 4:

There are several types of capital budgeting decisions and expanding the new branch, including:

1. Expansion Decision, for this case, SepuisLife manager must decide whether an asset is worth buying or whether it can increase SepuisLife production and sales capacity.
2. Cost Reduction Decision, for this decision, SepuisLife manager will determine whether some equipment it is worth to buy, in order to reduce SepuisLife production costs or not.
3. Equipment Selection Decision, for this case, SepuisLife manager will determine which equipment it is worth to buy, based on the level of to support company efficiency and effectiveness.

In this pandemic, many insurance companies use a Web site to sell their insurance product, but in this condition, many people use SepuisLife insurance but don't even know that SepuisLife has insurance to use, and other companies already do promotion marketing to invite people to use their Web site, but for marketing the application SepuisLife only, few respondents know about SepuisLife application.

6 Discussion

From the research result as we do from previous research, we discuss whether the results obtained are people who using insurance are they know about SepuisLife application, how many people choose the SepuisLife as insurance and how many people using insurance application as to buy/subscribe the insurances. From the result of our discussion, that many people don't know about SepuisLife has application can customer use to buy some SepuisLife insurance, and most of respondent said, insurance is needed in this pandemic; it is about from point 4–5 is 70%, and not agree or normal is 30%, and from the literature review, many author give solution to SepuisLife how to expand or increase the branch, and from that condition, SepuisLife still can expand their branch and only get information about how the insurance strategy during pandemic.

When we compare these responses to the research of "Feasibility of Investment Opening a New Branch of Beauty Clinic," which serves as previous research with similar topics and scope to our research. On the research of establishment of a new branch of Bver Beauty Clinic, it said opening up a new branch could also act as investment for the company not only the new clinic could attend as more centered on a workforce, which could also help to expand the company in general. So, in comparison, the answer of building and having a new branch is out of reach for SepuisLife Company because the general idea of opening a new branch for Bver Beauty Clinic as a new center of workforce is possible due to the face-to-face and work from office regulation still the main idea back then; for SepuisLife, it will be a waste of time regarding opening a new branch office because even with a new branch as another power house for the insurance sales, they still need to abide on the work from home regulation (WFH) due to the current pandemic, so the answers of having to implement their application as the center of the insurance sales process for SepuisLife company are more doable and preferable; lastly, the resemblance of the two is both could be taken as the company investment.

The marketing and customer awareness toward the companies new application (for SepuisLife) or New Beauty Clinic (for Bver Beauty Clinic) is still the deciding factor, here takes on the role of marketing which the surveys, and our research has answered on those topics which preferably during the pandemic customer would prefer an online transaction and try their best to avoid contact with other people, and having a better advertising campaign could solve the research question about "Would should SepuisLife during the pandemic and to make customer and potential customer know about their insurance sales application."

7   Conclusion

SepuisLife condition is needed support how to handle the crowded people who want to buy insurance, and from the SepuisLife system condition, SepuisLife still cannot do reengineering to the company to expand a new branch, but the conclusion from this condition is, SepuisLife don't to be focus to reengineering, but SepuisLife just need to they online marketing because most of SepuisLife customer does not know about SepuisLife has insurance application easy to use and simple, due to the customer don't know about that they just focus buy insurance directly to the office, so if the company focuses to the application, the density of customers who come to the company directly can be overcome by using an online application, this can also increase sales and prevent the spread of Covid-19 after this pandemic.

So, from the conclusion, we can get the answers for the research questions, it is:

1. The closest answer is no; most of the customers still don't know the existence of SepuisLife's application for them to access the insurance.
2. Yes, the current pandemic indeed affects the overall sales of the insurances.

3. Yes, the customer prefers an application to claim their insurance in the current pandemic.
4. SepuisLife Company needs to develop a new advertising campaign about their new application that has been developed, so their customer knows about its existence [8–11].

Acknowledgements Perkentangan adibakti company sponsored our research by giving the team a kickstarter almost 90% of the budget requirement that our planner has made in exchange they will be the first one to know the results of the research in hope to fix their own issue regarding their team workflow. The grant agreement term: January 30, 2022.

References

1. Lestari, H. (2017). Pengaruh Intellectual Capital Terhadap Kinerja Perusahaan Asuransi di Indonesia. *Jurnal Manajemen 4*(7), 491–493. http://journal.ecojoin.org/index.php/EJM/article/view/264
2. Oswari, T., & Suhendra, S. (2013). *Proyeksi Manajemen Risiko Operasional Pembukaan Kantor Cabang Perusahaan Asuransi di Indonesia.* http://toswari.staff.gunadarma.ac.id/Downloads/files/36751/Jurnal%20Asuransi%20an%2
3. Gepth, P. (2006). *Plant genetic resources conservation and utilization: The accomplishments and future of a societal insurance policy.* pp. 65–72.
4. Goodwin, S., & Celuch, K. (2002). Technology readiness in the e-insurance industry: An exploratory investigation and development of an agent technology e-consumption model. *Journal of Insurance Issues, 25*(2), 142–165. https://www.jstor.org/stable/41946210?readnow=1&refreqid=excelsior%3A83435ee4715fb1afc9671dfa1ff6f453&seq=1#page_scan_tab_contents
5. Mokoginta, A., Mananeke, L., & Jorie, R. (2019). *Implementation of marketing strategy using SWOT efforts to increase sales of insurance saving products in pt. Prudential Manado Branch.* https://doi.org/10.35794/emba.7.3.2019.25086
6. Djaelani, F. (2011). *Pertumbuhan Industri Asuransi Jiwa Di Indonesia: Suatu Kajian Dari Sisi Penawaran.* Indonesia: Yogyakarta, pp. 34–41
7. Feng, P. (2008). *Standard work from a lean theory perspective.* https://www.researchgate.net/publication/228425542_Standard_Work_From_a_Lean_Theory_Perspective
8. Antony, S. (2021). *Comparative analysis of the financial performance of Islamic insurance companies using the early warning system ratio before and after during covid-19.* https://journals.unihaz.ac.id/index.php/jaz/article/view/2448/1133
9. Bumulo, S. (2018). *Effect of trust and quality of service on insurance customer loyalty.* Yogyakarta, Indonesia.
10. Cain, C., & Haque, S. (2011) *Organizational workflow and its impact on work quality.* https://www.ncbi.nlm.nih.gov/books/NBK2638/
11. Spenche, M., & Zechka, R. (1978) *Insurance, information, and individual action.* pp. 335–343. https://doi.org/10.1016/B978-0-12-214850-7.50027-9

A Proposed System for Understanding the Consumer Opinion of a Product Using Sentiment Analysis

Eesha Martis, Rutuja Deo, Sejal Rastogi, Keshav Chhaparia, and Ameyaa Biwalkar

Abstract Sentiment analysis is essentially a process used to analyse or understand the underlying sentiment or feeling of the text in hand. This paper describes a software that leverages machine learning approaches such as support vector machine (SVM), random forest and logistic regression to perform opinion mining on Amazon consumer reviews. With an F1-score of 0.84, logistic regression and SVM prove to be better choices than random forest for a binary classifier. With online shopping becoming the new trend, this application is a one stop solution for businesses to track the popularity and sentiment of their products.

Keywords Sentiment analysis · Machine learning · Random forest · Logistic regression · Support vector machine · Amazon reviews

1 Introduction

Sentiment analysis or opinion mining [1], as the name suggests, helps one analyse the feeling or the state of mind of the author. That emotion can lie on either ends or between the positive–negative spectrums. Opinion mining is widely used in e-commerce, marketing, advertising, politics, market research and other fields of study.

In this era of e-commerce, where people largely purchase merchandise and avail different services online, they provide some kind of numeric ratings or feedbacks for other buyers to understand the utility of the purchased merchandise or availed service. Apart from a percentage of feedback given in the form of ratings, major feedback is generated via text. But, most of the times, it is challenging for humans to understand the textual feedback accurately as such responses are highly subjective. Looking at large volumes of such verbal information quantitatively is a task, which can be better done using techniques of sentiment analysis.

E. Martis (✉) · R. Deo · S. Rastogi · K. Chhaparia · A. Biwalkar
Computer Engineering Department, Mukesh Patel School of Technology, Management and Engineering, NMIMS University, Mumbai, India
e-mail: martiseesha@gmail.com

© The Author(s), under exclusive license to Springer Nature Singapore Pte Ltd. 2023
S. Shakya et al. (eds.), *Sentiment Analysis and Deep Learning*,
Advances in Intelligent Systems and Computing 1432,
https://doi.org/10.1007/978-981-19-5443-6_42

The 4 major categories of opinion mining are emotion detection, intent based, fine grained and aspect based [2]. Emotion detection analysis assists in perceiving and comprehending people's emotions such as anger, grief, excitement, irritation, fear, panic, anxiety and so on [3]. Intent analysis assists in determining the consumer's intent, i.e. whether the customer plans to purchase or is merely browsing. If the customer is interested in ordering, they can be tracked and targeted with offers; else time and resources can be saved by not advertising to them. Fine-grained sentiment analysis is useful for the study of reviews and ratings on a scale of 1–5. Whilst fine-grained analysis aids in determining the overall polarity of customer reviews, aspect-based analysis focuses on specific features of a product or service. For example, a phone's various aspects would be memory, speed, cost, aesthetic, etc.

The aim of this paper is to scrape online reviews and perform opinion mining on the scraped reviews in order to understand the overall customer opinion of the product. Binary classification is performed using logistic regression (LR), support vector machine (SVM) and random forest (RF), and the outcome is displayed using interactive visualisations on the web application.

This paper covers the entire work in 5 different sections. Section 2 provides a review of past literature on sentiment analysis to get a better understanding of this field. Section 3 covers the methods used to implement the system. It discusses the architecture, data flow in the system and the algorithms used. Section 4 this presents the results of the implemented framework. Section 5 concludes the paper and addresses the future endeavours in this area.

2 Background, Motivation and Objective

The reviews provided by the customers on e-commerce websites not only help the owners to improve their services accordingly but also help other customers to get an opinion of the products before buying them online. In order to analyse the customer feedback and the emotion associated to the reviews in the past, sentiment analysis has been performed on various data sets.

Paper [4] used various machine learning approaches such as Naïve Bayes and support vector machine alongside lexicon-based techniques to analyse the opinion of amazon products. The methodology included data collection, pre-processing, score generation followed by sentiment classification. NB with an accuracy of 98.17% performed better than SVM with an accuracy of 93.54%. The data set in [5] contained online product reviews and used k-means clustering. The data pre-processing techniques consisted of parts of speech tagging (POST) and negation phrase identification algorithm. Accuracy of 90.47% was achieved. Paper [6] implemented categorisation on sentence level as well as review level. They have also proposed methods on how to compute sentiment score, to identify negative phrases and feature vector generation. RF is the most effective amongst the various algorithms used. When reviews were categorised according to their individual star-scaled scores, dismal results were received. It is possible that it won't work well for reviews that are solely based on implicit feelings. F1-score of 0.8 was achieved. The authors of [7] used a combination of verbs, adjectives and adverbs to conduct sentiment analysis at the document level. Naive Bayes (NB), linear model and decision tree are some of the classifiers that have been used. Naïve Bayes performed best. Adjective-verb-adverb gave better results than any other combination of parts of speech tagging. Accuracy of 89.85% was achieved.

Paper [8] used Naïve Bayes, support vector machine and decision tree models to train and test the data. Using tenfold cross-validation, the accuracy gained was confirmed. SVM performed the best with accuracy of 81.77% whereas Naïve Bayes had a comparatively lower accuracy. The paper [9] examined the different types of approaches and the classifiers used for sentiment analysis. Traditional machine learning methods such as Naïve Bayes, SVM, decision tree and (KNN) K-nearest neighbour, deep learning based like GRU, LSTM and transformer based like—BERT and BERT large with UDA. The accuracy of BERT was a remarkable 95.22%. In paper [10], data pre-processing, analysis and classification are the main steps for the sentiment analysis model. Live reviews were fetched from Flipkart. Paper [11] investigated if a product's sentiment can be influenced by an artist's endorsement. Live reviews were fetched from YouTube and Twitter. Naïve Bayes rendered an accuracy of 84.50%.

Data obtained from OpinRank data set were used for performing sentiment analysis in [12]. The paper employed 4 algorithms, namely—Naïve Bayes multinomial, sequential minimal optimization, composite hypercubes on iterated random projections and compliment Naïve Bayes to build the model. The results showed a precision of 80.9% with Naïve Bayes multinomial outperforming the rest. In another study [13], the Twitter API was used to procure the tweets related to covid-19. Logistic regression, VADER and BERT have been used to classify the sentiment of the tweets. They obtained an F1-score of 0.92. Past works also tried and identified sentiments of viewers with respect to different cinemas. The authors of paper [14] aimed to find the best deep neural network to achieve a good classification on the movie reviews extracted from the IMDb data set. To distinguish and select a best neural network, three neural networks—convolution neural network (CNN), long short-term memory network (LSTM) and LSTM combined with CNN were used for feature extraction. Results indicated the suitability of applying CNN for sentiment analysis in comparison to other conventional methods and algorithms owing to its potential to recognise repetitive expressions in sentences.

The paper [15] used an augmented dictionary approach which means that only the domain-specific words are drawn from the review, and the polarity is assigned from the existing polarities for the words in the dictionary. The results showed an accuracy of 64.56%. The novelty of the paper was in the use of the augmented dictionary approach which outperformed other available methods. To present a thorough review of past literature, 30 such research papers were taken for the process in [16]. The outcome was to determine the best machine learning and feature extraction method on the postulation that the most admissible machine learning approach makes use of the most admissible feature extraction method. Out of the different machine learning algorithms—NB, and support vector machine were seen to be frequently used in past research, out of which SVM gave better accuracy most of the time. The feature extraction methods—n-gram and term presence have been frequently mentioned in literature as compared to other methods.

Paper [17] proposed a Fisher kernel method derived from a probability-based implicit opinion mining. The semantic analysis-based method can be deduced by using the Fisher function to evaluate the correlation between two entities on the derived framework and the mathematical framework. The obtained accuracy was 85.99%. The FK-SVM method overall performed better than PLSA-SVM. The use of neural networks was also seen in the past work. Paper [18] made use of convolutional neural networks (CNNs) for sentiment analysis of product reviews in Russian. Word2Vec was finalised for the vector input. Further, the paper did not make use of any heterogeneous features. The criticism on top-ranked products from the popular Russian e-commerce site was used to train the model. A F1-score of 75.45% was obtained. Training the model using reviews in the Russian dialect and emotion icons is what made the work unique. The improvement in the classification score can be credited to the use of emoticons.

In study [19], heterogeneous features like emoticons and SentiWordNet scores, negation and intensifier handling have been used as input for the sentiment analysis model which uses classifiers such as Naive Bayes and linear support vector machine to output the sentiment of the movie review. The results showed an accuracy of 89%. Naive Bayes algorithm performed the best. One drawback is that they have only tested basic performance metrics like accuracy, precision and recall. More advanced metrics like F1-score and roc plot should be used to get an accurate idea about performance of the algorithm. In paper [20], Twitter tweets for 5 products were collected using the Twitter API, and then, the WEKA software was used to calculate the sentiment score by applying classifiers like logistic regression, Naive Bayes and support vector machine. SVM with an accuracy of 72.6% performed far better than Naive Bayes and logistic regression on the data set. Paper [21] employed support vector machine on different data sets containing reviews of smart phones to analyse the underlying sentiments of the feedback given by the buyer. The paper highlighted the effectiveness of support vector machine owing to its robustness and accuracy of 90.99%.

Paper [22] analysed sentiments of user feedbacks on Windows phones and their classification using support vector machine. Further, the paper presented a comparison using different tokenisation and stemming techniques. The precision obtained with unigram, bigram and trigram was not up to the mark as frequency token was not adequate to be used as a distinguishing feature. However, use of n-gram proved to be a significant contributor to the rise in the value of accuracy. Maximum accuracy of 87.70% was achieved. Paper [23] assessed the opinions expressed over the reviews of online products from Amazon. The paper used four classifiers to analyse the reviews, i.e. logistic regression, Bernoulli Naive Bayes, random forest and multinomial Naive Bayes. The paper highlighted that the random forest with an accuracy of 93.17% performs best amongst all the algorithms. Using sentiment classification, the paper [24] investigated the use of machine learning approaches to detect positive and negative reviews collected from an e-commerce website for women's apparel.

The algorithms used for classification were SMO, Naïve Bayes, and JRip, J48. Out of the four algorithms used, SMO (SVM based) outperformed the best with an accuracy of 80.87%. In paper [25], sentiment analysis was performed on Amazon reviews data set making use of five classifiers, namely—support vector machine, random forest, Naïve Bayes, and ensemble method of bagging and boosting. Random forest, with an accuracy 87%, performed the best.

In paper [26], a sentiment analysis model was used to categorise product reviews into three classes. Using five classifying algorithms—support vector machine, K-nearest neighbour, NB, and maximum entropy, decision tree—the paper aimed to find the most potent algorithm. The accuracies obtained were—NB = 84.8%, SVM = 81.4%, maximum entropy = 84.8%. Paper [27] proposed a step-by-step approach to perform sentiment analysis on Yelp's review data and used machine learning algorithms like Gaussian Naive Bayes, combination of Gaussian Naive Bayes and AdaBoost, and KNN. The results presented the accuracies for the algorithms used— NB = 86.7%, AdaBoost = 74%, KNN = 73.2%. Gaussian Naive Bayes performed best with an accuracy of 86.7%. The combination of unigram and bigram features helped train the model well.

After conducting research in the concerned domain, logistic regression, Naïve Bayes and random forest seemed to be apt for classification of sentiments.

3 Methods

3.1 Architecture

Python is used for model building. The front end is built using react. The backend web application framework used for Nodejs is express. Material UI and framer-motion are used for designing. Beautiful Soup and the Requests Library are used to build the scraper. AWS amplify and Heroku are used to deploy the front end and back end, respectively.

Figure 1 illustrates the data flow through the framework of the web application. The model is initially trained using a data set of reviews with their already identified sentiment.

The user is initially provided with a dropdown where he can select a product. The selected product's url is sent to the back end which calls the scraper. Then, the scraper scrapes the given url and writes the data into a csv file. The three machine learning algorithms run on the csv file which predicts the sentiment of each review. Then, the csv file is converted into a json structure and is returned to the front end where the results are displayed. Figure 2 gives the UML diagram for the same.

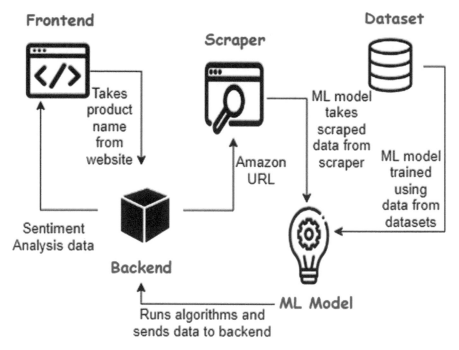

Fig. 1 Proposed data flow diagram

3.2 Algorithms

Logistic Regression

Logistic regression is essentially used for categorisation purposes. Unlike linear regression, logistic regression makes use of the sigmoid function to give a probabilistic output and hence is used to predict outputs for categorical variables [20]. The sigmoid function takes in the continuous and discrete inputs and restricts them to the range of 0–1, thus giving a probabilistic output. The maximum value that the function can give is 1 and the minimum 0. Hence, graphically, it forms an S curve called the "sigmoid function" [23].

SVM

In SVM, the main aim is to find a best suited separation boundary or a hyperplane. Initially, all points are plotted, and SVM tries to segregate them or generate a separation boundary. Cases of nonlinear data can be handled using a kernel function [4, 8, 20]. The best boundary is the one with the largest margin. This margin is the separation between the decision boundary and the support vectors, which are the points dominating the hyperplane's emplacement.

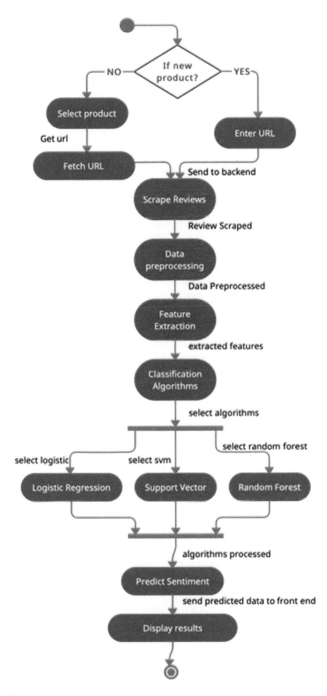

Fig. 2 UML diagram

RF

Random forest classifier is a machine learning algorithm utilised for the purpose of regression as well as classification. A forest comprises of many trees, and so, instead of making use of a single decision tree, it makes use of multiple such trees and improves the performance [23]. In this process, each decision tree provides some output, and the final result is decided on the basis of a majority vote [25]. A high accuracy can be achieved by increasing the number of trees in the classifier.

3.3 Implementation

Data Set

The training data set is taken from Kaggle [28]. Data set is a subset of Amazon Review 2018 data set which includes consumer reviews for category electronics. The data set contains 19,809 rows and 5 columns. A few rows of the data set can be seen in Fig. 3. After cleaning the data, a split of 80–20 is performed for training and testing, respectively.

Data Pre-processing

The process of cleaning and prepping text for classification is known as data pre-processing. It reduces the amount of noise and assists the algorithm to perform better [22]. This process includes the following steps:

- *Removing punctuations*: Punctuations become important for humans to understand the structure of any sentence but are futile to a computer. Thus, the punctuations are safely removed from the data by making use of string.punctuation in Python. The string.punctuation is a string already defined in Python which contains commonly used punctuation symbols [26].

	overall	vote	reviewTime	reviewText	summary
0	2	0	2010-02-10	Tech support is the worst	1265760000
1	2	0	2016-10-24	Screws were missing from the bracket and beaut...	Spend a little more and get much better.
2	1	0	2017-07-10	Trouble connecting and staying connected via b...	1499644800
3	4	5	2013-05-02	I purchased this unit for our RV to replace an...	Receiver Offers a Lot of Flexibility & Complexity
4	3	0	2013-01-04	It works. Nuff said but the review requires 1...	It's a cable

Fig. 3 Amazon review 2018 data set [28]

- *Removing Stop words*: Stop words are removed after tokenization. They are frequently used words that do not add anything to the context of the sentence, i.e. they are sentiment neutral [24, 26].
- *Stemming*: Stemming in layman terms is a way of cutting down a word into its simplest form. For every variation with every root word, more memory is required. Porter stemmer was initially used for stemming [26].
- *Lemmatization*: Lemmatization is a process akin to stemming with one distinguishing factor [26]. Lemmatization always returns a dictionary word and is therefore more precise, but it is also more computationally costly. The results show that lemmatization combined with POS tagging was more effective than lemmatization alone. The method of giving specific labels to each token in text to denote its part of speech is known as POS or part-of-speech tagging [21].
- *Vectorization*: Vectorization is the technique of encrypting text as integers in order to construct feature vectors. This procedure is used to translate text into an understandable format for the machine learning algorithms. The TFIDF vectoriser was used since it gave the best results [23].

Review Scraping

Review scraping is the process of scarping only the text or the content that is required or desired from the entire HTML web page. This data can be used for analysis later on.

Steps are as follows:

(a) Feed the website URL to the scraping algorithm.
(b) The scraper will then send an http request to the website. In Python, the requests library is particularly useful for this task.
(c) The website will send an http response.
(d) This response will be formatted with the help of the Beautiful Soup library.
(e) Then, using various built-in string functions in Python, the required reviews can be sliced.

Pagination will need to be used effectively to ensure the desired number of reviews getting scraped. Once the reviews get scraped, they can be inserted into a data frame in Python. The data frame can then be saved as a csv file for further pre-processing and analysis.

Front End

The predicted data are then converted into json which is sent back to the front end. Here, react is used to display all the results such as pie charts, bar charts and line charts. Sentiment highlights of a few random reviews are also shown. The code for the same is available on github [29].

Table 1 Performance metrics of algorithms

Algorithm	Precision	Recall	Accuracy	F1-score
LR	0.80	0.89	0.80	0.84
SVM	0.80	0.89	0.80	0.84
RF	0.77	0.91	0.79	0.83

4   Results and Discussions

4.1   Performance Comparison of Algorithms

Table 1 compares LR, SVM and RF using precision, recall, accuracy and F1-score. Out of the three selected classification algorithms, logistic regression and SVM perform the best as this is a binary classification problem. The data set contained ratings in the range of 1–5. To perform binary classification, the ratings are segregated into 2 classes 0 (ratings from 1, 2, 3) and 1 (ratings from 4, 5) due to which the data get imbalanced. Due to this reason, F1-score is a more reliable metric as compared to accuracy. With an F1-score of 0.84, logistic regression and SVM prove to be the best choices for a binary classifier.

The below mentioned formulae are used to mathematically compute the values of precision, recall, accuracy and F1-score.

$$Precision = TP/(TP + FP)$$

$$Recall = TP/(TP + FN)$$

$$Accuracy = TP + TN/(TP + TN + FP + FN)$$

$$F\ measure = 2 \times TP/(2 \times TP + FP + FN), \ where$$

TP = count of True positives; TN = count of True negatives; FP = count of false positives; FN = count of false negatives

4.2   UI

Once a particular product is selected, then the in-depth analysis of that product and its performance can be viewed using the three machine learning models implemented as be seen in Fig. 4. If the first classifier that is logistic regression is selected, then the sentiment of the product over the months can be analysed through the graph and the bar chart. According to logistic regression, there is a 28% difference between positive and negative sentiments. The first ten product reviews are also highlighted on the right side with their respective sentiments.

The next feature is a text box, which is provided for the user to pass some text and the sentiment of the text—positive (happy face emoticon) or negative (sad face emoticon) will be predicted algorithm wise, e.g. If "what an awesome product" has been entered, then all the algorithms rightly calculate it as a positive sentiment, as visible in Fig. 5.

Fig. 4 Sentiment analysis of Mi LED TV reviews scraped from Amazon

Fig. 5 Live sentiment analysis text box

5 Conclusion and Future Research Direction

In today's world, the users buy products that appeal to them or something they connect to. In a developing country with new business ventures, it is crucial to analyse the customer's opinion towards a merchandise or service. By understanding this sentiment, a business is more likely to cater to its users' needs better and in turn also boost its revenues.

Sentiment analysis plays a vital role for businesses trying to gauge the consumer's views on their brand, merchandises and offerings. Through the analysis, the traders understand the true opinion of their customers. The aim of this paper is to propose a system for understanding the consumer opinion of products listed on Amazon using sentiment analysis and machine learning classifiers such as SVM, random forest and logistic regression. With an F1-score of 0.84, logistic regression and SVM prove to be better choices than random forest for a binary classifier.

In future, the web application can be made more sophisticated by using aspect-based sentiment analysis and fake review detection. A product recommendation system can also be built as an extension, and the application can be generalised for any reviews like movie or food reviews. Considering certain quality measures like review helpfulness might also improve the model accuracy.

References

1. Kaur, H., & Mangat, V. (2017, February). A survey of sentiment analysis techniques. In *2017 International conference on I-SMAC (IoT in social, mobile, analytics and cloud) (I-SMAC)* (pp. 921–925). IEEE.
2. Sudhir, P., & Suresh, V. D. (2021). Comparative study of various approaches, applications and classifiers for sentiment analysis. *Global Transitions Proceedings*.
3. Kottursamy, K. (2021). A review on finding efficient approach to detect customer emotion analysis using deep learning analysis. *Journal of Trends in Computer Science and Smart Technology, 3*(2), 95–113.
4. Jagdale, R. S., Shirsat, V. S., & Deshmukh, S. N. (2019). Sentiment analysis on product reviews using machine learning techniques. In *Cognitive informatics and soft computing* (pp. 639–647). Springer.
5. Safrin, R., Sharmila, K. R., Subangi, T. S., & Vimal, E. A. (2017). Sentiment analysis on online product review. *International Research Journal of Engineering and Technology (IRJET), 4*(04).
6. Fang, X., & Zhan, J. (2015). Sentiment analysis using product review data. *Journal of Big Data, 2*(1), 1–14.
7. Sultana, N., Kumar, P., Patra, M. R., Chandra, S., & Alam, S. (2019). Sentiment analysis for product review. *International Journal On Soft Computing, 9*(7).
8. Singla, Z., Randhawa, S., & Jain, S. (2017, June). Sentiment analysis of customer product reviews using machine learning. In *2017 International conference on intelligent computing and control (I2C2)* (pp. 1–5). IEEE.
9. Ghosh, S., Hazra, A., & Raj, A. (2020). A comparative study of different classification techniques for sentiment analysis. *International Journal of Synthetic Emotions (IJSE), 11*(1), 49–57.

10. Singh, S. N., & Sarraf, T. (2020, January). Sentiment analysis of a product based on user reviews using random forests algorithm. In *2020 10th International conference on cloud computing, data science & engineering (Confluence)* (pp. 112–116). IEEE.

11. Ramdhani, S. L., Andreswari, R., & Hasibuan, M. A. (2018, November). Sentiment analysis of product reviews using naive bayes algorithm: A case study. In *2018 2nd East Indonesia Conference on Computer and Information Technology (EIConCIT)* (pp. 123–127). IEEE.

12. Zvarevashe, K., & Olugbara, O. O. (2018, March). A framework for sentiment analysis with opinion mining of hotel reviews. In *2018 Conference on Information Communications Technology and Society (ICTAS)* (pp. 1–4). IEEE.

13. Nair, A. J., Veena, G., & Vinayak, A. (2021, April). Comparative study of twitter sentiment on COVID-19 tweets. In *2021 5th International Conference on Computing Methodologies and Communication (ICCMC)* (pp. 1773–1778). IEEE.

14. Haque, M. R., Lima, S. A., & Mishu, S. Z. (2019, December). Performance analysis of different neural networks for sentiment analysis on IMDb movie reviews. In *2019 3rd International conference on electrical, computer & telecommunication engineering (ICECTE)* (pp. 161–164). IEEE.

15. Yadav, S., & Saleena, N. (2020, October). Sentiment analysis of reviews using an augmented dictionary approach. In *2020 5th International conference on computing, communication and security (ICCCS)* (pp. 1–5). IEEE.

16. Haberzettl, M., & Markscheffel, B. (2018). A literature analysis for the identification of machine learning and feature extraction methods for sentiment analysis. In *2018 Thirteenth international conference on digital information management ICDIM* (pp. 6–11). IEEE.

17. Han, K. X., Chiu, C. C., & Chien, W. (2019, October). The application of support vector machine (SVM) on the sentiment analysis of internet posts. In *2019 IEEE Eurasia conference on IOT, communication and engineering (ECICE)* (pp. 154–155). IEEE.

18. Smetanin, S., & Komarov, M. (2019, July). Sentiment analysis of product reviews in Russian using convolutional neural networks. In *2019 IEEE 21st conference on business informatics (CBI)* (Vol. 1, pp. 482–486). IEEE.

19. Bandana, R. (2018, May). Sentiment analysis of movie reviews using heterogeneous features. In *2018 2nd International conference on electronics, materials engineering & nano-technology (IEMENTech)* (pp. 1–4). IEEE.

20. Nafees, M., Dar, H., Lali, I. U., & Tiwana, S. (2018, November). Sentiment analysis of polarity in product reviews in social media. In *2018 14th International conference on emerging technologies (ICET)* (pp. 1–6). IEEE.

21. Kumari, U., Sharma, A. K., & Soni, D. (2017, August). Sentiment analysis of smart phone product review using SVM classification technique. In *2017 International conference on energy, communication, data analytics and soft computing (ICECDS)* (pp. 1469–1474). IEEE.

22. Hidayah, I., Permanasari, A. E., & Wijayanti, N. W. (2019, July). Sentiment Analysis on Product Review using Support Vector Machine (SVM). In *2019 5th International conference on science and technology (ICST)* (Vol. 1, pp. 1–4). IEEE.

23. Shah, B. K., Jaiswal, A. K., Shroff, A., Dixit, A. K., Kushwaha, O. N., & Shah, N. K. (2021, January). Sentiments detection for amazon product review. In *2021 International conference on computer communication and informatics (ICCCI)* (pp. 1–6). IEEE.

24. Noor, A., & Islam, M. (2019, July). Sentiment analysis for women's e-commerce reviews using machine learning algorithms. In *2019 10th International conference on computing, communication and networking technologies (ICCCNT)* (pp. 1–6). IEEE.

25. Alrehili, A., & Albalawi, K. (2019, April). Sentiment analysis of customer reviews using ensemble method. In *2019 International conference on computer and information sciences (ICCIS)* (pp. 1–6). IEEE.

26. Salem, M. A., & Maghari, A. Y. (2020, December). Sentiment analysis of mobile phone products reviews using classification algorithms. In *2020 International conference on promising electronic technologies (ICPET)* (pp. 84–88). IEEE.

27. Faisol, H., Djajadinata, K., & Muljono, M. (2020, September). Sentiment analysis of yelp review. In *2020 International seminar on application for technology of information and communication (iSemantic)* (pp. 179–184). IEEE.

28. https://www.kaggle.com/magdawjcicka/amazon-reviews-2018-electronics?select=electronics_sample.csv
29. https://github.com/keshav263/sentinic

Security and Privacy in Social Network

**Abhinav Agarwal, Himanshu Arora, Shilpi Mishra, Gayatri Rawat,
Rishika Gupta, Nomisha Rajawat, and Khushbu Agarwal**

Abstract For many people, social media has become an essential part of daily life. While many people began by exchanging data in the form of text and images in the media sphere, others moved on to sharing test papers, coursework, and masterclasses in the academic domain and e-learning materials, marketing, and a performance of the business clientele in the amusement sphere as well as jokes, music, and recordings in the entertainment sphere. Even the tiniest of Internet users would prefer long-range social media to the current Internet culture because of its widespread use. Sharing personal information on social media may be fun, but it also demands a great deal of security and safety. Data about customers should be kept private if it is to be kept private.

Keywords Security · Social media · Data privacy · Security threats · Social engineering

1 Introduction

Social media's significance has grown to mind-boggling proportions. Personal communication networks such as Facebook, Instagram, Napster, and Myspace have exploded in popularity over the last several years, reaching over two billion members. Almost every PC-savvy people has at least one social media account, and they spend significant hours per day on social media each day. Social media platforms may be thought of as Web apps. Individuals may use precision strike interpersonal communication endeavors for a variety of reasons, including connecting with new contacts,

A. Agarwal
Department of CSE, Amity University, Jaipur, Rajasthan, India

H. Arora
Department of CSE, Arya Institute of Engineering and Technology, Jaipur, Rajasthan, India

S. Mishra · G. Rawat (✉) · R. Gupta · N. Rajawat · K. Agarwal
Department of CSE, Arya College of Engineering and Research Center, Jaipur, Rajasthan, India
e-mail: gayatri28rawat@gmail.com

© The Author(s), under exclusive license to Springer Nature Singapore Pte Ltd. 2023   569
S. Shakya et al. (eds.), *Sentiment Analysis and Deep Learning*,
Advances in Intelligent Systems and Computing 1432,
https://doi.org/10.1007/978-981-19-5443-6_43

reconnecting with previous companions, maintaining existing connections, establishing or advancing a business or endeavor, participating in discussions about a specific topic or simply having a good getting acquainted, and working more closely with other users. Attackers attack the victim's system to get their personal information by changing the files of local hosts on the system. The easy prey for the hackers to steal information is those accounts that always share their personal data publicly without any concern [1].

Social media is a medium of expression among data owners (data generators) and observers (end users) for digital communications that utilize OSN to form virtual communities [1]. A social media platform is a graph of relationships between individuals, organizations, and their group events. These people, organizations, and groups, for example, form the nodes of the graph, while the interactions between the users, organizations, and groups from the edges. An OSN is a Web-based platform that enables users to develop online platforms or interactions with others who share their ideas, hobbies, activities, and/or real-world connections [2]. In today's Internet environment, a plethora of different forms of social networking programs are free. Several of the most prevalent aspects of online communities are [2, 3].

Several organizations, for example, Twitter and Instagram, have a broad clientele, while others have specified inclinations in mind. LinkedIn, for example, has positioned itself as an expert network management site—profiles include résumé information, and meetings are organized to share questions and opinions with colleagues in related disciplines. However, MySpace is well-known for its emphasis on entertainment and other forms of entertainment. Additionally, there are interpersonal contact perks that have been specifically designed to reunite former classmates [4].

2 Social Network and Public Information

The great majority of people use social media to exchange information and remain connected with people they know. The core characteristic of multimedia is a friend finder, which enables users to search for people they know and then build their online platform [5, 6]. The majority of social media users reveal a great deal of personal information in their informal organization space. This data include segment information, contact information, notes, photographs, and recordings, among other things [7, 8]. Numerous individuals readily release their details without hesitation. As a result, social media platforms have amassed a massive amount of sensitive data. Additionally, users of informal organizations will often have a higher level of trust in other Internet users [9]. They will generally respond successfully to partner demands and believe the information sent to them by companions. Given social media's massive user base and database, as well as its inherent vulnerability, lengthy communication skills platforms have emerged as new magnets for digital miscreants [10, 11].

There are a plethora of social media platforms to choose from as shown in Fig. 1. The main point of social communications may shift over time since most

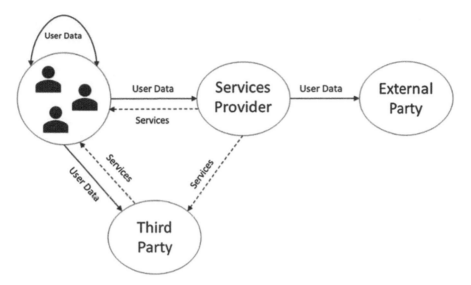

Fig. 1 Flow of users' data in social media objects

interpersonal groups integrate components from several platforms. Various systems can benefit from a wide range of security and defense recommendations. People's networks clients can create detailed online profiles and engage with other users, with an emphasis on social relationships, thanks to these technologies. Customers often share their gender identity, religion, hobbies, educational background, and employment information with other approved clients using these platforms, as well as documents and links to media, images, and recordings. People and apps that have not been approved as contacts can also be given access to selected data throughout these stages [12].

These sorts of virtual communities are designed to allow customers to publish brief announcements so that they may communicate with other customers quickly. Twitter, for example, focuses its services on providing quick, brief updates. The function of this layer is to quickly and publicly share information; however, privacy options may deny access to alerts [13]. We can build the virtual community by identifying the stakeholders and defining the goal and purpose of the virtual community. The second thing is to get a community platform and build a profile. After having a profile, one can develop the rules and can set up the community. At last, we can promote our community. These were the steps that can be used to build a virtual community using GPS-enabled mobile phones, location networks are becoming more prominent. If you are looking for a way to keep track of where you are at all times, these solutions can help you do so. An update made on one of these systems might be pushed out to the user's other informal groups as well.

Networks for distributing content: They are intended to be platforms for digital media distribution, such as music or images. These sites become unstructured organizations as well as data centers when they offer the ability to create personal brands, build up relationships, and connect with other customers via comments [14]. It can be understood as the process of exchanging, producing, and adding information to the different platforms of social media.

Mutually beneficial networks: Several networks were based on a common interest or geared at a certain group of people. A subset of people, such as those with roughly similar diversions, proposed conceptual models, political beliefs, ethnic principles, rigid viewpoints, sexual orientations, or other characterizing preferences are targeted by these systems, which combine the best features of various kinds of business associations.

3   Previous Workings and Findings

In light of the many dangers to user privacy and security in the SN environment, this review article examines a variety of data leakage scenarios involving SPs, users, and third parties, as well as external data links. On the other hand, this article examines information security focused on the social engineering concerns like phony accounts, identity theft, and spear phishing. Relevant research reveals a wide range of initiatives to address security and privacy vulnerabilities in the context of an SN environment, including both. This study presents and discusses a variety of dangers to information privacy and security, some of which are already accessible in surveys, as indicated in Table 1. However, the collection and formulation of these categories in single research are what make this review paper significant. Traditional Internet security risks like spam, malware, and DDoS assaults have been the focus of earlier research; thus, they are not discussed in this analysis.

The search parameters of this research article are to obtain relevant publications written in English and issued from 2009 to 2020. Scholarly articles, IEEE Xplore and ScienceDirect, are among the scientific databases and Internet-based articles that have been used in the search. Using the keywords listed in Table 1, we were able to do our search.

Table 1 Previous findings and research provided by other authors

Ref	Year	Social engineering threats		
		Identity theft	Fake accounts	Spear phishing
[18]	2018	Partial discussed	✓	Partial discussed
[22]	2013	–	–	–
[13]	2017	Partial discussed	✓	Partial discussed
[21]	2010	✓	✓	Partial discussed
[17]	2015	Partial discussed	–	–
[19]	2017	Partial discussed	✓	✓

4   Threats from Social Engineering in Social Networks

There is a wide range of dangers to the SN environment, but social engineering by attackers plays a major part in endangering information security. Through deception and persuasion, social engineering assaults attempt to get sensitive information from their victims [15]. As a result of the capacity to combine social and technological techniques, social engineering assaults are more complicated than other security threats.

By using the high value of user information that is available in SNs, social engineering attackers may get victims' information and execute their assaults [16].

However, the risk rises considerably when the victim's location is divulged and a physical crime is attempted [17]. Phishing, vishing, baiting, shoulder surfing, dumpster diving, and watering holes are some of the most common forms of social engineering in the SN environment.

4.1   Identity Theft

The next paragraph discusses a more risky fraud tactic, known as identity theft, which follows the discussion of phony accounts in the preceding section. Using a genuine person's identity may be accomplished either by cloning the victim or by infiltrating their account. On the same social network as the victim, the attacker utilizes publicly published information of a user to create a new profile with a nearly identical name and picture and then tries to convince the victim's friends that the new page is really a new profile for the same person (the victim) [18]. When a user's information from one social network is used to create a new profile on another, the victim is tricked into thinking they are the same person. Getting to know the victim's friends and family opens the door to learning more about the victim and his or her life [19].

4.2   Fake Accounts

It is an SN profile with a fictitious identity. Attackers are drawn to SN platforms because of the ease with which they may join, without having to reveal their true names and at no additional expense. Fake profiles are often recommended by social networking sites (SNSs) to other legitimate users, but many of those users do not care to add those recommended fake profiles because they do not want their personal information shared with strangers, despite the fact that such behavior allows them to appear popular and has a large network [20]. Humans or social bots might be the perpetrators of an assault in this or any other online community. "Sybil attacks" refer to malicious acts carried out via the creation of huge numbers of false social network

(SN) profiles in order to propagate spam, malware, misinformation, or tamper with online voting results [21].

4.3 Spear Phishing

To avoid being detected, spear-phishing targets particular individuals inside an organization, as opposed to the general public as is the case with regular phishing. Sophisticated network attacks, like spear phishing, go at a small set of SN users who have been pre-selected based on their individual characteristics. In this case, the SNs are used as a data source for an organization's staff and then targeted by messages on the SN platform or separate emails, asking them to open a URL that leads them to a fake Web site of that organization, where the victims may provide more sensitive data such as their login credentials. Individual and organizational information security will be compromised as a result of this incident [21].

To prevent phishing attacks, you can't use firewalls or complex authentication mechanisms like PINs to protect yourself since either the attacker utilizes the SN platform to send messages or the SN serves as the source of the victim's personal information.

5   General Security Measures

When providing information on the social network, a customer expects authorized contacts to be able to see it. Among the permitted data collectors [9], who can access it are as follows:

1. Advertisers are interested in local data, so they may target their ads to people most likely to be interested in the item.
2. Outside programmers combine data to create applications, such as Web games that integrate social media.
3. Character hackers who get personal data from clients or others.
4. Online criminals who aim to deceive or irritate people or infect systems with malware.

With the growing usage of social media, many users have unwittingly exposed themselves to risks to their safety and security. The main categorization comprises typical threats, including safety and security threats to users on Facebook and netizens who do not utilize informal networks [20]. The following classification includes contemporary hazards, that is, threats that are unique to the social networking world and use the social networking framework to jeopardize client privacy.

Malware is harmful software designed to disrupt a computer's operation and steal a user's credentials and personal data. Ransomware in social networking sites uses the social network model to spread among customers and their friends. With these skills, malware may sometimes mimic the client and transmit infected information to the client's Internet friends [21].

Phishing Attacks: Spear phishing is a type of social deception that impersonates a trusted third party. A current study found that social media consumers are more susceptible to phishing attacks owing to their trusting character [22].

Online scams, Internet fraud, often known as digital extortion, refer to using the Internet to deceive or abuse others. Earlier, extortionists used traditional in-person groupings, including weekly group sessions, to gradually build solid connections with their future victims.

Swindlers: A fake profile is a pre-programmed profile that mimics human behavior in OSNs. Generally, fake profiles may be used to obtain users' confidential data from online networks. The chatbots can collect a client's data and send companion requests to other OSN clients who accept them [16].

6 Security Practices for Social Media

Information on small networks can be used for many applications other than those given by the customer. Listed below are some helpful ideas to help clients minimize security risks when using interpersonal organizations. These suggestions are not fully effective. The long-distance conversation has its own set of obstacles. Mindfulness, alertness, and skepticism are arguably the most solid defenses [17].

In the registration process, social networks commonly ask another consumer for an email passcode to access the customer's email directory. The group promises to join the new user with some other users on the network. Some unofficial groups collect a plaintiff's email connections and then repeatedly invite individuals to participate. These communications may appear to have been from the initial client [17, 18].

Tips for Using Social Media:

For doing the authentication methods on the SMN make sure that our privacy encryption is tuned to "Custom Only." Alternatively, use the "Custom" options to achieve the highest level of protection. Keep an eye on modifications to social media service agreements and cybersecurity. Caution while clicking on truncated links. When in doubt, use a URL expander [17]. When you exit a social networking site, delete all cookies, particularly streak goodies. Remember that anything that happens on a network is viewed by persons, not the demographic profile. Don't publish negative comments about any business, clientele, products, or services. Keep an eye out for photos on social media. Don't be afraid to un-tag yourself and ask for material removal. Avoid making travel plans too far in advance. Users should be careful while publishing any form of location or using geotagging features since criminals may use it to track their whereabouts. Similarly, avoid sharing the daily routine [18].

Use caution when using third-party apps. Maintain a strategic distance from them completely for maximum protection and stability. When considering using one, check the platform's security measures and terms of service. If you get a request to connect with someone and you don't know who they are, don't accept it. Consider phoning, emailing, or presenting an inquiry to which only your contact may respond [13].

If an outsider approaches you for a connection, you should decline. If you accept the request, utilize security settings to limit what data are shown to others and avoid providing personal data such as your current location. Be wary of monetary requests, even from known contacts [16].

Avoid possible danger if you are a victim of stalking, incitement, or domestic abuse. If the account's security is compromised, notify the network and the other contacts.

Keep pruning the companion's list as after a while, it is easy to forget the connections, and therefore, sharing of data can be harmful.

7 Conclusion

Vast Web users spend more time on social media than they do on any other online activity, according to a recent study. On different social media platforms, we appreciate the chance to communicate with people through the exchange of experiences, images, and videos with our followers. In their shady underbelly, social networks are filled with hackers, criminals, and cyberstalking, who are all likely to exploit the platform provided by social media to recruit new victims for their schemes. In this essay, we will analyze the factors that weaken social media network members and put their identities, security, and financial well-being at risk both in the digital and actual worlds.

The acceptance of our ideas and the education of themselves, and also their relatives, about online dangers, is a motivating factor for social network members. As protectors, we cannot afford to be naive; we must be able to distinguish the charms of unstructured societies and be aware of the hazards that lurk behind the surface. Every effort is being made to inform and educate our children about possible dangers, as well as to train them to avoid interacting with strangers, whether in the actual or digital world.

References

1. Zephoria. (2017). The top 20 valuable Facebook statistics, Zephoria2017. https://zephoria.com/top-15-valuablefacebook-statistics/
2. Twitter. (2017). Twitter usage/company facts, Twitter2017. https://about.twitter.com/company
3. Protalinski, E. (2012). 56% of employers check applicants' Facebook, LinkedIn, Twitter. http://www.zdnet.com/article/56-of-employers-check-applicants-facebookLinkedIn-twitter/
4. Lotan, G., Graeff, E., Ananny, M., Gaffney, D., Pearce, I., et al. (2011). The arab spring—the revolutions were tweeted: Information flows during the 2011 Tunisian and Egyptian revolutions. *International Journal Communication, 5*(31), 32–48.

5. Mishra, S., Singh, D., Pant, D., & Rawat, A. (2022). Secure data communication using information hiding and encryption algorithms. In *2022 Second International Conference on Artificial Intelligence and Smart Energy (ICAIS)*, pp. 1448–1452.
6. Arora, H., Soni, G. K., Kushwaha, R. K., & Prasoon, P. (2021). Digital image security based on the hybrid model of image hiding and encryption. In *2021 6th International Conference on Communication and Electronics Systems (ICCES)*, pp. 1153–1157.
7. Soni, G. K., Rawat, A., Jain, S., & Sharma, S. K. (2020). A pixel-based digital medical images protection using genetic algorithm with LSB watermark technique. In *Springer Smart Systems and IoT: Innovations in Computing*, pp. 483–492.
8. Soni, G. K., Arora, H., & Jain, B. (2020). A novel image encryption technique using arnold transform and asymmetric RSA algorithm. In *Springer International Conference on Artificial Intelligence: Advances and Applications 2019 Algorithm for Intelligence System*, pp. 83–90.
9. Kumar, M., Soni, M., Shekhawat, A. R. S., & Rawat, A. (2022). Enhanced digital image and text data security using hybrid model of LSB steganography and AES cryptography technique. In *2022 Second International Conference on Artificial Intelligence and Smart Energy (ICAIS)*, pp. 1453–1457.
10. Mail, D. (2011). Bank worker fired for Facebook post comparing her 7-an-hour wage to Lloyds boss's 4000-an-hour salary.
11. Vladlena, B., Saridakis, G., Tennakoon, H., & Ezingeard, J.-N. (2015). The role of security notices and online consumer behavior: An empirical study of social networking users. *International Journal of Human Computer Studies, 80*, 36–44.
12. Shibchurn, J., & Yan, X. (2015). Information disclosure on social networking sites: An intrinsic extrinsic motivation perspective. *Computers in Human Behavior, 44*, 103–117.
13. Adewole, K. S., Anuar, N. B., Kamsin, A., Varathan, K. D., & Razak, S. A. (2017). Malicious accounts: Dark of the social networks. *Journal of Network and Computer Applications, 79*, 41–67.
14. Facebook's new terms of service: we can do anything we want with your content. forever. Consumerist, 2009. http://consumerist.com/2009/02/15/facebooksnew-terms-of-service-we-can-do-anything we want-with-your-content-forever/
15. Krishnamurthy, B., & Wills, C. E. (2008). Characterizing privacy in online social networks. In *Proceedings of the First Workshop on Online Social Networks*, pp. 37–42.
16. Fire, M., Goldschmidt, R., & Elovici, Y. (2014). Online social networks: Threats and solutions. *IEEE Communication Survey Tutorials, 16*, 2019–2036.
17. Deliri, S., & Albanese, M. (2015). *Data management in pervasive systems* (pp. 195–209). Springer.
18. Ali, S., Islam, N., Rauf, A., Din, I. U., Guizani, M., & Rodrigues, J. J. P. C. (2018). Privacy and security issues in online social networks. *Future Internet, 10*, 114.
19. Rathore, S., Sharma, P. K., Loia, V., Jeong, Y. S., & Park, J. H. (2017). Social network security: Issues, challenges, threats, and solutions. *Information Sciences, 421*, 43–69.
20. *International Conference on Innovation, Management and Technology Research, Malaysia*, September 22–23, 2013, pp. 191–197.
21. Timm, C., & Perez, R. (2010). *Seven deadliest social network attacks.* Syngress Publishing.
22. Smith, O. (2013). Facebook terms and conditions: why you don't own your online life. http://www.telegraph.co.uk/technology/socialmedia/9780565/Facebookterms-and-conditions why-you-dont-own-your-online-life.html

Understanding Images of Tourist Destinations Based on the Extract of Central Sentences from Reviews Using BERT and LexRank

Da Hee Kim, Kang Woo Lee, Ji Won Lim, Myeong Seon Kim, and Soon-Goo Hong

Abstract The reputation of tourist attractions depends greatly on tourist reviews. Tracking the online evaluation of tourists is essential to grasp the reputation of a tourist attraction and provide better tourism services. We used text summarization techniques on online reviews of tourist attractions to investigate their representative images. Additionally, we also investigated the changes in the image of tourist attractions by season and year. Our research helps identify the image of tourist attractions and provide better resources to those who perform tasks related to tourism policy.

Keywords Tourist destinations image · Tourism destination online reputation · Online reputation · Extract representative sentence · Text summarization

1 Introduction

Online reviews have a huge influence on tourists' decisions on where and when they travel. Tourist reviews are also an indicator of an indicator of a destination's post-visit destination image (DI) because tourists write reviews of their experience based on their post-trip image of the destination [1].

The reputation of a tourist attraction depends heavily on the tourist reviews. Positive reviews will encourage people to visit, while negative reviews will deter their

D. H. Kim · M. S. Kim
Department of Computer Engineering, Dong-A University, 37, Nakdong-daero 550beon-gil, Saha-gu, Busan, Republic of Korea

D. H. Kim · K. W. Lee · J. W. Lim · M. S. Kim
Smart Governance Research Center, Dong-a University, 225 Gudeok-ro, Seo-gu, Busan, Republic of Korea
e-mail: ji1e@naver.com

S.-G. Hong (✉)
Department of Management Information System, Dong-a University, 225 Gudeok-ro, Seo-gu, Busan, Republic of Korea
e-mail: shong@dau.ac.kr

decision to visit the tourist destination. Therefore, tracking tourists' online evaluations is essential to the reputation of tourist destinations because online reviews directly affect tourists' visits. It is crucial to maintain and improve the reputation of tourist destinations for those who perform tasks related to the tourism policies to provide high-quality tourism services.

However, tons of reviews and comments shared by tourists can be overwhelming. Individually, analyzing a massive number of reviews is extremely time-consuming and laborious. Thus, effective data mining is needed to minimize the required hard work and effort. Valuable information can only be acquired by unveiling patterns through text mining and machine learning algorithms.

Text summarization is a method to reduce workload by extracting numerous reviews into several representative sentences. Text summarization is very fast and accurate in the tourism field to analyze large-scale data on tourist attractions, obtaining valuable knowledge without much effort. Moreover, knowing the evaluation of tourist attractions improves the quality of the tourism industry [2–4].

We extract representative sentences for tourist attractions in Busan, Korea, using LexRank, one of the techniques of text summarization, and BERT algorithms. Additionally, we investigate images by tourist destination, year, and season.

2 Extract Central Sentences

2.1 System Architecture

Figure 1 shows the architecture of the central sentence extraction system implemented in this study. It consists of a dataset, data preprocessing, and central sentence extraction process. A dataset was constructed based the tourist destination review data crawled by TripAdvisor. Next, the data preprocessing process was carried out through text cleaning and sentence split. In the central sentence extraction process, after embedding using 'Ko-Sentence-BERT-SKTBERT', cosine similarity and centrality score are calculated to extract the most central sentence in the data.

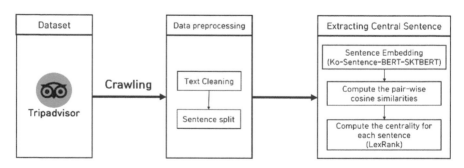

Fig. 1 System architecture

2.2 Data Crawling

The dataset in this paper was collected from TripAdvisor, a tourist information Web site [5]. The data crawled from the Web site contained tourists' evaluation of Busan, the second-largest city in South Korea. The data comprise tourist destination names, languages, dates, reviews, and ratings. We collected 27,752 reviews of tourist attractions in Busan, dated from August 2007 to May 2021; the data are written in Korean. The rating ranges from 1 (very negative) to 5 (very positive). The tourist destinations include 329 attractive tourist sites such as Haeundae, Gamcheon Culture Village, Haedong Yonggungsa Temple, among others. A review consists of an average of four or five sentences, more than fifteen sentences, and each sentence had an average of 15 words.

2.3 Data Preprocessing

- Text cleaning

  In the case of TripAdvisor's review data, there are many special characters owing to it being an online review. Unnecessary special characters or symbols such as '!, ?, ^, (,)' were removed, and spacing was corrected in the analysis process of the review data. Space correction was performed using the Python package for automatic Korean word spacing [6].
- Sentence split

  A tourist review is typically composed of multiple evaluations of a tourist attraction, and these evaluations are usually expressed in multiple sentences. Therefore, review data were split into sentences to accurately extract such evaluations. We performed sentence separation work using 'Korean Sentence Splitter (KSS)', a Python Korean sentence separator [7].

2.4 Embedding Sentence Using Ko-Sentence-BERT-SKTBERT

Embedding is a significant problem in natural language processing because computers must understand and process text efficiently. Language model pre-training is effective for improving many natural language processing tasks. The language model widely used for embedding is 'Bidirectional Encoder Representations from Transformers (BERT)'. The fine-tuning method in the existing pre-trained language model uses the unidirectional language model in the pre-transforming process. The unidirectional language model entails the context of the sentence being previously configured to move from left-to-right. This structure has the disadvantage of not understanding the context from both sides because it can only be attached to previous tokens and not the current tokens. BERT eased these unidirectional constraints by applying bidirectional encoder presentation. BERT is more suitable for sentence-level tasks compared with the existing pre-trained language model. BERT exhibits state-of-the-art performance on sentence-pair regression tasks like semantic textual similarity (STS) [8].

BERT supports multilingualism, but there is a limit to understanding the irregular characteristics of the Korean language. To alleviate this problem 'KoBERT' was introduced. 'KoBERT' was developed to overcome the performance limitations of the existing BERT model in the Korean language, exhibiting 2.6% better performance than BERT in Korean data [9]. The preprocessed data are embedded using the 'Ko-Sentence-BERT-SKTBERT' model, noticeably improving the performance of sentence embedding in the KoBERT model [10].

There are two types of models for evaluating the trained data in the 'Ko-Sentence-BERT-SKTBERT' model: 'training_nli' and 'training_sts'; 'training_nli' is only used for training NLI, and 'training_sts' is only for STS. Natural language inference (NLI) and semantic textual similarity (STS) are key tasks in natural language understanding (NLU). A model with NLI receives a pair of sentences, a premise, and a hypothesis, and it classifies their relationship into one out of three categories: entailment, contradiction, and neutral. Comparatively, STS is used to evaluate the semantic similarity between the two sentences and how well the model understands the intimacy of the two sentences in terms of meaning [11]. In this study, the similarity between several sentences is determined by embedding the 'training_sts' model.

2.5 Calculating the Centrality of Each Sentence

It is crucial to conceptualize the centrality for extracting a central sentence from multi-documents. The extracted central sentence would be different depending on how centrality is calculated. When calculating the centrality of a sentence, it is not reasonable to obtain it by degree centrality of a sentence alone. As all the connections to a sentence are not equally important, it is necessary to consider the degree centrality of the adjacent node and the importance of the adjacent nodes. PageRank is an algorithm that considers both the degree centrality and the importance of adjacent nodes [12]. The existing PageRank is an algorithm that represents a link to a Web page on the Internet in a direct graph. However, in the case of sentences, the similarity between sentences is represented as an undirected graph [13]. LexRank calculated the similarity between sentences using an undirected graph. LexRank assumes that the sentences similar to many of the other sentences in a cluster are more central (or salient) to the topic.

Since not all of the connections to a sentence are not equally important, it is necessary to consider not only the degree centrality of the adjacent node but also the importance of the adjacent nodes. PageRank is an algorithm that considers both the degree centrality and importance of adjacent nodes [12]. The existing PageRank is an algorithm that represents a link to a Web page on the Internet in a directed graph. However, in the case of sentences, the similarity between sentences is represented as an undirected graph [13]. LexRank calculated the similarity between sentences using an undirected graph. LexRank assumed that the sentences similar to many of the other sentences in a cluster are more central (or salient) to the topic.

A short description of how to calculate the centrality of each sentence in Lexrank is as follows. First, the idf-modified-cosine between sentences is calculated using (1), and a graph is constructed [13].

$$
\text{idf - modified - cosine}(x, y) = \frac{\sum_{w \in x, y} \text{tf}_{w,x} \text{tf}_{w,y} (\text{idf}_w)^2}{\sqrt{\sum_{x_i \in x} (\text{tf}_{x_i,x} \text{idf} x_i)^2} \times \sqrt{\sum_{y_i \in y} (\text{tf}_{y_i,y} \text{idf} y_i)^2}} \quad (1)
$$

Next, the centrality score of each sentence is calculated using the 'power method', one of the methods of obtaining 'stationary distribution' in the Markov chain.

Depending on whether the threshold is applied to LexRank, the algorithm is divided into continuous LexRank and classic LexRank. The classical LexRank limits the connectivity of nodes using the threshold value. In continuous LexRank, the connectivity of nodes is accomplished through the similarity of idf-modified-cosine without applying any threshold [13]. As the classic LexRank causes information loss due to a threshold, we used a continuous LexRank model without any threshold.

The central sentence extraction algorithm calculates the centrality score of each tourist destination review data and returns the sentence with the largest score. The result of embedding **Corpus C** into the 'Ko-Sentence-BERT-SKTBERT' model is **Corpus_Embedding E**. Calculates cosine similarity between the results of embedding. Next, use LexRank's 'degree_centrality_score' function to calculate the centrality score for each sentence. As mentioned, the threshold is not used in LexRank. Extract the most central sentence of the sentences based on the obtained *centrality_score*.

Algorithm 1 Central Sentence Extraction Algorithm

1: **Input: Corpus C, Corpus_Embedding E**
2: *cos_scores = cosineSimilarity(E, E);*
3: **function** *degree_centrality_scores (cos_scores)*
4: **return** *centrality_score*
5: **end function**
6: *centrality_score = degree_centrality_scores (cos_scores*, threshold = None*)*
7: *most_central_sentence_indices = sort_by_index(-centrality_score)*
8: **return** *C[most_central_sentence_indices]*

---

3 Results

3.1 Extracting Representative Sentences by Tourist Destination

The results of central sentence extraction are presented according to tourist destinations, years of visit, and seasons of visit. The tourist destinations include Haeundae, Gamcheon Culture Village, and Haedong Yonggungsa Temple—three representative tourist destinations in Busan.

Initially, the central sentence extraction process was done for Haeundae beach data. Haeundae beach, located in Busan, Korea, is adjacent to many other tourist attractions, such as Dongbaekseom, Dalmaji Pass, and it offers several facilities, including hotels and restaurants [14].

Among the online reviews of Haeundae, the five most central sentences were extracted using the algorithm. In Fig. 2, the most central sentence is 'Haeundae in Busan. It is so good to look at the sea of Haeundae on a nice day. It is so refreshing and good to come out around Dongbaekseom'. There was also a review of Haeundae beach being called the best beach as well as the best coast, but most of the reviews, referred to Dongbaekseom, located near Haeundae. Hence, the centrality score is high for the reviews related to Dongbaekseom. Therefore, the sentences related to Dongbaekseeom were extracted as representative sentences. Haeundae is generally evaluated in terms of 'the clean and beautiful sea'. Also, Haeundae is appreciated for having modern and tall buildings equipped with facilities such as hotels, restaurants, galleries, and shopping centers.

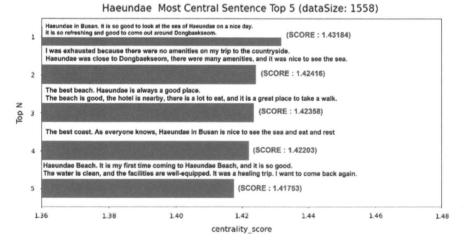

Fig. 2 Haeundae online reputation top 5. Multiple English sentences are generated from a single Korean sentence via Korean-to-English translation

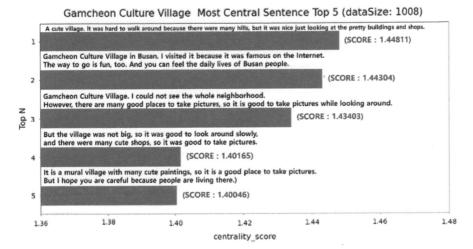

Fig. 3 Gamcheon culture village online reputation top 5

Figure 3 shows the five most central sentences concerning Gamcheon Culture Village. Gamcheon Culture Village was built when refugees settled in the Korean War. As refugees flocked, a village was inevitably formed on the hillside. However, since the early 1970s, the population has seen a sharp decrease. Revitalization efforts included decorating the exterior of the buildings with cute paintings and murals, eventually becoming the symbol of this village in later years [15]. The most central sentiment sentence of Gamcheon culture village is 'A cute village. It was hard to walk around because there were many hills, but it was nice just looking at the pretty build ings and shops'. Most of the sentences selected as central in the tourist destination were extracted, such as those that are good for taking pictures or village with many cute pictures, which generally come to mind when thinking of Gamcheon Culture Village.

Figure 4 shows the five most central sentences concerning Haedong Yonggungsa Temple. The temple is introduced as the most famous temple in Busan with beautiful scenery, according to Wikipedia [16]. The most central sentiment sentence of Haedong Yonggungsa Temple is 'It is a beautiful temple where I went so well. I wondered if there was a place like this. The scenery is so beautiful because there is a temple in front of the sea. Moreover, this is where my heart gets quiet, and I want to visit again'. The Haedong Yonggungsa Temple is located near the sea, making it is a beautiful temple with good scenery. Although many people visit the temple, it was evaluated as a temple worth it. Consequently, all five extracted sentences speak in a similar context.

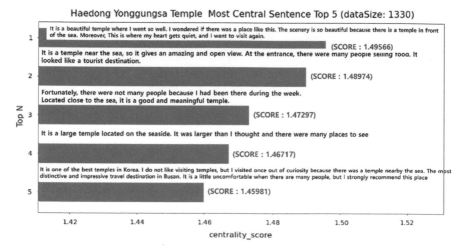

Fig. 4 Haedong Yonggungsa temple online reputation top 5

3.2 Extracting Representative Sentences from Tourist Destinations by year

The central sentences of the reviews of Haeundae beach were extracted year-by-year to investigate the change in the general image of the beach over time. The result is presented in Fig. 5. Their periods were divided 'prior-to-2018', 2018, 2019, and 'post-2020'. The review data size of 'prior-to-2018' and 'post-2020' was quite small and was therefore grouped. In particular, the review data in 2020 and 2021 were particularly small due to the decrease in the number of tourists after the COVID-19 outbreak.

In general, Haeundae beach is known as 'the best beach surrounded by hotels and restaurants, and close to other tourist sites, but crowded'. The central sentences, extracted from reviews—prior-to-2018, 2018, and 2019—describe that Haeundae is the best beach (in prior-to-2018 and 2018 reviews), and that facilities, including hotels and restaurants, are well equipped (in prior-to-2018, 2018, and 2019). Unlike prior-to-2018, 2018, and 2019, reviews of the Haeundae Light Festival were extracted in post-2020.

3.3 Extracting Representative Sentences from Tourist Destinations by Seasons

Figure 6 shows that the most central sentences of Haeundae were extracted from the reviews of 4 different seasons—spring (from March to May), summer (June to August), autumn (September to November), and winter (December to February).

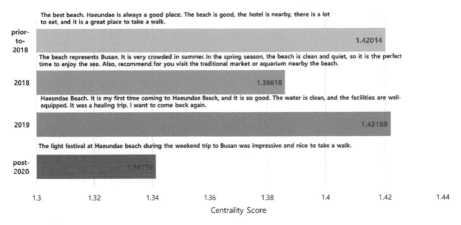

Fig. 5 Haeundae most central sentence by year

Haeundae beach has multiple facets that change with different seasons. Summer is the busy season for Haeundae beach when many tourists visit. The beach in spring is quiet and clean, while it is romantic in autumn. However, it presents a lonely atmosphere in winter.

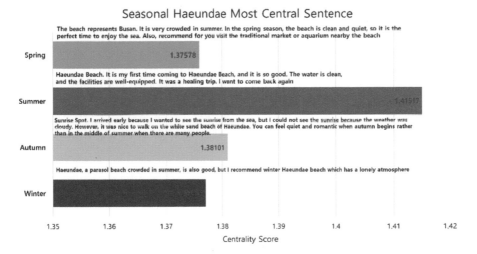

Fig. 6 Seasonal Haeundae most central sentence

4 Conclusion and Discussion

Using text summarization techniques—LexRank and BERT algorithms—we extracted representative sentences about tourist attractions in Busan, Korea, to investigate the destination images. Additionally, tourist reviews by destination, year, and season were analyzed. First, the central sentence was extracted according to tourism destinations. The destination image (DI) of Haeundae, Gamcheon Culture Village and Haedong Yonggung Temple was similar to the generally assumed images. Next, Haeundae's annual central sentence was extracted. The reviews of prior-to-2018, 2018, and 2019 described Haeundae as the 'best beach', a generally thought of image of Haeundae. However, Haeundae's review extracted in post-2020 extracted an unusual sentence related to the Haeundae Light Festival. This extraction was unique because access to Haeundae beach was restricted post-2020 due to COVID-19 quarantine rules. As a result, reviews related to the sea were rare post-2020. Before COVID-19, light festivals were not popular, but data on Haeundae's light festivals were extracted as an extraordinary result because it became impossible to visit the beach. Finally, the central sentence of Haeundae by season was extracted. In the case of Haeundae, summer is the busy season. However, in the case of autumn or winter, unlike the existing Haeundae, there is a romantic and lonely atmosphere, indicating various image facets experienced in the same tourist destination.

Therefore, it is crucial to identify, maintain, and improve the reputation of tourist attractions for those who perform tasks related to the tourism policy. This study can easily grasp the image of a tourist destination and identify potential improvements and characteristics of the tourist destination. In addition, through the annual analysis of tourist destinations, tourism policy managers can track changes in the perception of tourist destinations by year and grasp the image of tourist destinations by season. By grasping the changes in the image of tourist destinations according to the season, it is possible to understand the optimal season of the tourist destination and help marketing.

Acknowledgements This work was supported by the Ministry of Education of the Republic of Korea and the National Research Foundation of Korea (NRF-2018S1A3A2075240).

References

1. González-Rodríguez, M. R., Martínez-Torres, R., & Toral, S. (2016). Post-visit and pre-visit tourist destination image through eWOM sentiment analysis and perceived helpfulness. *International Journal of Contemporary Hospitality Management.*
2. Prameswari, P., Sujandari, I., Zulkarnain, & Laoh, E. Mining online reviews in Indenosia's priority tourist destinations using sentiment analysis and text summarization approach.
3. Iinuma, S., Nanba, H., & Takezawa, T. Automatic summarization of multiple travel blog entries focusing on travelers' behavior.

4. Premakumara, N., Shiranthika, C., Welideniya, P., Bandara, C., Prasad, I., & Sumathipala, S. (2019) Application of summarization and sentiment analysis in the tourism domain. In *International Conference for Convergence in Technology.*

5. TripAdvisor. https://www.tripadvisor.co.kr/. Last accessed February 11, 2022.

6. PyKoSpacing. https://github.com/haven-jeon/PyKoSpacing. Last accessed February 11, 2022.

7. KSS: Korean Sentence Splitter. https://github.com/likejazz/korean-sentence-splitte. Last accessed February 11, 2022.

8. Devlin, J., Chang, M. W., Lee, K., & Toutanova, K. (2018) BERT: Pre-training of deep bidirectional transformer for language understanding.

9. KoBERT. https://sktelecom.github.io/project/kobert/. Last accessed February 11, 2022.

10. Ko-Sentence-BERT-SKTBERT. https://github.com/BM-K/KoSentenceBERT_SKT. Last accessed February 11, 2022.

11. Ham, J., Choe, Y., Park, K., Choi, I., & Soh, H. (2020). KorNLI and KorSTS: New benchmark datasets for Korean natural language understanding.

12. Page, L., Brin, S., & Motwani, R. (1998). The pagerank citation ranking: Bringing order to the web, technical report. Stanford University.

13. Erkan, G., & Radev, D. R. (2004). Lexrank: Graph-based lexical centrality as salience in text summarization. *Journal of Artificial Intelligence Research, 22,* 457–479.

14. Haeundae. https://en.wikipedia.org/wiki/Haeundae_District. Last accessed February 11, 2022.

15. Gamcheon Culture Village. https://en.wikipedia.org/wiki/Gamcheon_Culture_Village. Last accessed February 11, 2022.

16. Haedong Yonggung Temple. https://en.wikipedia.org/wiki/Haedong_Yonggungsa. Last accessed February 11, 2022.

Comparative Study of Machine Learning and Deep Learning for Fungi Classification

Nandakishor Velu, Sanjay Rojar Utharia Muthu,
Nitheesh Kumar Narasimmalu, and Madheswari Kanmani

Abstract The aim of this paper is to compare and contrast between deep learning (DL) and various machine learning (ML) algorithms for fungi classification. The Danish Fungi data set provided by Kaggle, for this study. Only, 10 classes from the provided data set were extracted which consists of 1775 images. In this work, the used machine learning techniques are decision tree (DT), Naive Bayes (NB), K-nearest neighbour (KNN) and random forest tree (RFT) and achieved accuracies of 25, 28, 29 and 33%, respectively. The reason for low accuracies for the machine learning algorithms is because machine learning algorithms are usually used for numerical data and not suitable for images. Deep learning model using Keras was used to achieve an accuracy of 75.82%. On comparing the quantitative metrics like precision, recall, f1-scores, it can be concluded that deep learning algorithms are much better than machine learning algorithms.

Keywords Fungi classification · Deep learning · Machine learning · Decision tree · Naive Bayes · k-nearest neighbour · Random forest · Convolutional neural networks · Accuracy metrics

N. Velu (✉) · S. R. U. Muthu · N. K. Narasimmalu · M. Kanmani
Department of Computer Science and Engineering, Sri Sivasubramaniya Nadar College of
Engineering, Chennai, India
e-mail: nandakishor2010608@ssn.edu.in

S. R. U. Muthu
e-mail: sanjayrojar2010085@ssn.edu.in

N. K. Narasimmalu
e-mail: nitheesh2010343@ssn.edu.in

M. Kanmani
e-mail: madheswarik@ssn.edu.in

© The Author(s), under exclusive license to Springer Nature Singapore Pte Ltd. 2023   591
S. Shakya et al. (eds.), *Sentiment Analysis and Deep Learning*,
Advances in Intelligent Systems and Computing 1432,
https://doi.org/10.1007/978-981-19-5443-6_45

1 Introduction

Fungi are a group of microorganisms like mould, yeast, rusts, mildew and mushrooms that grow freely in soil or water. Unlike plants, fungi do not photosynthesise and absorb food from dissolved molecules. Certain fungi have a relationship that is symbiotic with other organisms. Fungi have great significance in the carbon cycle and are an active initiator of fermentation. Fungi have a very important impact in medicine and the environment. Penicillin and many antibiotics are synthesised using fungi. Each fungi has their own importance to the environment, and in medicine industry, some fungi are also dangerous and cause harm to humans and other animals. In some cases, fungi might even cause fatal damage depending on the kind of fungi, so it is all the more reason for classifying fungi correctly and precisely, to find out which fungi are useful and which are not. Fungi are identified by noting the morphological characteristics of mushrooms such as their size, colour, presence or absence of volva, stipe, ring, scales, reticulum, zonation, striation, warts, cap, areolae and gills. Transverse sections of the sporocarps are prepared. Mycelial growth and morphology of dried spores are examined microscopically, to identify and classify different fungi classes.

Classification helps in identifying images and organising information. It helps the researchers order useful biodiversity data. Deep learning is used to process unstructured data, and its algorithms are used to determine the most important features to distinguish between objects. Researchers used deep learning and neural networks to classify images with their ID. Since fungi are highly important in the industrial, agricultural and environmental sectors, research has emerged to classify fungi. This project focusses on classifying fungi using deep learning techniques and machine learning techniques. Concepts of deep learning, like convolution neural networks, were applied to distinguish between various fungi by their attributes and associate them into 10 different classes. Rest of the paper is organised as follows: Sect. 1 deals with introduction; Sect. 2 deals with literary survey which gives an overview on the previously published work; Sect. 3 explains the proposed work; Sect. 4 discusses the result obtained; Sect. 5 deals with the conclusion and scope for future works, and finally, contains the references.

1.1 Background, Motivation and Objective

The motivation for our work is to determine which is better for image classification, deep learning or machine learning. Our objective is to prove why deep learning is more suitable for image classification. Machine learning algorithms are way faster than deep learning and suitable for well-structured data but for deep learning is more advanced and more accurate on unstructured data like image classification, but the one down side is that deep learning models are usually take large computational time and power than their counterparts.

2 Related Works

Alessia et al. [1] have developed the random forest model for the earliest diagnosis of Alzheimer disease, and the authors have been successful in doing it [2–6]. The authors were able to predict Alzheimer disease with the huge data set the authors had. The authors considered all limitations of random forest algorithms to predict the early progression of multi-cognitive impairment to Alzheimer disease.

Kong et al. [7] have also used the random forest model classification for the gene expression data. Sparsity of effective features with unknown correlation structures in gene expression brought more risks to handle this classification task. Whilst using a random forest model, Yunchuan was able to overcome these risks. Working on the gene data set with this model gave him relatively higher classification accuracy compared to other models.

Andy and Christopher [8] developed random forest model classification on a compound's quantitative or categorical biological activity based on a quantitative description of the compound's molecular structure. The authors developed using the bootstrap of training data set and random features of the data. The authors used six data sets so that the model could predict the data accurately. Prediction is made by the aggregation of the average prediction that is the majority vote. The authors were able to conclude that random forest model classification was better suited for this problem statement.

Zhang et al. [9] used random forest model classification for transmission line images. During classification, output of each base classifier has the same sort of error. To overcome it, this model is developed by the idea of deleting inferior. The authors obtained higher accuracy and stability in the three data sets.

Xiang and Junhao [10] modulated random forest model classification for predicting employee turnover. The data set is critical that not only sustainability of work, it also has the continuity of enterprises. Now, this model focuses on forecasting employee turnover. It offers an analytic model that predicts employee turnover more accurately than any other model, and it reduces employee turnover intention.

Deepika and Ranjan [11] found a major problem in the plants which led to less productivity and huge loss. So, the authors focussed on a model to overcome the problem that was a random forest classification model. By this model and the various data set, the authors used, the authors were able to project the disease of the crop at the earliest stage. The authors were able to achieve an accuracy of 80.68%. Thus, the authors proposed a random forest model to predict the disease in the plants.

Radko and Ayyub [12] approached a random forest classification model using copula-based variants. Based on this model, the authors were able to classify a label-valued outcome. Their model is enabled to select only the relevant features connected by a linear prototype. The authors were even able to execute the study with Covid-19 and for diabetes data sets. It helped them to predict the nonlinear data accurately.

Yuita and Fitri [13] generated a random forest classification model to explore foods. People found it difficult to enjoy the local food when the authors travel places. So, the authors tried to automatically identify food images. The authors extracted the colour moments RGB, LAB and HSV spaces of different food images and classified it with different texture and colour using class estimators. The authors were successful enough to predict accurately with an accuracy of 93.5% using 500 estimators.

Ebrahimi et al. [14] proposed a detection method based on images that use the SVM classification method to identify pests. The authors developed a technique using image processing to detect thrips in a strawberry greenhouse. SVM with difference kernel function was used for identification. Using the SVM method with intensified colour and region index, the authors obtained an average error percentage of less than 2.25%.

Mathur et al. [15] proposed a supervised classification analysis that uses SVM for training and classification users. The authors used a one-shot approach that had an accuracy of 92.00% and binary classification-based approaches that yielded an efficiency of 89.22 and 91.33%.

Inglada [16] developed a classification algorithm for the recognition of man-made objects which were not clearly seen in lower resolutions using a supervised learning approach. An algorithm was developed using SVM that uses geometric image features to identify classes of man-made objects that had a precision rate of 80%.

Bolbol et al. [17] developed a solution for analysing and obtaining the mode of transportation using SVM classification without user inputs or assumptions. It first uses speed and acceleration to discriminate between variables and then SVM classification. It has been tested using GPS data with a precision percentage of 88%.

Kumari [18] proposed a method of using brain scan images (using MRI) to detect abnormalities. The authors came up with an algorithm to identify abnormalities and tumours in the brain to differentiate between patients. The authors used feature extraction with GLCM and SVM classification and obtained an accuracy of 98%.

Li and Guo [19] proposed two active learning strategies to integrate into multilabel SVM classification. The authors used a label cardinality inconsistency strategy and a max-margin prediction uncertainty strategy. Their results showed the effectiveness of instance selection strategies.

Mavroforakis and Theodoridis [20] proposed SVM classification using the geometric framework. The authors use the "reduced convex hull" to solve the problem. It is used efficiently to solve non-separable problems.

Baten et al. [21] developed an algorithm using SVM classification and probabilistic parameters to identify the splice sites. MM1 was used in the first stage, and SVM classification was used in the second stage. This method was highly accurate, efficient, and fast machine learning is an artificial intelligence technique that makes the computer systems learn and make predictions based on the available data, previous history and examples. Mohammad.

Ashraf et al. [22] neural networks (NN), decision tree (DT) and K-nearest neighbour (KNN) to classify the mushrooms into two categories such as poisonous and non-poisonous. The authors extracted the features such as Eigen, histogram, and parametric features from the data set. The authors applied different machine learning models for classification and achieved high accuracy 94% using KNN.

Babu et al. [23] used support vector machine and Naïve Bayes algorithms for classification of mushrooms. The authors achieved better results for SVM.

Al-mejibli et al. [24] built a mobile application called Mushroom Diagnosis Assistance System with the aim of realising safety whilst gathering mushrooms. The authors applied decision tree and Naïve Bayes classifiers to classify.

Beniwal et al. [25] applied Naïve Bayes to classify several mushroom diseases. The authors used metrics such as accuracy, kappa statistics and mean absolute error to predict the quality of the classification algorithms. Finally, the authors find that Bayes net gives the lowest mean absolute error and highest accuracy and then Naïve Bayes.

The authors in [26] applied the Weka tool that has many machine learning techniques that can be used for pre-processing, analysing, classification and predicting the given images and predicting the given data. The authors used the Wrapper method and Filter method in the Weka tool to select the best attributes for the mushroom classification. Decision tree is applied on the selected key attributes to classify the images.

Agung [27] also used eka tool for feature selection and decision tree (C4.5), Naïve Bayes and support vector machine (SVM) for image classification.

Karunakaran [28] applied deep learning algorithm to classify domain generation algorithms. The neural network developed by the author is a 3-layered neural network, namely input, hidden and output layer. The authors extracted special features and additional knowledge base feature extraction which improvised the accuracy of prediction.

Jacob [29] applied capsule network (Caps-Net) architecture for text classification. The author compared the results of single task learning and the multi-task learning and the CNN-based MTL, RNN-based MTL and Caps-Net-based MTL and concluded that the Caps-Net-based gated MTL provided the better results compared to the other methods in text classification.

2.1 Limitation of Related Works

Random forest algorithms are faster to train but are slower for predictions. Upon usage of a random forest classifier, to obtain high accuracy, the number of trees must be greater which thus increases the time to predict in real-time applications. So, for an application which needs high run-time performance, random forest is not preferred. Also, a random forest classifier requires high computational powers along with resources as it generates many trees to combine their outputs.

Support vector machine (SVM) does not suit for larger data sets as the time taken to train increases to a point where training and usage of the model becomes infeasible due to constraints. SVM classifiers classify data based on the location of the feature points with respect to classifying hyperplane. There are high possibilities for SVM models trained on imbalanced data set to predict the data points on the decision boundaries of the hyperplane as negative. Also, there are no probabilistic reasons for the prediction.

KNN algorithm does not generate model from the input data set instead the whole data are considered to be the model for this algorithm. Hence, for large data sets, KNN algorithms are inefficient. The cost of computing the distance between the new point and all of the existing points to make prediction is extremely high if the data set is huge. Also, there is a need of feature scaling in all dimensions on the data set before using KNN, without which the model may produce incorrect predictions.

One of the implicit assumption of the Naive Bayes classifier is the conditional independence of features (all features are mutually independent), which is mostly not the case in real-world applications. This drawback limits the use of the model in real-world applications. Also, the algorithm faces a "zero-frequency problem" where the model assigns a zero probability when it encounters a feature/category variable in the test set which wasn't encountered by the model during training with the train data set.

3 Materials and Methods

3.1 Data Set

The fungi data set consists of 295,938 training images which belong to 1604 species. The data were collected mainly in Denmark. The test data set contains 59,420 observations with 118,676 images. Along with the data set, two metadata csv files have been provided which includes meta information on each image. From this, 1775 images belonging to 10 classes have been derived. The data set was downloaded from Kaggle.

3.2 Proposed Work

The workflow begins by loading the data set given by Kaggle, from which only 10 classes were selected, which is used for both training the model as well as testing the trained model. The input images are then pre-processed for model training. Due to the lack of image for some classes, data generators are used to produce more images to improve the quality of the model. A sequential model is constructed using various layers. This is then trained using the input images. The trained model is then used to predict the test images from the provided test data set, which returns the predicted output Class Id. The metrics are calculated and compared with machine learning models, which are trained using sklearn. The proposed workflow is depicted in Fig. 1.

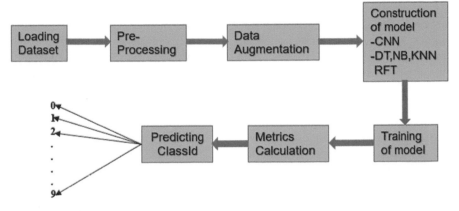

Fig. 1 Workflow

3.2.1 Data Pre-Processing and Data Augmentation

Image pre-processing is the step done to format images even before the images are used for model training. In this process, images are resized, enhanced, converted to grey scale and oriented for proper utilisation in the model. This is done by using histogram equalisation and greyscale numbers. Histogram equalisation is a technique that helps to adjust the contrast of an image.

The cdf is a sum of all probabilities lying in its domain and defined by:

$$\mathrm{cdf}(x) = \sum k = -\infty x \mathrm{P}(k) \tag{1}$$

Cumulative distribution function cdf(x) is used to evaluate the probability of a real-value variable x is less than or equal to x.cdf(x) is the summation of $P(k)$ probability of limits between 0 to k.

Technique to perform histogram equalisation:

$$\mathrm{Sk} = (L - 1)\mathrm{cdf}(x) \tag{2}$$

Histogram equalisation (Sk) is used to contrast images in image processing.

Sk is the product of total number of grey levels L and cumulative distribution function cdf(x).

Grey scale is to evaluate the value of each pixel through intensity of light in the image. Grey scale is the average value of red R, green G and blue B from the pixel matrix of the image.

$$\mathrm{Grey\ scale} = (R + G + B)/3 \tag{3}$$

Data augmentation is the process of creating more relevant data from the existing data which is achieved by using image data generators. Data augmentation is very helpful when the size of the data set is less. Data augmentation helps in generating more relevant data for a large data set. It generates more images with the existing images. The various methods used for data augmentation are flip—it flips the image vertically or horizontally, scaling—scaling an image is used to zoom in or out based on the type of scaling used, translation—translation shifts the image to the left or right, and rotation—it is used to rotate the image by some degree between (0–360) degree. The main reason for the necessity of data augmentation is because some of the classes have smaller training sample sizes compared to other classes. For example—the ClassId 4 has 436 images whereas the ClassId 7 has only 32. The graph in Fig. 2 shows the relation between the number of images and the classId.

The number of images is augmented 4 times to generate more images. The number of images before and after augmentation is given in Table 1.

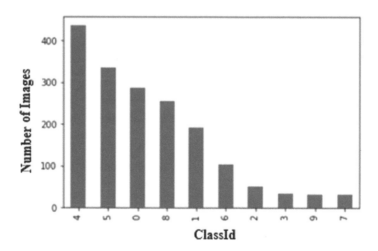

Fig. 2 Graph depicting relation between classId and number of images

Table 1 Number of images before and after augmentation

ClassId	Before augmentation	After augmentation
0	288	1152
1	192	768
2	52	208
3	35	140
4	436	1744
5	336	1344
6	105	420
7	32	128
8	256	1024
9	33	132

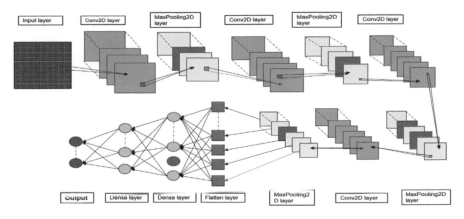

Fig. 3 Sequential model

3.2.2 Building Model

For classifying fungi into classes, 1 deep learning model and 4 machine learning models were trained, whose accuracy is calculated and compared to determine which algorithm performs better. The model used for classification is CNN; the main reason for choosing behind CNN is feature engineering is not required. Feature selection is necessary before training the CNN model. CNN performs well and gives better accuracy. It covers local and global components. It also learns different parts from images.

CNN Model—DL

A sequential model is a method to build a model using Keras which allows construction of the model layer by layer, where each layer has one input tensor and one output tensor. Each layer is stacked together in a sequence to form the model; hence, it is called a sequential model. Some of the advantages are, it is easier to track progress, construct, and understand. Sequential model is depicted in Fig. 3.

Decision Tree Model—ML

Decision tree is developed to split a node into a number of sub-nodes. This split helps in most homogeneous sub-nodes. The algorithm selection is based on the kind of target variables the data set possesses. This is depicted in Fig. 4.

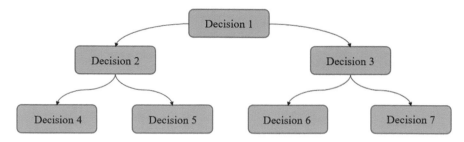

Fig. 4 Decision tree

Naive Bayes Model—ML

The Naive Bayes classifier gives the conditional probability. Bayes theorem used to calculate the conditional probability of event A, given that event B has happened.

$$P(A|B) = \frac{P(B|A)P(A)}{P(B)} \tag{4}$$

This classifier predicts probabilities for each data belonging to a particular class. It then comes to a point that the class with the highest probability is taken to be the most like class.

KNN Model—ML

The *K*-nearest neighbour algorithm is one of the simple and highly supervised machine learning algorithms to solve problems based on classification and regression. This model works by finding distance between a given query and all the taken examples in the data set near to the given query. This is depicted in Fig. 5.

Decision Tree Model—ML

Random forest classification deals with various number of decision trees on the given data set and predicts the accuracy of that data set from the average values to improve the prediction. It is based on the majority votes of each tree.

Higher accuracy is obtained when greater numbers of trees are used to predict. This even protects from overfitting.

Random forest trees work in different phases, i.e. creation of the model by combining various numbers of decision trees, and then, the prediction for each tree created in the first phase is proceeded. This is depicted in Fig. 6.

Fig. 5 K-nearest neighbour

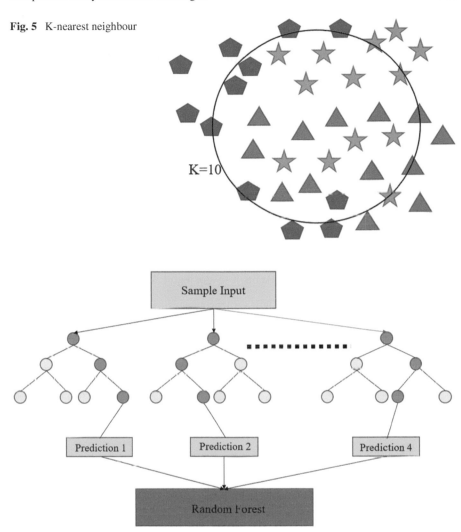

Fig. 6 Random forest

3.3 *Quantitative Metrics*

Quantitative metrics are a way to calculate the accuracy of a model and decide which model is better suited. There are three popular metrics provided by sklearn, they are mentioned below:-

3.3.1 Precision

Precision metrics quantify the number of correct positive predictions made. The following equation is used to calculate precision

$$\text{Precision} = \frac{\text{True Positive}}{\text{True Positive} + \text{False Positive}} \tag{5}$$

Precision is the ratio of number of true positives to the sum of number of true positives and number of false positives. This formula is used when the cost of a false positive is very high, and the cost of a false negative is low.

3.3.2 Recall

Recall metrics quantify the number of correct positive predictions made out of all positive predictions that could have been made. The following equation is used to calculate recall-

$$\text{Recall} = \frac{\text{True Positive}}{\text{True Positive} + \text{False Negative}} \tag{6}$$

Recall is the ratio of number of true positives to the sum of number of true positives and number of false negatives. This formula is important where false negatives (Overlooked cases) are more costly than false positives (False alarms).

3.3.3 F1 Score

F1 score is the mean between precision and recall. The following equation is used to calculate F1 score.

$$\text{F1 score} = 2 * \frac{\text{Precision*Recall}}{\text{Precision} + \text{Recall}} \tag{7}$$

F1 score is double the ratio of the product of precision and recall to the sum of precision and recall. This equation takes the harmonic mean of precision and recall and converts it into one metric, which is used to compare the performance of models.

4 Results and Discussion

Jupiter notebook (Python) was used for training the model and classifying the images into classId. The train data set was split into 80 and 20% for validation. The sequential

Table 2 Comparison of metrics

	CNN	Decision tree	Naïve Bayes'	KNN	Random forest tree
Accuracy	75.82%	25%	28%	29%	33%
Precision macro avg	–	0.19	0.27	0.25	0.19
Precision weighted avg	–	0.26	0.35	0.51	0.56
Recall macro avg	–	0.19	0.23	0.22	0.17
Recall weighted avg	–	0.25	0.28	0.29	0.33

model was trained using 100 epochs and attained an accuracy of 75.8%, whereas all the ML models returned less accuracy compared to the CNN model.

The reason for low accuracy for the machine learning model is the closeness of the species, and the number of images for each class differs a lot. Machine learning models are not great at classifying images because of less feature extraction from the images, especially when all the classes are very closely connected, and there are no distinct feature that separates them apart. The proposed workflow describes the steps used to classify fungi into 10 classes, which starts with pre-processing the images using histogram equalisation, which is followed by data augmentation, this step is essential for generating new data set from existing data set that is not commonly used. The metric used in the model is the "accuracy" metric for CNN and precision, recall and f1 for ML models. The metrics for various methods are depicted in Table 2.

The predicted images with their respective classes are shown in Fig. 7.

5 Conclusion and Future Work

This paper discusses how different machine learning and deep learning models are used to classify fungi. On implementing and in-depth researching, it is concluded that deep learning is way better than machine learning algorithms in distinguishing images. This paper establishes that CNN is better in classifying images than classical machine learning algorithms, but there is always a scope of improvement. There are many different deep learning algorithms which might give a higher accuracy than CNN, and there is always room to groom in the field of deep learning.

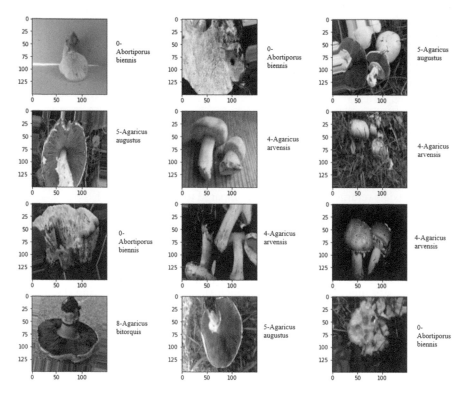

Fig. 7 Predicted images

References

1. Sarica, A., Cerasa, A., & Quattrone, A. Random forest algorithm for the classification of neuroimaging data in alzheimer's disease: A systematic review 9. https://doi.org/10.3389/fnagi. 2017.00329
2. Kanmani, M., & Narasimhan, V. An optimal weighted averaging fusion strategy for remotely sensed images. *Multidimensional Systems and Signal Processing, 30*(4), 1911–1935.
3. Kanmani, M., & Narasimhan, V., Optimal fusion aided face recognition from visible and thermal face images. *Multimedia Tools and Applications, 79*(25), 17859–17883.
4. Madheswari, K., Venkateswaran, N., & Ganeshkumar, N. Entropy optimized contrast enhancement for gray scale images. *International Journal of Applied Engineering Research.*
5. Madheswari, K., & Venkateswaran, N. Particle swarm optimization aided weighted averaging fusion strategy for CT and MRI medical images. *International Journal of Biomedical Engineering and Technology.*
6. Nathan, S. S., Kanmani, S., Kumar, S., & Kanmani, M. (2018) AP/CSE, published a paper titled, Survey on digital age- smarter cradle system for enhanced parenting. *International Journal of Applied Engineering Research* 13(10), 8187–8193. ISSN 0973-4562.
7. Kong, Y., & Yu, T. (2018). A deep neural network model using random forest to extract feature representation for gene expression data classification. *Scientific Reports, 8*, 16477.
8. Svetnik, V., Liaw, A., Christopher Tong, J., Culberson, C., Sheridan, R. P., & Feuston, B. P. (2003). A classification and regression tool for compound classification and QSAR modelling. *Journal Chemical Information of Computer Science, 43*(6), 1947–1958.

9. Bingzhen, Z., Xiaoming, Q., Hemeng, Y., & Zhubo, Z. (2020). A random forest classification model for transmission line image processing. In *2020 15th International Conference on Computer Science & Education (ICCSE)*. https://doi.org/10.1109/ICCSE49874.2020.9201900

10. Gao, X., Wen, J., & Zhang, C. (2019) An improved random forest algorithm for predicting employee turnover *2019,* Article ID 4140707. https://doi.org/10.1155/2019/4140707

11. Chauhan, D, Walia, R., Singh, C., Deivakani, M., & Kumbhkar, M. (2021). Detection of maize disease using random forest classification algorithm *12*(9). https://doi.org/10.17762/turcomat.v12i9.3141

12. Mesiar, R., & Sheikhi, A. (2021). Nonlinear random forest classification, a copula-based approach. *Applied Science, 11*(15), 7140. https://doi.org/10.3390/app11157140

13. Sari, Y. A., Utaminingrum, F., Adinugroho, S., Dewi, R. K., Adikara, P. P., Wihandika, R. C., Mutrofin, S., & Izzah, A. (2019). Indonesian traditional food image identification using random forest classifier based on color and texture features. In *2019 International Conference on Sustainable Information Engineering and Technology (SIET)*. https://doi.org/10.1109/SIET48054.2019.8986058

14. Ebrahimi, M. A., Khoshtaghaza, M. H., Minaei, S., & Jamshidi, B. (2017). Vision-based pest detection based on SVM classification method. *Computers and Electronics in Agriculture, 137*, 52–58. ISSN 0168-1699.

15. Mathur, A., & Foody, G. M. (2008). Multiclass and Binary SVM Classification: Implications for Training and Classification Users. *IEEE Geoscience and Remote Sensing Letters, 5*(2), 241–245. https://doi.org/10.1109/LGRS.2008.915597

16. Inglada, J. (2007). Automatic recognition of man-made objects in high resolution optical remote sensing images by SVM classification of geometric image features. *ISPRS Journal of Photogrammetry and Remote Sensing, 62*(3), 236–248. ISSN 0924-2716.

17. Bolbol, A., Cheng, T., Tsapakis, I., & Haworth, J. Inferring hybrid transportation modes from sparse GPS data using a moving window SVM classification. *Computers, Environment and Urban Systems, 36*(62012), 526–537. ISSN 0198-9715.

18. Kumari, R. (2013). SVM classification an approach on detecting abnormality in brain MRI images. *International Journal of Engineering Research and Applications (IJERA), 3*(4), 1686–1690. ISSN: 2248-9622.

19. Li, X., & Guo, Y. (2013). Active learning with multi-label SVM classification. IjCAI. http://citeseerx.ist.psu.edu/viewdoc/download?doi=10.1.1.417.4976&rep=rep1&type=pdf

20. Mavroforakis, M. E., & Theodoridis, S. (May 2006). A geometric approach to Support Vector Machine (SVM) classification. *IEEE Transactions on Neural Networks, 17*(3), 671–682. https://doi.org/10.1109/TNN.2006.873281

21. Baten, A., Chang, B., Halgamuge, S., et al. (2006) Splice site identification using probabilistic parameters and SVM classification. *BMC Bioinformatics, 7,* S15. https://doi.org/10.1186/1471-2105-7-S5-S15

22. Ottom, M. A., Alawad, N. A., Nahar, & K. M. O. (2019). Classification of mushroom fungi using machine learning techniques. *International Journal of Advanced Trends in Computer Science and Engineering, 8*(5).

23. Babu, P., Thommandru, R., Swapna, K., & Nilima, E. (2014). Development of Mushroom Expert System Based on SVM Classifier and Naive Bayes Classifier. *International Journal of Computer Science and Mobile Computing, 3*(4), 1328–1335.

24. Al-Mejibli, & Hamed Abd, D. (2017). Mushroom diagnosis assistance system based on machine learning by using mobile devices Intisar Shadeed AlMejibli University of Information Technology and Communications Dhafar Hamed Abd Al-Maaref University College, 9(2), 103–113. https://doi.org/10.29304/jqcm.2017.9.2.319

25. Beniwal, S., & Das, B. (2015). Mushroom Classification Using Data Mining Techniques. *International Journal of Pharma and Bio Sciences, 6*(1), 1170–1176.

26. Vanitha, V., Ahil, M. N., & Rajathi, N. (2020). Classification of mushrooms to detect their edibility based on key attributes. *Bioscience Biotechnology Resources Communication, 13*(11), 37–41.

27. Wibowo, A., Rahayu, Y., Riyanto, A., & Hidayatulloh, T. (2018). Classification algorithm for edible mushroom identification. In *2018 International Conference on Information and Communications Technology (ICOIACT)*.
28. Karunakaran, P. (2020). Deep learning approach to DGA classification for effective cyber security. *Journal of Ubiquitous Computing and Communication Technologies (UCCT), 2*(04), 203–213.
29. Jacob, I. J. (2020). Performance evaluation of caps-net based multitask learning architecture for text classification. *Journal of Artificial Intelligence, 2*(01), 1–10.

Improved Tweet Sentiment Analysis by Features Weight Optimize by GWO and Classify by XG-Boost

Pankaj Kumar(ID) and **Monika**(ID)

Abstract The emergence of Twitter as a significant social media network and its strong interest among sentiment analysis experts are big talking points. Despite its recent focus, state-of-the-art Twitter sentiment analysis methods are known to perform very badly with claimed classification accuracies that are frequently around 70%, which negatively impacts the ability to use the sentiment information generated from the techniques. This study looks at the difficulties involved with Twitter, research the literature to find out whether the solutions were addressed challenges. We perform a benchmark assessment to evaluate the state-of-the-art in Twitter sentiment analysis. In this paper analysis of sentiment 140 dataset and classify by XG-Boost with feature optimization by XG-Boost and GWO. Result shows that in proposed approach accuracy improves 2–3%, precision 4%, recall 5%, and F-sore3%.

Keywords Twitter sentiment analysis · Emotions · Opinions · Machine learning · Support vector machine · Natural language processing

1 Introduction

Abstract feelings and convictions, such as feeling, appraisal, and pitch, have a huge impact on human behavior. Because expressing other's opinions is built into every individual naturally and presents us as 'social creatures', our decisions are heavily influenced by others' perceptions of the world. Social media platforms such as Facebook, YouTube, and Twitter have exploded in popularity in recent years [1–3]. Sentiment Analysis is another name for Opinion Mining, which falls under the data mining and machine learning categories. Sentiment analysis and Opinion mining approaches must begin with people's data for the research of a distinct kind of field such as economy, politics, or biology, among other things [4]. Each day, massive amounts of data relating to distinctive single entities are stored in digital formats.

P. Kumar (✉) · Monika
Department of Information Technology Engineering, UIET, Panjab University, Chandigarh, India
e-mail: pankajchauhan.jnv@gmail.com

© The Author(s), under exclusive license to Springer Nature Singapore Pte Ltd. 2023   607
S. Shakya et al. (eds.), *Sentiment Analysis and Deep Learning*,
Advances in Intelligent Systems and Computing 1432,
https://doi.org/10.1007/978-981-19-5443-6_46

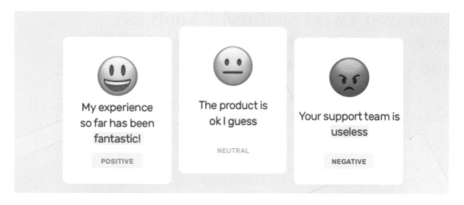

Fig. 1 Sentiment analysis

As a result, the field's rapid expansion coexists alongside other social media-related content like blogs, forums, consumer reviews, Twitter, and social networking sites.

Sentiment analysis [5] entails categorizing data into positive, or excellent sense, or negative, or terrible sense, or neutral, or non-effective, categories. The task of identifying polarity (negative, neutral, positive) as well as emotions and feelings (angry, pleased, sad, etc.), immediacy (not urgent, urgent), and sometimes even intentions is called sentiment analysis] (interested versus not interested) as shown in Fig. 1. The amount of information on social media is unorganized due to free text formatting and easy access to microblogging platforms [5]. Whenever it comes to a conclusion or producing a final product, it is critical to seek out the opinions of others. From many people's experiences and reviews, which provide valuable information. With the application of opinion mining that fulfills individual preferences; significant work has already been done inside the promotion and demanding domains [6]. For improved performance, a proper training set is necessary, as well as an accurate database for good textual analysis.

2 Related Work

There is a lot of research being done in the field of sentiment analysis. We give a thorough assessment of features using several supervised learning methods, as Jonnathan and Alexandre [1] did with an important issue commencing of twenty-two databases of tweets. The researchers look at a large number of meta-features that have been studied in recent publications, as well as a large number of pre-trained word embedding models. Antonakaki et al. [7] shows a map of current Twitter research issues, focusing on three primary areas: social graph structure and features, sentiment analysis, and dangers such as spam, bots, fake news, and hate speech. Pandya and Mehta [8] uses diverse ML and Lexicon investigation approaches to reveal the aftereffects of testing. The outcomes are analyzed in order to conduct an empirical assessment and verify the current composition estimation. In this way, future investigators will be better able to comprehend current beginnings in the design of potential examination. Manguri et al. [9]. Throughout this study, twitter data was extracted from the Tweet social media platform that use the Python programming language and the Tweepy module, and SA was performed utilizing Python TextBlob package. Following the sentiment analysis, a graphic structure of the database has been presented. Joshi and Gupta [4] use a hybrid technique to tackle the challenge of Twitter sentiment analysis, using a SentiWordNet (SWN)-based feature vector as input to the classification model SVM. The main goal is to increase classification performance by handling lexical modifier negation throughout SWN score calculation. Jaiswal and Kumar [2] give a review of the literature in which they collect, examine, interpret, and assess attempts and patterns in a very well manner in order to discover research gaps that define the coupling's future potential. Tiwari et al. [3]. This project is based on the concept of TSA, which is a method for determining what a person thinks about a certain tweet they sent. Any individual can reach a conclusion after learning about other people's perspectives on a topic. NLP has been used in the current work to do these types of analyses. As a result, this study paper explains the many methodologies used as well as the procedure followed all throughout the whole process. The fourth year of the 'Sentiment Analysis in Twitter Task' is discussed by Nakov et al. [5]. Task 4 of SemEval-2016 consists of five subtasks, three of which differ significantly from prior editions. The first two subtasks are repeats from previous years, and they require you to forecast overall sentiment as well as sentiment toward a specific topic in a tweet. Two variations of the basic sentiment classification in Twitter problem are the focus of the three new subtasks. In this study, Idrissi and Rehioui [10] proposed new clustering algorithms for twitter sentiment analysis. Tripathi [11] performed Nepali COVID-19 tweet sentiment analysis using Long Short-Term Memory [LTSM] SVM and Bernoulli Naïve Bayes. Panadian [12] uses Deep Learning techniques for comparison and performance evolution of sentiment analysis. Meena et al. [13] performed sentiment analysis for cancer disease and treatment combination using google Book Search Engine and Pumbed, with the use NB algorithm. Khatun et al. [14] performed lexicon analysis on Amazon book review data, on the basis of user review comments and rating the quality of books and author is determined.

3   Proposed Work

This work proposes a sentiment analysis technique that improves the text features sparsity using an optimization approach. The optimization approach reduces the weight of sparse features and ignores these features by the classifier. After reduction of features learn by ensemble learner [15] using XG-Boost.

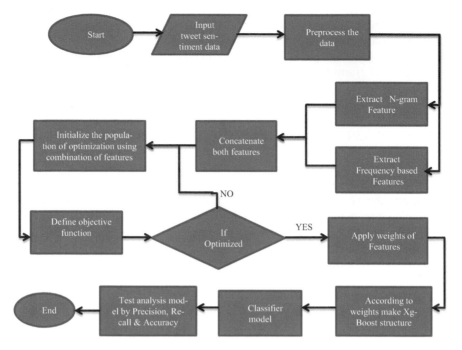

Fig. 2 Process flow for the proposed method

3.1 Dataset

Sentiment 140 [16] dataset with 1.6 million tweets can be collected from "https://www.kaggle.com/datasets/kazanova/sentiment140". The training dataset contains 1,600,000 tweets, i.e., includes 800,000 positive and 800,000 negative tweets, there is no neutral tweet. Testing is performed using 1,280,000 tweets data.

3.2 Proposed Methodology Flowchart

See Fig. 2.

3.3 Proposed Methodology

Collect the tweet dataset "sentiment140" [16] from Kaggle website published by Kazanova. Pre-process tweet text by tokenization and stemming process as shown in Fig. 2

After pre-processing the text change in features for features uses frequency and n-gram features; both features are concatenated. After that, features weight was optimized by AI-based optimization and reduced the sparsity in features. According to features weight learning by structure-based XG-Boost classifier and make classifier model classifies tweets sentiments. Analysis of the precision, recall, and accuracy of classifier models.

$$X(t+1) = \frac{x_1 + x_2 + x_3}{3} \tag{1}$$

$$D = |C * X(t) - X(t)| \tag{2}$$

$$X(t+1) = D`.e^{bl}.\cos 2\pi l + X * t \tag{3}$$

3.4 Proposed Methodology Algorithm

Step-1: Initialize population X_i $(I = 1,2,3……, n)$.
Step-2: Initialize a, A, and C.
Step-3: Calculate fitness of each search agent
$\qquad\qquad X_a =$ the best
$\qquad\qquad X_p =$ the second best
$\qquad X_\delta =$ the third best
Step-4: While $(t < \max no. of iterations)$
$\quad$ for each search agent.
$\quad$ update a, A, C, 1 and p
$\quad$ if 1 $(p < 0.5)$
$\quad$ if 2 $(|A| < 1)$
$\quad$ Update the position current search by Eq. (1)
$\qquad\quad else\, if\, 2\,(|A| \geq 1)$
$\quad$ Update the position current search by Eq. (2)
$\quad$ $end\, if\, 2$
$\quad$ $else\, if\, 1(p \geq 0.5)$
$\quad$ Update the position current search by Eq. (3)
$\quad$ $end\, if\, 1$
$\quad$ $end\, for$
$\qquad$ $check\, if\, any\, search\, agent\, goes\, beyond\, the\, search\, space\, and\, ammend\, it$
$\qquad\quad calculate\, fitness\, of\, each\, search\, agent$
$\qquad\quad update\, X_a, X_p, \& X_\delta$
$\qquad\quad t = t + 1$
$\quad$ $end\, while$
Step-5: $return X_a.$

GWO is inspired by the spotted hyena's encircling, attacking, hunting, and searching behaviors. The encircling property is simulated in the following manner [2, 5, 9].

4 Results

In experiments use sentiment 140 dataset and compare proposed approach with existing ensemble learning SVM-based approach.

Table 1 shows the metrics values (in percentage) for the precision, recall, and F-score for Ensemble Learning [17], PLSA-SVM and Optimize-XG-Boost.

Figure 4 shows experiment done and comparison of accuracy, precision, recall, and F-score for Ensemble learning, PLSA-SVM and Optimize-XG-Boost. Proposed approach accuracy improves, in proposed approach use XG-Boost with optimize feature. So its improve accuracy, precision, recall, and F-score. which shows in Fig. 3.

Table 1 Accuracy, precision, recall, and F-score for various approaches

Approaches	Accuracy	Precision	Recall	F-score
Ensemble learning	92	83.2	85.67	84.13
PLSA-SVM	75.81	75.79	76.63	73.23
Optimize-XTG-Boost	96	89.34	90.12	93.23

```
from nltk.stem.porter import PorterStemmer
def cleanup_tweets(tweet_df):
    # remove handle
    tweet_df['clean_tweet'] = tweet_df['TweetText'].str.replace("@", "")
    # remove links
    tweet_df['clean_tweet'] = tweet_df['clean_tweet'].str.replace(r"http\S+", "")
    # remove punctuations and special characters
    tweet_df['clean_tweet'] = tweet_df['clean_tweet'].str.replace("[^a-zA-Z]", " ")
    # remove stop words
    tweet_df['clean_tweet'] = tweet_df['clean_tweet'].apply(lambda text : remove_stopwords(text.lower()))
    # split text and tokenize
    tweet_df['clean_tweet'] = tweet_df['clean_tweet'].apply(lambda x: x.split())
    # let's apply stemmer
    stemmer = PorterStemmer()
    tweet_df['clean_tweet'] = tweet_df['clean_tweet'].apply(lambda x: [stemmer.stem(i) for i in x])
    # stitch back words
    tweet_df['clean_tweet'] = tweet_df['clean_tweet'].apply(lambda x: ' '.join([w for w in x]))
    # remove small words
    tweet_df['clean_tweet'] = tweet_df['clean_tweet'].apply(lambda x: ' '.join([w for w in x.split() if len(w)>3]))
```

Fig. 3 Tokenization and stemming process

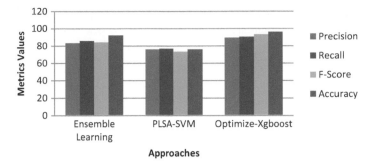

Fig. 4 Comparison for precision, recall, *F*-score, and accuracy for various approaches

5 Conclusion

Twitter provides a one-of-a-kind platform for developing and applying research and search theories, as well as technology attitudes. The study provided in this article suggests a novel approach to analyze Twitter sentiment data. To expose the feeling, the perspective statements have been removed (Adjectives combined with the verbs and Tweets of adverbs). The corpus technique was used in this study. Finding a dictionary-based strategy for defining the semantic methodology and finding semantic adjective viewpoint. There are two main types of words: adverbs and verbs. There are two types of words: adverbs and verbs. A linear equation provided in the code was used to calculate the tweet emotion. Emotions are also on the rise. It is clear that this is still a work—in—progress. The prototype being examined is, in fact, a rudimentary prototype. The preliminary data indicate that it is a motivational technique. In proposed approach accuracy improve 2–3%, precision 4%, recall 5%, and *F*-sore 3%.

References

1. Carvalho, J., & Plastino, A. (2021). On the evaluation and combination of state-of-the-art features in Twitter sentiment analysis. *Artificial Intelligence Review, 54*(3), 1887–1936.
2. Kumar, A., & Jaiswal, A. (2020). Systematic literature review of sentiment analysis on Twitter using soft computing techniques. *Concurrency and Computation: Practice and Experience, 32*(1), e5107.
3. Tiwari, S., Verma, A., Garg, P., & Bansal, D. (2020). Social media sentiment analysis on Twitter datasets. In *Proceedings of 6th International Conference on Advanced Computing and Communication Systems*.
4. Gupta, I., & Joshi, N. (2020). Enhanced twitter sentiment analysis using hybrid approach and by accounting local contextual semantic. *Journal of Intelligent Systems, 29*(1), 1611–1625.
5. Nakov, P., Ritter, A., Rosenthal, S., Sebastiani, F., & Stoyanov, V. (2019). arXiv preprint arXiv: 1912.01973.
6. Wagh, B., Shinde, J., & Kale, A. (2020). A Twitter sentiment analysis using NLTK and machine learning techniques. *International Journal of Emerging Research in Management and Technology, 6*(12).

7. Antonakaki, D., Fragopoulou, P., & Ioannidis, S. (2021). A survey of Twitter research: Data model, graph structure, sentiment analysis and attacks. *Expert Systems with Applications, 164,* 114006.
8. Mehta, P., & Pandya, S. (2020). A review on sentiment analysis methodologies, practices and applications. *International Journal of Scientific and Technology Research, 9*(2), 601–609.
9. Manguri, K. H., Ramadhan, R. N., & Amin, P. R. M. (2020). Twitter sentiment analysis on worldwide COVID-19 outbreaks. *Kurdistan Journal of Applied Research,* 54–65.
10. Rehioui, H., & Idrissi, A. (2019). New clustering algorithms for twitter sentiment analysis. *IEEE Systems Journal, 14*(1), 530–537.
11. Tripathi, M. (2021). Sentiment analysis of Nepali COVID19 Tweets using NB, SVM AND LSTM. *Artificial Intelligence and Capsule Networks Journal, 3*(3), 151–168.
12. Pandian, A. P. (2021). Performance evaluation and comparison using deep learning techniques in sentiment analysis. *Journal of Soft Computing Paradigm (JSCP), 3*(02), 123–134.
13. Meena, R., Thulasi Bai. V., & Omana. J. (2019). Sentiment analysis on Tweets for a disease and treatment combination. In *Proceedings of International Conference On Computational Vision and Bio Inspired Computing* (pp. 1283–1293). Cham: Springer.
14. Khatun, F., Chowdhury, S. M, Zerin, Tumpa, N., Rabby, S. S. K., Akhter, H., & Sheikh, A. (2019). Sentiment analysis of amazon book review data using lexicon based analysis. In *International Conference On Computational Vision and Bio Inspired Computing,* pp. 1303–1309. Cham: Springer.
15. Alsaeedi, A., & Khan, M. Z. (2019). A study on sentiment analysis techniques of Twitter data. *International Journal of Advanced Computer Science and Applications, 10*(2), 361–374.
16. Go, A., Bhayani, R., & Huang, L. (2009). Twitter sentiment classification using distant supervision. In *CS224N Project Report, Stanford (*Vol, 1, p. 12).
17. Fouad, M., Tarek, M., Gharib, F., & Abdulfattah, S. M. (2018) Efficient Twitter sentiment analysis system with feature selection and classifier ensemble. In *Proceedings of International Conference on Advanced Machine Learning Technologies and Applications.* Springer.

Toward Smartphone Energy Optimization Through Users Operating Pattern Discovery

S. Pandikumar, S. Bharani Sethupandian, B. Arivazhagan, and P. Rajeswari

Abstract A smartphone has evolved into an indispensable device, with the majority of people treating it as their third hand. Mobile phone users must be studied to make better smartphones, improve the user experience, and produce generalizable and repeatable results. Screen-on (active mode) and screen-off (inactive mode) are the two states of every mobile phone, depending on how it is used. The majority of people believe they use their phone 24 h a day, although the active usage time (screen-on) of the phone is extremely short. This paper focuses on smartphone usage patterns and user attitudes toward smartphone energy optimization. The overall trial result demonstrates that the average user uses their smartphone actively for 2.96 h each day, as well as other fascinating data about smartphone usage. This study shows that the average smartphone is inactive for the majority of the time, and this result leads to the optimization of energy consumption in the inactive mode. The findings of this paper will make a significant impact on extending the battery life of smartphones.

Keywords Screen-off · Inactive mode · Smartphone · Usage pattern · Energy optimization · Battery drain

1 Introduction

The smartphone market has been expanding at a breakneck pace. Despite such a high rate of smartphone adoption, the user experience has been and will continue to be significantly limited by the phone's battery life. Accessing the internet via 3G or wireless fidelity while running several interactive apps and background services is a

S. Pandikumar (✉)
Department of Computer Science, The American College, Madurai, India
e-mail: spandikumar@gmail.com

S. Bharani Sethupandian · P. Rajeswari
Department of Computer Science, Sri Krishna Arts and Science College, Coimbatore, India

B. Arivazhagan
Department of Computer Science, Erode Arts and Science College, Erode, India

© The Author(s), under exclusive license to Springer Nature Singapore Pte Ltd. 2023   615
S. Shakya et al. (eds.), *Sentiment Analysis and Deep Learning*,
Advances in Intelligent Systems and Computing 1432,
https://doi.org/10.1007/978-981-19-5443-6_47

significant source of smartphone energy consumption [1, 2]. As a result of roughly 40% of the radio network energy being spent in the screen-off state, screen states are playing an essential role in energy loss [3].

Normally, smartphone's screen is having two statuses called screen-on and screen-off; in 3G and 4G network, the IP data are sent for both screen-off and screen-on states. During the inactivity of user equipment (UE), the consumption of radio energy is between 30 and 40%. There are plenty of researches done previously regarding the behavior of always-on apps during screen-off [4, 5]. This paper analyzes various usage patterns of the smartphone by the user and found the average usage time of the phone. As a result of these discoveries, researchers are concentrating their efforts on reducing energy consumption in the inactive mode. The paper moves toward focus to answer the following questions:

- How much time the user actively spends in smartphone?
- How much time the users switch between screen states?
- What is the average screen-on time per day?

The following is a breakdown of the paper's structure. The literature review and background studies are the emphasis of Sect. 2. The basic experimental setup of the trace is discussed in Sect. 3. Section 4 delves into the performance analysis of smartphone user behavior and events. In Sect. 5, the conclusion and future directions are discussed. The detailed result and methodology are discussed in the paper [6].

2   Literature Study

Hossein Falaki et al. [7] argue that their findings show that the average number of interactions per day for a user ranges from 10 to 200; the average interaction length ranges from 10 to 250 s; the average number of applications used ranges from 10 to 90; and the average amount of traffic per day ranges from 1 to 1000 MB, with 10 to 90% of that traffic being exchanged during interactive use. Longer interactions are more common in games and mapping applications [7]. Another finding of the study is that many short connections consume more energy than a few long interactions, because those interactions affect the state of important components including network data modules, CPUs, and memory. Variations in how people engage with their smartphones lead to variations in energy use and data usage [7]. Physical user-app interactions (such as taps and scrolling) might afford a rich foundation for assessing mobile usage, according to the authors of the study [8]. Using a large dataset of 82,760,000 interaction events from eighty-six individuals over an 8-week period, the researchers used attribute embedding and unsupervised algorithm to extract important interfaces within group of mobile phone usage sessions. These clusters are not reflected in high-level aspects like session time, unlock state, or app updates. More research into the utility of leveraging user-app interaction behavior as the foundation for the aforementioned use cases is encouraged by the findings.

The present [9] investigative and preventive power-saving tactics based on user groups are given a new dimension in this research. The relationship between user group and power consumption is quite valuable, and it may be used by marketing groups of mobile phone operators/producers as well as network builders to size radio interfaces in mobile networks. The research [10–12] presents an in-depth investigation on user behavior-based energy consumption on residential cyber-physical, construction, communication devices. For the Android environment, a search-based energy testing technique was developed by Jabbarvand et al. [13]. The experimental results utilizing real-world apps show that it is not only capable of testing app energy behavior effectively and efficiently, but also that it outperforms previous techniques by detecting a broader and more diverse range of energy flaws. The researches [14–17] suggest various techniques of energy measurements and profiling techniques.

3 Experimental Setup

The experimental data was gathered from fifty individuals of various ages, categories, and professional backgrounds. For a month, traces were collected from Android 5 + phones. Packet payloads, user input events, RSRQ value, and power usage log are all included in these traces; however, this article focuses solely on user interactions. In the future articles, the remaining data will be evaluated.

Users are recommended to uninstall unneeded apps during the test. The following apps are available to users for network traffic monitoring and other uses. WhatsApp, Facebook, Instagram, Cricbuzz, Twitter, YouTube, Gmail, Google Play Store, and Browser are some of the most popular social media platforms. The gaming apps have been left out on purpose.

The test participants include a diverse group of professionals with a variety of smartphone models (Table 1). The logging procedure focuses solely on user interactions and is used to assess how the user uses his or her smartphone.

4 Performance Evaluation

The outcome of a 30-day usage pattern features a multitude of quantitative findings and data. Obviously, each user has a unique thinking and usage pattern, which is determined by age and the nature of the job. Traces were acquired from a variety of experts (Table 1) and age groups for this study. Because middle-aged people are the hungriest for cellphone usage, the average age is 32–35. The trace's lowest age is 18 and maximum age is 65. The users are drawn from a wide range of professions and practically cover every major topic. The mechanic and taxi driver domains each have one user, whereas the remaining areas have at least two individuals (Table 1).

Table 1 Category of users based on professional

Category	Total participants
UG students	3
PG students	2
Homemaker	4
Politician	2
Asst. professor	3
Developer	4
Web developer	2
Business	4
Homeo	2
Animator	1
Research scholar	2
Unemployed	2
Sales rep	4
Entrepreneur	3
Mechanic	1
Taxi driver	1
Office clerk	2
Shop owner	2
Medial rep	2
ECE engineer	1
Civil engineer	1
Insurance agent	1
Retired professor	1
Total	50

Marketing professionals use their phones regularly during business hours, according to the trace study. The top three phone users are marketing professionals, academics, and homemakers, with average usages of 4.0, 3.0, and 3.54 h, respectively (Figs. 1 and 2, Table 2).

The user $U4$ (medical rep) holds maximum individual day usage as well as maximum average usage among the users (Fig. 2). He uses his smartphone up to 11.21 h in one particular day (Fig. 5), and his minimum usage is 2.17 h in the entire trace period because that day was Sunday.

Not only the user $U4$, even the users $U6$, $U7$, $U8$, and $U10$ are having maximum average usage that are 5.33, 5.66, 5.31, and 5.12 h, respectively. The maximum usages no need to have the highest screen events because most of the users use their phone with long active duration without switching between the screen events.

The interesting fact is the user $U4$ having the highest screen event that is 12148 and daily average is 404. At the same time, the user $U10$ is a politician having average

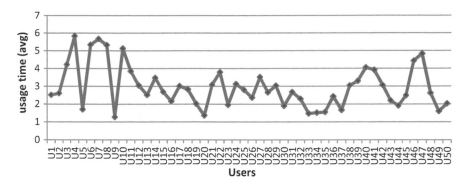

Fig. 1 Average usage of mobile phone (screen-on duration) in hours

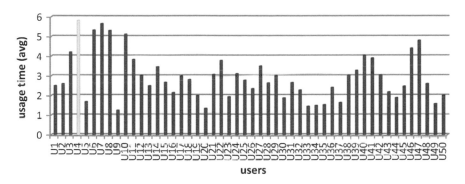

Fig. 2 Highest usage of mobile phone (screen-on duration) in hours

usage of 5.12 h, but he has the lowest total screen events 1208 in the entire trace because he does not check the phone very often, but once he unlocks the phone, he uses long period that is average of 20–30 min because he has 2 smartphones for communication and mostly uses for voice calls and occasionally uses social media.

The user $U9$ (homemaker) is having the average minimum usage 1.26 h among the entire users. This user's maximum and minimum usage of the trace is 2.1 and 0.52, respectively. This user normally uses phone for voice calls and use WhatsApp and Gmail. $U20$ (Asst. professor) comes under the second minimum usage of the phone that is 1.33 h averagely. The lowest and highest usage of $U20$ is 0.56 and 2.14 h in a day not only the usage this user does not check the phone often, she has 2nd lowest total screen event 1328 in the entire trace and having average 44 events per day.

The user $U2$ (UG student) is the lowest minimum usage of the day (Fig. 4). She used her phone for 0.41 h in a particular day, and her standard deviation is 2.12 (Fig. 3). Not only $U2$, the users $U1$, $U5$, $U9$, and $U20$ come under the day minimum usages, their day usage are 0.52, 0.57, 0.52, and 0.56, respectively.

Table 2 Average usage time and screen events during the log period for 25 users

Id	Δ^{α}	Δ^{β}	σ_{usage}	Δ_{Low}	Δ_{High}	Δ_{Tot_Events}	Δ_{Avg_Events}	σ_{events}
$U1$	2.6	1.50	1.94	0.53	7.4	2885	97	86.82
$U2$	2.61	2.13	1.76	0.42	5.18	10,399	347	110.21
$U3$	4.22	4.22	1.42	2.46	6.55	2915	97	15.29
$U4$	5.83	5.39	2.87	2.18	11.22	12,149	405	160.98
$U5$	1.70	1.21	1.08	0.58	3.35	2272	77	11.52
$U6$	5.34	5.39	1.07	4.15	7.17	9198	308	45.29
$U7$	5.67	5.38	1.44	3.54	8.05	1884	62	21.54
$U8$	5.29	4.99	1.38	3.44	7.14	4312	143	15.92
$U9$	1.24	1.45	0.56	0.53	2.11	1507	51	4.92
$U10$	5.13	5.23	2.15	2.16	8.06	1207	39	8.53
$U11$	3.83	2.14	2.91	1.13	7.31	3033	102	17.51
$U12$	3.02	2.49	1.77	1.00	5.46	4756	156	28.43
$U13$	2.60	2.29	1.00	1.36	4.22	4303	142	24.14
$U14$	3.36	3.35	0.89	2.21	4.47	4227	140	29.08
$U15$	2.67	2.80	0.90	1.14	3.56	3111	104	18.22
$U16$	2.18	2.16	0.71	1.10	3.49	2961	98	11.20
$U17$	3.01	3.39	0.59	2.16	3.56	2393	80	7.32
$U18$	2.83	3.46	1.12	1.13	4.08	2119	72	9.22
$U19$	2.20	2.20	1.00	1.10	3.61	1544	53	8.31
$U20$	1.36	1.46	0.50	0.66	2.24	1330	45	5.42
$U21$	3.08	3.55	1.19	1.24	4.55	4288	144	39.10
$U22$	3.80	4.55	1.34	1.19	5.18	6944	234	65.10
$U23$	1.96	1.45	1.14	1.35	4.42	2210	72	17.84
$U24$	3.10	3.11	0.80	2.31	4.30	2915	96	18.80
$U25$	2.80	3.16	1.18	1.20	4.58	2576	84	21.00

Δ^{α} Mean of screen-on duration in hours, Δ^{β} Median of screen-on duration in hours
σ_{usage} Standard deviation of screen usage, Δ_{Low} Minimum screen-on duration in hours
Δ_{High} Maximum screen-on duration in hours, σ_{events} Standard deviation of screen usage.

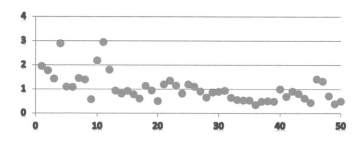

Fig. 3 Standard deviation of mobile phone usage (screen-on duration)

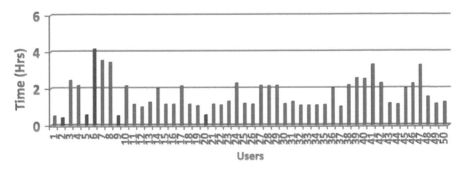

Fig. 4 User-wise minimum screen-on duration

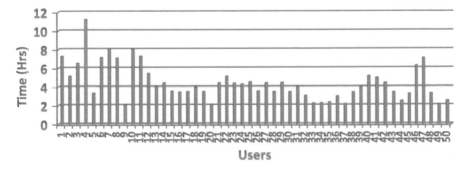

Fig. 5 User-wise maximum screen-on duration

According to the consistency of the user behavior, the user $U36$ (web developer) having low standard deviation 0.32 (Fig. 3) (Table 2), and she uses her phone with consistent interval in all days. The average usage of $U36$ is 2.41 h, and the median is 2.45 h.

The second consistent user is $U49$ (homemaker). She used her phone for less duration than $U36$, but her average usage is 1.58 h per day, the median is 1.51 h, and 0.36 is the standard deviation. At the same time, $U11$ (PG student) having 2.92 is the standard deviation (Fig. 3) because he uses his phone inconsistently depending on his work schedule, his usage low is 1.14 h, usage high is 7.3 h (Fig. 3), and his average usage is 3.83 h per day. The second inconsistent user is $U4$ (medical rep), he uses his phone less time during holidays that is why he has high standard deviation of 2.86 (Fig. 3), his recorded usage low is 2.17 h, and usage high is 11.21 h (Fig. 5).

The study of screen interaction is more fascinating because some users use their phone for lengthy periods of time with few screen events, while others, particularly those under the age of 30, check their phone (unlock) frequently without receiving any notification (Figs. 6 and 7).

The user $U2$ (UG student) has 10,397 total screen events in a month with average of 346 screen events per day, but the average usage of phone is just 2.6 h. This data is almost near to the maximum usage level 5.82 h per day. At the same time, the user

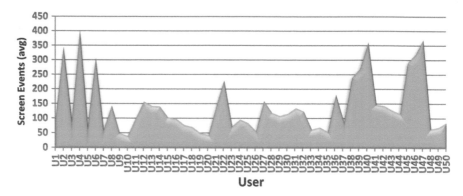

Fig. 6 Average screen events of all users

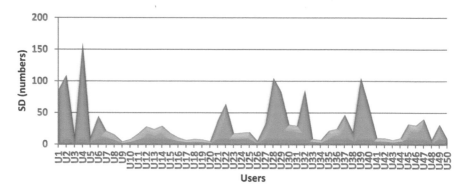

Fig. 7 Standard deviation of screen events

U10 (politician) has low phone interaction but high phone usage. The user U2 does not check the status throughout the day, but she unlocks often in particular hours like evening 4–7 pm and 9–10 pm.

The study of the trace reveals that smartphone usage is significantly dependent on the user's career and mindset as in Fig. 8. The users of IT, marketing, academic business, and homemakers are having the average highest usage of their smartphone in the trace (Figs. 9 and 10). Equation 1 derives that the average usage of the smartphone is 2.94 h in a day. The average usage of various activities was calculated by Eq. (1), let 'n' be total number of users and U_i denote each users in the trace.

$$\text{Usage}_{\text{Avg}} = \frac{1}{n}\sum_{i=1}^{n} U_i = \frac{1}{n}(U_1 + U_2 \ldots + U_n) \tag{1}$$

A typical user actively uses their phone by averagely 2.54 h per day and 21.06 h that is inactive. Most of the time the smartphones are not used by the user, but then there is a process on the inactive mode.

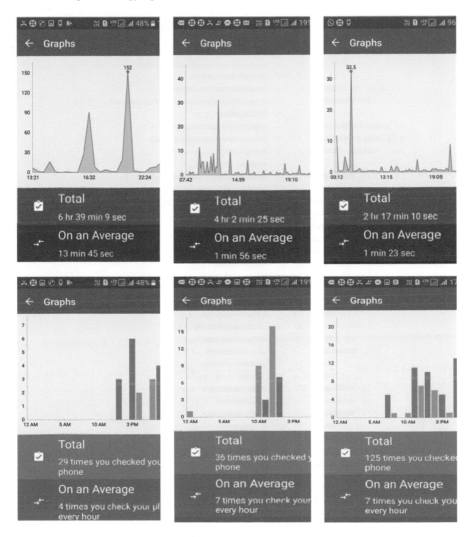

Fig. 8 Mobile usage screen samples of U4

Figures 11 and 12 compare marketing and academic people because there are seven users in the trace and they are the maximum mobile users in this study.

5 Conclusion

The active and idle states of a common smartphone are examined in this study's experimental results. For 30 days, experimental data was collected from 50 users. Participants were chosen from a variety of professions and were advised to download

Usage minimum   Usage maximum

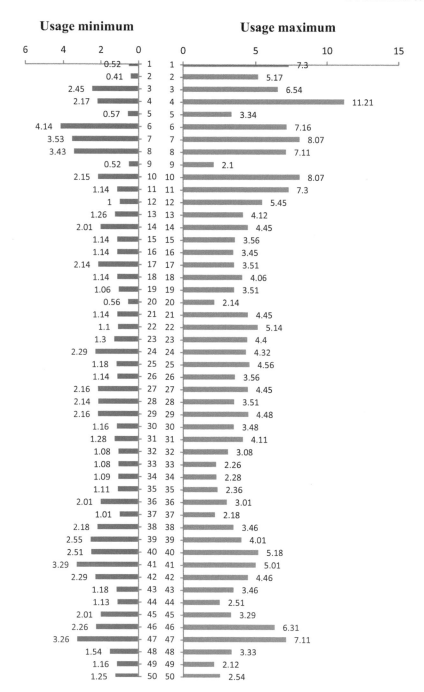

Fig. 9 Comparison of all users usage minimum and maximum

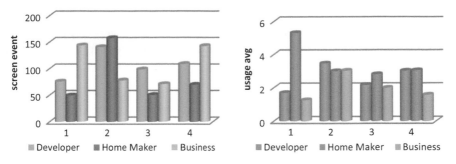

Fig. 10 Screen event and usage comparison of developer, homemaker, and business people

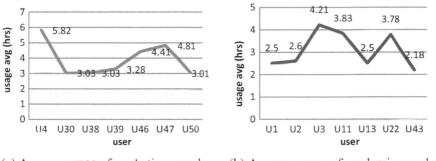

(a) Average usage of marketing people   (b) Average usage of academic people

Fig. 11 Average usage of marketing and academic people

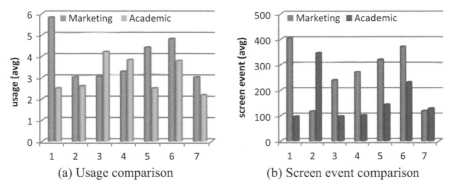

(a) Usage comparison   (b) Screen event comparison

Fig. 12 Average usage and screen event comparison of marketing and academic people

communicative apps. The trace reveals some intriguing facts, particularly regarding phone interactions and total smartphone usage. People in white-collar jobs are more likely to use their smartphones than those in blue-collar jobs. On average, a smartphone is utilized for 2.54 h a day, with the rest of the time spent by the user dormant.

This study highlights the research topic of how much energy and resources the smartphone consumes while it is not in use, and it leads to optimize energy consumption. In the future, the research proposes a novel methodology to reduce energy consumption of smartphone.

Conflict of Interest   The authors declare that they have no conflict of interest.

References

1. Pihkola, H., Hongisto, M., Apilo, O., & Lasanen, M. (2018). Evaluating the energy consumption of mobile data transfer—from technology development to consumer behaviour and life cycle thinking. *Sustainability., 10*(7), 2494. https://doi.org/10.3390/su10072494
2. Tawalbeh, M., Eardley, A., & Tawalbeh, L. (2016). Studying the energy consumption in mobile devices. *Procedia Computer Science, 94,* 183–189. https://doi.org/10.1016/j.procs.2016.08.028
3. SelimIckin.'2 et al., (2013). QoE-based energy reduction by controlling the 3G cellular data traffic on the smartphone. In *22nd ITC Specialist Seminar on Energy Efficient and Green Networking (SSEEGN)*, pp. 13–18.
4. Nika, A., Zhu, Y., Ding, N., Jindal, A., Hu, Y., Zhou, X., Zhao, B., & Zheng, H. Energy and performance of smartphone radio bundling in outdoor environments, pp. 809–819. https://doi.org/10.1145/2736277.2741635
5. Li, X., Zheng, C., Zhang, C., Li, S., Guo, L., & Xu, J. (2017). Understanding usage behaviors of mobile apps by identifying app package in network traffic. In *2017 IEEE 9th International Conference on Communication Software and Networks (ICCSN)*, pp. 1037–1041. https://doi.org/10.1109/ICCSN.2017.8230268
6. Pandikumar, S., & Sumathi, M. (2019) Energy efficient algorithm for high speed packet data transfer on smartphone environment. *International Journal of Engineering and Advanced Technology, 8*(6).
7. Falaki, H., Mahajan, R., Kandula, S., Lymberopoulos, D., Govindan, R., & Estrin, D. (2010). Diversity in smartphone usage. In *MobiSys'10—Proceedings of the 8th International Conference on Mobile Systems, Applications, and Services*, pp. 179–194. https://doi.org/10.1145/1814433.1814453
8. Friedrichs, B., Turner, L. D., & Allen, S. M. (2021). Discovering types of smartphone usage sessions from user-app interactions. In *2021 IEEE International Conference on Pervasive Computing and Communications Workshops and other Affiliated Events (PerCom Workshops)*, pp. 459–464. https://doi.org/10.1109/PerComWorkshops51409.2021.9431034
9. Almasri. & Sameh, A. (2019). Saving energy on smartphones by categorizing users. In *2019 2nd IEEE Middle East and North Africa COMMunications Conference (MENACOMM)*, pp. 1–4. https://doi.org/10.1109/MENACOMM46666.2019.8988558
10. Aksanli. & Rosing, T. S. (2020). Human behavior aware energy management in residential cyber-physical systems. In *IEEE Transactions on Emerging Topics in Computing*, vol. 8, no. 1, pp. 45–57. https://doi.org/10.1109/TETC.2017.2680322
11. Chuang, M. et al. (2021). Research on user electricity consumption behavior and energy consumption modeling in big data environment. In *2021 IEEE 2nd International Conference on Big Data, Artificial Intelligence and Internet of Things Engineering (ICBAIE)*, pp. 220–224. https://doi.org/10.1109/ICBAIE52039.2021.9389990
12. Qian, W., & Gechter, F. (2018). Modeling smartphone energy consumption based on user behavior data. In *2018 International Conference on Computational Science and Computational Intelligence (CSCI)*, pp. 788–793. https://doi.org/10.1109/CSCI46756.00158

13. João Bazzo, J., de Melo Pires, R., Javier Ortega, A., Portela Salehi, N., & Thiago Marreiros Santos, P. (2020). UE current consumption on carrier aggregation in LTE-A systems. In *2020 International Conference on Information and Communication Technology Convergence (ICTC)*, pp. 600–602. https://doi.org/10.1109/ICTC49870.2020.9289502
14. de Sousa, J. O., Fonseca, G. C., de Oliveira, J. W. A., & Santos, R. N. (2021) A novel non-invasive method to measure power consumption on smartphones. In *2021 Brazilian Power Electronics Conference (COBEP)*, pp. 1–6. https://doi.org/10.1109/COBEP53665.2021.968 4072
15. Almasri., & Sameh, A. (2019). Rating Google-Play Apps' energy consumption on android smartphones. In *2019 2nd IEEE Middle East and North Africa COMMunications Conference (MENACOMM)*, pp. 1–6. https://doi.org/10.1109/MENACOMM46666.2019.8988554
16. Niemann, C., Ewert, H., Puttnies, M., Rethfeldt, D. T., & Danielis, P. (2020). Modeling energy consumption for task-offloading decisions on mobile and embedded devices. In *2020 IEEE 2nd Global Conference on Life Sciences and Technologies (LifeTech)*, pp. 400–404. https://doi. org/10.1109/LifeTech48969.2020.1570618809
17. Ahmadoh, & Tawalbeh, L. A. (2018) Power consumption experimental analysis in smart phones. In *2018 Third International Conference on Fog and Mobile Edge Computing (FMEC)*, pp. 295–299. https://doi.org/10.1109/FMEC.2018.8364082

Efficient Novel Approach Prediction of Survival Time on Pancreatic Cancer Using Machine Learning Paradigms Toward Big Data Analytics

P. Santosh Reddy and M. Chandra Sekhar

Abstract Pancreatic cancer that is usually found in the pancreas is stomach-related enzyme cells and hormone-producing cells, which occur when cells begin to grow out of control in an abnormal manner. Pancreatic adenocarcinoma is the most dangerous malignancy and the fourth highest deadliest cancer that causes death. The prognosis for those diagnosed with pancreatic cancer is still extremely poor with the little chance of recovery. Since it is surrounded by diverse organs and fatty tissues in the middle, identification and separation are ineffectual. Clinical patient data have had beneficial and useful in revealing the success rate of cancer diagnosis. Models built with patient level data also have the potential to yield different research insights in predicting and diagnosing the best course of treatment with improved outlook for individual patients. Yet, there is a need for an efficient and impactful technique, which predicts the results effectively and correctly. Hence, with the idea of automated meta-learning approach that learns to predict the finest results help in improving the performance. This approach is then used to determine the survival rate for the patient thereby increasing the chances of beating the cancer. The proposed approach reveals the use of machine learning algorithms, mainly by using appropriate feature selection and prediction techniques, which increases the accuracy of models employed in predicting patient survival. The main objective of this paper is to build the machine learning models by using clinical cancer patient data for predicting the survival rate of patient with pancreatic cancer and to understand the factors influencing the outcome.

Keywords Pancreatic · HDFS · WEKA

P. Santosh Reddy (✉)
Department of Computer Science and Engineering, Presidency University, Bengaluru, India
e-mail: santosh.reddyp@presidencyuniversity.in; santoshreddy@bnmit.in

Department of Computer Science and Engineering, B. N. M. Institute of Technology, Bengaluru, India

M. Chandra Sekhar
Department of CSE, Presidency University, Bengaluru, India
e-mail: mchandrasekhar@presidencyuniversity.in

© The Author(s), under exclusive license to Springer Nature Singapore Pte Ltd. 2023   629
S. Shakya et al. (eds.), *Sentiment Analysis and Deep Learning*,
Advances in Intelligent Systems and Computing 1432,
https://doi.org/10.1007/978-981-19-5443-6_48

1 Introduction

Adenocarcinoma, the results in lining of ducts in the pancreas, is said to be an aggressive and lethal type of cancers. The treatment for the carcinogenic disease is very hard to detect it in the initial stage but for the patients for whom the diagnosis is carried at an earlier stage have curable outcomes. However, the diagnosis of the cancer stage at the right time plays a very crucial role in determining the survival rate. For example, for a localized cancer, the rate of survival is shown to be four times its average. Attributes and factors like age, gender, smoking, obesity raise the risk of having pancreatic cancer and also affect prognosis. With the growth of digital healthcare centers, efforts have been made for implementing new techniques in making cancer disease-free. The data have been powerful and direct in patient treatment and at the same time have given equal importance to evaluate the best treatment to the patient. Patient-level data are analyzed by the models using machine learning techniques that have ability to determine the best type of cure. The potency of this paper is to apply machine learning techniques to patient-level cancer data and analyze and understand various factors like medical history, severity of the disease, and provide the best treatment [6].

Pancreatic cancer can have debilitating side effects and is quite difficult to detect at the early stage. For example, in a whipped procedure—an operation carried out to remove the head of the pancreas, the surgical treatment may take around eight hours to perform, but the recovery time will take several months. But, if the cancerous tissues have expanded into other organs or arteries of the patient, then the survival of the patient is under threat. In such a scenario, it is more convenient and sensible to help the patient improve the quality of life while living with the disease. For these reasons, it is wise to opt for surgical treatment only if the cell can be removed. Therefore, our main aim is to develop machine learning models and predict the survival time along with the factors influencing the outcome.

2 Literature Survey

In AI, different training methods have been developed, such as recovery, reinforcement, and stacking (see below). Because they encapsulate the opportunities of a set of AI models to develop the ultimate insightful model, these procedures are also known as cadaveric techniques. Malignant growth has lately been studied using AI troupe approaches [4]. Other topics addressed by AI and its application to forecast cancer growth of the pancreas are detected in this work. Given these group paradigms, they are typically characterized by a comparable AI technique using a variety of patchy real-time empowerment datasets. Because of the clear ordering, most votes for expectation stacking are cast across scientific paradigms and across paradigm-averaged forecasts at the expense of a set of numerical targets. In this prototypical, for any focus arranged informational junctures that are fruitlessly served by previous patterns, grouping of paradigms is established on a regular basis through a training method. Democracies that are balanced between scientific perspectives set normal growth expectations. The agreement provides for the development of concepts on which the accretion must take place using various forms of AI techniques. The only ideas now in use are zero-level paradigms. The outcomes of the zero-level paradigms are regarded contributions to a second layer of learning called zero-level paradigms, which is the entire throughput employed in analysis for real-time forecasting.

3 Background

3.1 Machine Learning

Machine learning is a booming technology, which is growing progressively in predicting outcomes efficiently and accurately with human interruption. In medical science, machine learning has advanced right from identification and diagnosis of the disease to suggesting drugs for therapy. Machine learning covers various aspects like decision tree induction, naïve Bayes probabilistic classifiers, and support vector machines, discriminant analysis. Major attention is drawn to supervise learning, which the present paper highlights, each data instance used for learning (training) consists of two portions; it labels the data which are then tested on unlabeled data, and a supervised classification or regression is done to display the output. Basically, classification is used for predicting a finite set of values, whereas regression is used for predicting a continuous range of values.

Machine learning has found to be beneficial in detecting pancreatic cancer [3] and has helped in efficient analysis. Various supervised machine learning algorithms have proved to show outstanding results for diagnosis of pancreatic cancer, determine the rate of survival, and improve the standard of life when combined with logistic regression.

The predictions have different probabilistic and statistical approached for each individual. Hence, a method that is favorable for one patient need not be beneficial to the other patient. The latter fact implies that the overall predictive performance can be improved using advanced computational techniques such that it reliably predict what machine learning method provides the best performance for the sufferer. Hence, this is the main motive that is captured in the present paper.

3.2 Meta-Learning Approach

Meta-learning as the name suggests is learning from its own learning process. It serves as 'knowledge for completion' where the learner in which the learner gets an understanding of the knowledge and learns how to apply it. Some of the meta-learning approaches are bagging, boosting, and stacking.

In bagging [11], which is also known as bootstrap aggregator, the models in the ensemble are typically generates extra data in the training dataset. This is done by generating random samples of the dataset over which learning is to be applied. The bagging prediction is done by casting vote conducted through the learned models, and final prediction goes by the maximum votes. In boosting [12], learning happens in a sequential manner, usually by the same learning technique, with each model focusing on the fact that each instance of data is learnt by the previous model. Weighted voting is how the learnt models do the prediction. The combined models are usually the second-level learners. In this context, the individual models are named as level 0 models which is the base model. The outputs of base model are fed as inputs to a level 1 model, known as the metamodel, whose output is used for prediction.

4   Proposed Methodology

The proposed model is an ensemble meta-learning approach which is simply mixing or combining two models. One approach is to combine boosting and stacking, so that the whole dataset is trained and there is complete learning. However, it differs from classical bagging, boosting, and stacking and is characterized by its adoption of a new prediction target. This approach which learning model is best suitable to make the correct and accurate prediction. Hence, the selected model is used and implemented.

Presenting the details of the various models and then choosing the best meta-learning model are the potency of this paper.

4.1   Clinical Dataset Modeling

A group of experts from the University of Massachusetts created and developed a clinical database containing retrospective records of 60 patients for treating pancreatic adeno. Each patient record includes details preliminary outlook, medical history of the family, diagnostic tests, tumor pathology, treatment suggested, surgical proceedings, and span of survival. The attributes are divided into three major categories: 111 pre-operative attributes, 78 peri-operative attributes, and the target attribute.

Pre-operative and all-parameters forecast parameters: If the subsets of attribute consist of only 111 pre-operative attributes, then the dataset is named as pre-operative dataset, while if both the pre-operative and peri-operative attributes of 189 non-target attributes are considered, then the dataset is called an all attribute predictive attribute as shown in Table 1.

Table 1 Pre-operative attributes	Classification patient	Parameter numbers	Specification
	Physicians	10	Biographical, physicians
	Croaker	19	Symptoms at diagnosis
	Chronicle	24	Health record
	Fluid	7	Laboratory analysis accuracy
	Tomography	20	Diagnosis image details
	Proctoscopy	22	Proctoscopy info
	Preliminary outlook	09	Physicians pre-surgical evaluation
	Total	111	

Prediction target attribute: The analysis is survival time which is nothing but the time from diagnosis to the time of demise that is named as the prediction target attributes.

Survival discretization: The final target attribute for survival prediction is listed below:

A 9-month split is defined when one target class splits for less than 9 months and the other target class splits for 9 or greater than 9 months.

A 6-month split is defined when one target class splits for less than 6 months and the other target class splits for 6 or greater than 6 months.

A 6–12-month split is defined when one target class splits for less than 6 months, the other target class between 6–12 months, and another target class is split as over 12 months.

4.2 Studying to Forecast the Perfect Model

4.2.1 Training the Model Selection Meta-Learner

Let i0 denote original input. When a predictive model is fed to this input, the trained model predicts a probabilistic distribution of all the target values. Consider an example of 6–12-month survival split, the level 0 models will output the probabilistic distribution of survival rate for less than 6 months, more than 12 months, and in between 6 -12 months.

Table 2 Generation of the level 1 dataset for model

Instance	Class	$P_{ANN}(+)$	$P_{NB}(+)$	Selected model
$N^{[1]}$	Add	0.21	0.71	ANN
$N^{[2]}$	Sub	0.39	0.49	NB
$N^{[3]}$	Add	0.82	0.71	ANN
$N^{[4]}$	Sub	0.81	0.77	NB

Table 3 Resulting level 1 dataset $i1$ based on Table 2

Instance	Target class
$N^{[1]}$	ANN
$N^{[2]}$	Bayes
$N^{[3]}$	ANN
$N^{[4]}$	Bayes

4.2.2 Training Procedure

The training follows a two-stage approach. With the use of cross-validation, $i1$ is constructed from the original dataset. The learning algorithms are tagged as l0…lo and l1.Relabeling each training instance using level 0 technique l0 outputs the highest probability. This level technique is the top predictor. In the second stage, the learning algorithm is applied to construct l1 over i1 which is the set of new dataset. Finally level zero types are built over full dataset $i0$.

Example 1. Example to delineate the development of the dataset to prepare the level 1 model is shown in Table 2.

Instances $N^{[1]}$, $N^{[2]}$, $N^{[3]}$, $N^{[4]}$ correspond to a tranquil instance f. The first column takes input attribute vector $a^{(I)}$ and can have two classes, ' $+$ ' or '$-$'. For the $+$ class shown in 3rd column and 4th column, the probability for level 0 is reflected for ANN and NB classifiers. The model selected shows whether it belongs to ANN class or NB class.

The Table 3 shows a new instance that is added to level dataset i_1. The new instance contains the input attribute vector $a^{(I)}$ together with the model selected as the target class. The result is displayed in Table 3.

5 Evaluation

5.1 Preprocessing and Predictive Methods

Attribute and feature selection: The models are proposed and evaluated using machine learning algorithms wherein the subsets of attributes are selected using attribute selection techniques. It has been proved that feature selection enhances the performance of classification methods. We analyze the gain ratio, principle component analysis for dimensionality reduction, relief, and support vector machines and look for the optimal approach.

Machine learning techniques for level 0 and level 1 classifiers: The popular classifiers for impactful and effective detection of pancreatic cancers are artificial neural networks (ANN), Bayesian networks (BN), decision trees, naive Bayes networks (NB). Along with these classifiers, we propose the use of support vector machines (SVM) which plays the role of an attribute selector as well as a classifier.

For each dataset, we find the best combination of feature selection and machine learning algorithm. We evaluate the model by using zero and logistic regression (LR) that set the benchmark for with the comparison of each of these models that are made.

Experimental protocol:

1. **When selection of level 0 classifier is considered.** Machine learning techniques are evaluated by using and not using feature selection, and results are computed using tenfold cross-validation. All the features were taken into consideration and tested by varying the attributes; it was found that with the use of feature selection, the accuracy of machine learning performed well. Out of all the models, the top three best models are selected.
2. **When selection of level 1 classifier is considered.** As the upper three-stage zero versions are recognized, we experimentally established for which one of the models in level 1 would blend for the best accuracy.

6 Outcome

The outcome of the evaluated model for pre-operative datasets is displayed below.

6.1 Parameter Selection

The above Fig. 1 considers the pre-operative attributes and observes the behaviors that occur when the attributes are varied. The pre-operative attribute is used with a 6-month lifespan target fragmented along with SVM parameter selection. We observe that the NB or naïve Bayes classification accuracy is higher when small attributes training is taken into account, while BN or Bayes network classifiers work well when large attributes are taken into account with little or no selection. The ROC curve plotted below is a comparison between the performance of logistic regression (LR) and random survival time prediction as shown in Fig. 2.

Dataset for 9-month split: The pre-operative dataset for a 9-month split is done into two classes. It is found that the highest accuracy is obtained when gain ratio attribute selection is used in conjunction with logistic regression or SVM classifier

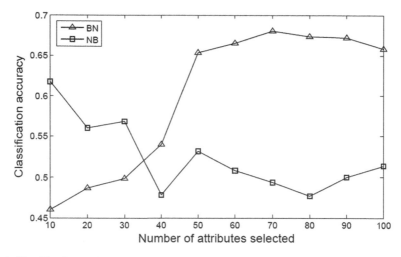

Fig. 1 Classification performance of Bayesian procedures for various levels of parameter choice

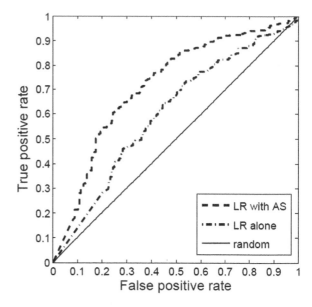

Fig. 2 Categorization execution of LR with and without gain ratio parameter assortment

which is 65.5 percent for both and relief attribute selection for BN classifier that gave an accuracy of 65.6%. This shows how our proposed model slightly overtops the standard meta-learning approaches as well.

Dataset for a 6-month discrete: Discreting a lifespan time at six periods is shown in Table 5 which shows an increase in classification accuracies.

Table 4 Classification accuracy: 9-month split

ML algorithms	Selected attributes	No parameters reliability	Reliability
Logistic regression	Gain ratio	70.00	0.655
Support vector machine	Gain ratio	80.00	0.655
Bayes network	Relief	100.00	0.653
Performing model selection meta-learner			
Level 1: naïve Bayes Level 0: LR, SVM	None gain ratio	111.0	0.673

Table 5 Classification accuracy: 6-month split

ML algorithms	Selected attributes	No parameters reliability	Reliability ML + AS
LR	Gain ratio	40.00	0.702
SVM	Gain ratio	26.00	0.598
Best performing model selection meta-learner			
Stage 1: logistic regression Level 0: SVM_LR	PCA gain ratio	12.00	0.675

Dataset for a split between 6–12 months: Splitting survival time between 6–12 months is shown in Table 6, and it is observed that the accuracy in Table 6 is slightly less than the accuracy in Tables 4 and 5 due to the fact that it has larger number of classes.

Table 6 Classification accuracy 6–12-month categorization

ML algorithms	Selected attributes	No parameters reliability	Reliability ML + AS
Bayes network	Relief	20	0.527
ANN	Gain ratio	50	0.518
SVM	Relief	72	0.387
Performance model selection			
Level 1: naïve Bayes Level 0: ANN, SVM	None Gain ratio relief	111	0.533

Table 7 Categorization accuracy: pre- and peri-operative parameters, 6-month split

P_{ANN}(>Six_Months)	P_{NB}(> Six_Months)	P_{Actual}(> Six_Months)	Accurate_Model(s)
0.25	0.77	Zero	ANN
0.10	0.59	Zero	ANN
1.94	0.11	Zero	Bayes
0.07	0.98	One	Bayes
0.02	0.46	One	Bayes
0.5	0.49	One	Bayes
0.04	0.12	Zero	Together
0.99	0.84	One	Together
1.99	0.059	One	Together
0.89	0.49	One	Together
1.00	0.84	One	Together
0.87	0.77	One	Together
0.991	0.60	One	Together
1.09	0.82	Zero	–
0.82	0.79	Zero	–
0.88	0.89	Zero	–

Pre- and peri-operative parameters, 6-month split: In the pre- and peri-operative attributes for a 6-month split, it is observed that meta-classifier model approach for all-attribute dataset is 75.2% which is greater than logistic regression 61.3%. For bagging, boosting, and stacking, the accuracies are 74.5, 67, and 72.5% from Table 7.

Info of prototype selection: In the research work that is carried out by the meta-learning classifier, the models taken into consideration are naïve Bayes classifiers and artificial neural networks which are the top two best level 0 models.

For a level 1 classifier, we have associated decision tree and SVM for feature selection.

7 Conclusion

The main goal of this paper is to accurately detect the pancreatic cancer using the best-suited method. This is mainly done with the help of meta-learning, which creates a model that foresees the best model to produce the most accurate results. For such an example, the manner that different tactics produce conflicting forecasts from time to time inspired this approach. The model carefully selects which model gives an optimal predictive performance when experimented on each of the individual indicators. Taking note on the patients's resistance time based on characteristics such as segment data, first symptoms, and symptomatic test results, the current evaluation is done. The proposed learning approach proved to be more beneficial that individual insight approaches. It is found that the research work conducted promises to find the good accuracy and can hence be developed to detect and diagnose the patient with pancreatic cancer and suggest the best treatment.

In the future, we intend to try out deep transfer learning techniques using DenseNet, VGG-19 and compare the effectiveness of our proposed method with it.

References

1. Pei, H., Li, L., Fridley, B. L., Jenkins, G. D., Kalari, K. R., Lingle, W., Petersen, G., Lou, Z., & Wang, L. (2009). FKBP51 affects cancer cell response to chemotherapy by negatively regulating Akt. *Cancer Cell, 16*(3), 259–266.
2. Altekruse, S. F., Kosary, C. L., Krapcho, M., Neyman, N., Aminou, R., Waldron, W., Ruhl, J., Howlader, N., Tatalovich, Z., Cho, H., Mariotto, A., Eisner, M. P., Lewis, D. R., Cronin, K., Feuer, E. J., Stinchcomb, D. G., & Edwards, B. K. (Eds.) (2009). Seer cancer statistics review, 1975–2007, national cancer institute. bethesda, md, http://Seer.Cancer.Gov/Csr/1975_2007/; based on november 2009 seer data submission, posted to the seer web site 2010.
3. Bektas, B., & Babur, S. (2016). Machine learning based performance development for diagnosis of breast cancer. *Medical Technologies National Congress (TIPTEKNO), 2016*, 1–4.
4. Huang, G., Liu, Z., Weinberger, K. Q., & van der Maaten, L. (2018). Densely connected convolutional networks. arxiv 2016," arXiv preprint arXiv:1608.06993, Vol. 1608.
5. McGuigan, A., Kelly, P., Turkington, R. C., Jones, C., Coleman, H. G., & McCain, R. S. (2018). Pancreatic cancer: A review of clinical diagnosis, epidemiology, treatment and outcomes. *World Journal of Gastroenterology, 24*(43), 4846.
6. Shah, A. P., Mourad, M. M., & Bramhall, S. R. (2018). Acute pancreatitis: Current perspectives on diagnosis and management. *Journal of Inflammation Research, 11*, 77.
7. Hayward, J., Alvarez, S. A., Ruiz, C., Sullivan, M., Tseng, J. & Whalen, G. (2010). Machine learning of clinical performance in a pancreatic cancer database. In Kim, S. (Ed.), *Special issue on data mining approaches to the study of disease genes and proteins, artificial intelligence in medicine* (Vol. 49, no. 3, pp. 187–195).
8. Floyd, S., Ruiz, C., Alvarez, S. A., Tseng, J., & Whalen, G. (2010). Model selection meta-Learning for the prognosis of pancreatic cancer, full paper. third international conference on health informatics (HEALTHINF 2010). In *Conjunction with the Third International Joint Conference on Biomedical Engineering Systems and Technologies (BIOSTEC 2010)* (pp. 29–37). Valencia, Spain.
9. Qu, Y., Adam, B. L., Yasui, Y., Ward, M. D., Cazares, L. H., Schellhammer, P. F., Feng, Z., Semmes, O. J., & Wright, G. L. (2002). Boosted decision tree analysis of surface-enhanced laser desorption/ionization mass spectral serum profiles discriminates prostate cancer from non-cancer patients. *Clinical Chemistry, 48*, 1835–1843.
10. Lin, T.-Y., Dollar, P., Girshick, R., He, K., Hariharan, B., & Belongie, S. (2017). Feature pyramid networks for object detection. In *Proceedings of the IEEE Conference on Computer Vision and Pattern Recognition* (pp. 2117–2125).
11. Wang, X., Girshick, R., Gupta, A., & He, K. (2018). Non-local neural networks. In *Proceedings of the IEEE Conference on Computer Vision and Pattern Recognition* (pp. 7794–7803).
12. Ren, S., He, K., Girshick, R., & Sun, J. (2015). Faster r-cnn: Towards realtime object detection with region proposal networks. In *Advances in neural information processing systems* (pp. 91–99).

An Artificial Intelligence-Based Approach to Recognize Faces: A Perspective from Pandemic

Paramita Bhattacharjee, Ajitesh Moy Ghosh, and Tanaya Das

Abstract Face recognition technology comprises methods that analyze human faces to determine their identity. Like other biometric systems, it uses for surveillance connected with public cameras without anyone's consent. Nowadays, numerous challenges associated with recognizing faces, masked faces have also been added due to pandemic situation. Hence, in this paper, the authors concentrated on the problems of detecting masked and unmasked faces with the help of artificial intelligences-based technique, i.e., convolution neural network. It comprises methodology to develop an AI-based face recognition model using deep neural network concept. The proposed methodology includes several steps of capturing both image and video using computer vision model, i.e., OpenCV with the help of training and testing datasets. This research work helps to identify an individual regardless of masked and unmasked faces based on the recognition percentage.

Keywords Face recognition · Face detection · Convolution neural network · Computer vision · Deep neural network

1 Introduction

Biometric authentication technology has been a significant industry trend for years, and it will be even more due to artificial intelligence (AI)-based innovations. It uses automated methods for verifying or recognizing a person's identity based on physiological characteristics such as fingerprint, face pattern, handwriting, and voice. The term "facial recognition technology" refers to identify persons in real time [1]. It maps facial feature from a photograph or video using biometrics to establish a match with the person's real identity. Since the earliest pioneering works on facial recognition started 40 years ago, steady and continuing progress in popularity and innovation have been made in the development of biometric technology [2]. Face

P. Bhattacharjee · A. Moy Ghosh · T. Das (✉)
Department of Computer Science and Engineering, Adamas University, Kolkata, West Bengal, India
e-mail: tanayadas.das23@gmail.com

© The Author(s), under exclusive license to Springer Nature Singapore Pte Ltd. 2023
S. Shakya et al. (eds.), *Sentiment Analysis and Deep Learning*,
Advances in Intelligent Systems and Computing 1432,
https://doi.org/10.1007/978-981-19-5443-6_49

recognition difficulties represent substantial hurdles for face-based authentication solutions [3]. In recent years, COVID-19 has impacted the opinion on the use of biometric technology worldwide [4]. The pandemic situation accelerates the global adoption of facial recognition technology for masked faces.

Face recognition and detection have faced challenges due to factors like occlusion, pose variation, facial expression, aging, and resolution either in still image or in video. Another addition to the list of challenges is use of face mask worldwide during this pandemic situation. Currently, due to pandemic, people cannot be forced to open their mask in public while visiting proctored location for identity verification. Due to use of masked faces, this kind of situation may result in identity theft as well as unauthorized access to personal information. Hence, in this paper, the authors have focused on dealing with the challenge of face detection for masked faces. This research work recognizes identity irrespective of masked and unmasked faces with recognition percentage of the person's identity.

Section 2 describes the related literatures on face recognition and detection systems. Section 3 and its subsection discussed the proposed architecture and dataset used to train the neural network model to detect person's identity. Section 4 describes the result analysis and discussion with the help of recognition rate. Section 5 finally concludes the paper and the future work of this study.

2 Related Study

The sectors like finance [5], marketing [6], education [7], web service [8], surveillance [9], etc., have security issues where face recognition can be used to provide security. Nowadays, home security system uses face recognition with door access management to identify fraud and avoid robbery using AI-based technique like deep neural network, Internet of Things (IoT) to deal with the security problems [10, 11]. Nowadays, face recognition system uses IoT devices and Raspberry Pi as the principal controller to send a notification to the user at the time of any security concern [10]. In recent time, closed-circuit television (CCTV) surveillance systems are another motivating application to recognize masked and unmasked faces at the time of emergency that rely heavily on humans to monitor screens [12]. Globally, the use of closed-circuit television (CCTV) for monitoring has increased to unprecedented levels. Mobile banking is another sector where biometric face recognition is highly utilized. In this period, globally some places exist like Nigeria where the availability of mobile phones in remote places where banks or ATMs are not easily available. To resolve this critical situation, biometric system has become a primary focus in order to provide secure mobile transactions between clients and bank servers [13].

Biometric mask face recognition has become mandatory for identifying the person's identity without compromising the security concern during present pandemic situation.

Some recent literatures have discussed about face mask detection techniques like triplet [14], machine learning [15], support vector machines [16], computer vision [17], etc. A recent literature offers a mask detection framework that adapts the attention mappings feature refinement using feed-forward convolution neural network to enable effective attention of the feed-forward convolution neural network [18].

From the above literatures, the authors have found that most face recognition system are giving solutions for the unmasked faces. Some of the recent research works are detecting masked faces only. Also, the authors have found a research gap of detecting both masked and unmasked faces with accuracy percentage of recognition. Hence, in this paper, the authors have proposed an approach that detect a person's identity in both masked and unmasked condition and comparing the accuracy level in both the cases using cv2 for image capturing and train DNN model by feeding datasets. It provides a trained deep learning model which predicts the face recognition and detection. In the future, this research work can be utilized with some application-based tool to develop a biometric system. It can also be transferred to IoT platform for larger utilization.

3 Proposed Architecture

As shown in Fig. 1, the proposed architecture is based on cloud where the datasets are stored in cloud-based environment, and the training and test datasets are processed through machine learning algorithm to recognize masked and unmasked faces to recognize person's identity. The subsequent section discussed the stepwise procedure of the proposed architecture in the local environment where the datasets are stored in local machine.

3.1 Stepwise Methodology for Local System

Facial mask detection task and masked face recognition task are two closely connected and different applications for the current popular face masks. The face mask detection task must determine whether or not a person is wearing a mask as needed. As shown in Fig. 2, the purpose of the masked face recognition task is to determine the unique identification of a person wearing a mask. The dataset requirements for each task are different. The former just requires masked face picture samples, whereas the latter requires a dataset that includes both masked and unmasked face photographs of the same subject. Face recognition datasets are more harder to create than other datasets.

The proposed model describes the way to create the dataset by capturing huge number of images with masked faces as well as unmasked faces and saved as .jpg files. It explains about the storage of .jpg files as the dataset. The next process describes the creation of a face detection model using deep neural network. It

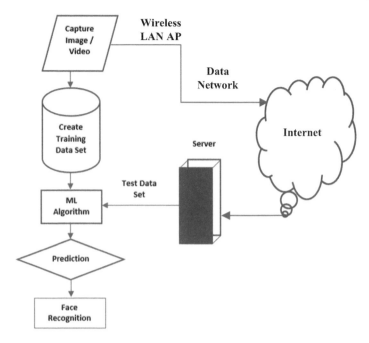

Fig. 1 Cloud-based proposed architecture

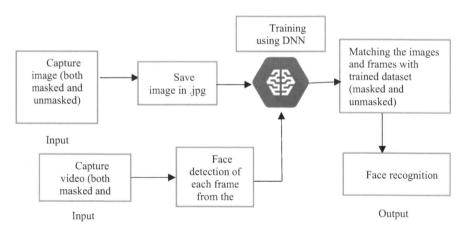

Fig. 2 Stepwise proposed methodology

includes OpenFace model, used for the purpose to train dataset using method "open-face_nn4.small2.v1.t7." After the above process, it describes the way to capture the video dataset of both masked and unmasked faces. Next, the methodology includes the checking of the trained dataset with the video file, frame by frame to check all the faces that appears in it. It detects faces and captures the portion of the face within a square. Finally, the methodology compares all those face frames with the dataset, and detects the face when matches, and labels the square frame with the recognized name.

3.2 Algorithm Steps

Step 1:   Capture image and video through webcam to create dataset
            image = read image
Step 2:   Save images in .jpg and videos in .mp4 format
            dataset = save image/video
Step 3:   Train the model using the created dataset
            train = model training with dataset
Step 4:   If the face recognition confidence is greater than equal to 0.8

      a.   Localize the face in the frame and create a rectangle around it.
      b.   Show the name and accuracy percentage above the rectangle.

else

print unknown image

Image capturing using cv2 and train DNN model by feeding datasets. It provides a trained deep learning model which predicts the face recognition and detection.

The above algorithm has been used as it provides good accuracy with minimal hardware requirement.

3.3 Proposed Dataset

The authors used both masked and unmasked datasets for the above-mentioned proposed methodology.

The dataset was created by capturing numerous images of the test subjects with mask in their face along with bare face images. To make sure of the detection and recognition capacity, amount of the images was huge. These images are then saved in .jpg format and used as training dataset. The testing dataset used is saved in .jpg as well as .mp4 format.

The glimpse of the proposed datasets is furnished as shown in Figs. 3 and 4.

Fig. 3 Masked dataset

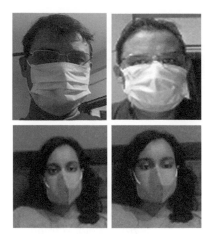

Fig. 4 Unmasked dataset

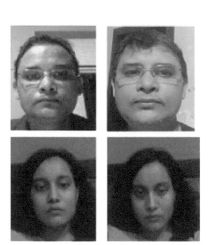

4 Results Analysis and Discussion

After training and testing, the accuracy percentage of our proposed model is shown in Fig. 5. The highest accuracy percentage is also captured and stored in .txt file for each face for both masked and unmasked as shown in the same figure. The authors also captured the average accuracy percentage for both masked and unmasked datasets as given in Table 1. The graphical representation of average accuracy percentage for both masked and unmasked face is shown in Fig. 6.

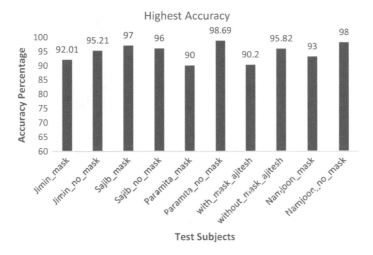

Fig. 5 Highest accuracy achieved for each face

Table 1 Average accuracy percentage with mask and without mask

Name	Average % of accuracy (with mask)	Average % of accuracy (without mask)
Jimin	89.89	90.29
Sajib	95.98	94.44
Namjoon	91.71	93.40
Paramita	88	97.86
Ajitesh	87.75	94

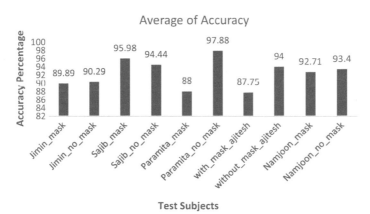

Fig. 6 Graphical representation of average accuracy percentage

The accuracy percentage was determined by calculating the average of the recognition rate of each test subject using the following formula:

$$\text{Average accuracy \%} = \frac{\text{Sum of the accuracy \% found for a test subject}}{\text{Total number of accuracy \% found for the test subject}}$$

The proposed method provides high recognition rate with less processing power unlike various other models. The GPU requirement is low and can work with huge amount of data.

5 Conclusion and Future Scope

Face recognition is a study in the science of biometrics. Still, some of the algorithms are computationally expensive to be useful for real-time processing. It is likely to change as computer technology improves. In this research work, the authors have focused on the significant issues in face recognition system when the face is hidden by a mask along with a solution. The solution provides high recognition rate with less processing power and low GPU requirement. Though the system provides good accuracy when it comes to detection of masked as well as unmasked faces, factors like illumination [19], low resolution image [20] can affect the results. Thus, our future scope is to make the system efficient enough to deal with the above-mentioned factors. Also, this proposed approach can be enhanced by incorporating some artificial intelligence-based applications like attendance tracking system using face recognition technology.

References

1. Gupta, Y., Prasad, A., Touti, S., Sachdev, K., Jaiswal, V., & Naranje, V. (2021). Realtime face recognition: A survey. In *2021 International Conference on Computational Intelligence and Knowledge Economy (ICCIKE)*. https://doi.org/10.1109/iccike51210.2021.9410792
2. Jayaraman, U., Gupta, P., Gupta, S., Arora, G., & Tiwari, K. (2020). Recent development in face recognition. *Neurocomputing, 408*, 231–245. https://doi.org/10.1016/j.neucom.2019.08.110
3. Lahasan, B., Lutfi, S. L., & San-Segundo, R. (2017). A survey on techniques to handle face recognition challenges: Occlusion, single sample per subject and expression. *Artificial Intelligence Review, 52*(2), 949–979. https://doi.org/10.1007/s10462-017-9578-y
4. Damer, N., Grebe, J. H., Chen, C., Boutros, F., Kirchbuchner, F., & Kuijper, A. (2020). The effect of wearing a mask on face recognition performance: An exploratory study. In *2020 International Conference of the Biometrics Special Interest Group (BIOSIG)* (pp. 1–6).
5. Misra, A., Dev, R. K., & Rajasekaran, M. M. (2020). Secured payment system using face recognition technique. In *4th International Conference on the Science and Engineering of Materials (ICoSEM2019)*. https://doi.org/10.1063/5.0028457

6. Spivak, I., Krepych, S., Faifura, V., & Spivak, S. (2019). Methods and tools of face recognition for the marketing decision making. In *2019 IEEE International Scientific-Practical Conference Problems of Infocommunications, Science and Technology (PIC S&T)*. https://doi.org/10.1109/picst47496.2019.9061229

7. Balmik, A., Kumar, A., & Nandy, A. (2021). Efficient face recognition system for education sectors in COVID-19 pandemic. In *2021 12th International Conference on Computing Communication and Networking Technologies (ICCCNT)*. https://doi.org/10.1109/icccnt51525.2021.9579523

8. Jayaprakash, C., & Maheswari, V. (2016). Face recognition alert mechanism and idealness for monitoring user motion by web services. In *2016 10th International Conference on Intelligent Systems and Control (ISCO)*. https://doi.org/10.1109/isco.2016.7727073

9. Wang, Y., Bao, T., Ding, C., & Zhu, M. (2017). Face recognition in real-world surveillance videos with Deep Learning Method. In *2017 2nd International Conference on Image, Vision and Computing (ICIVC)*. https://doi.org/10.1109/icivc.2017.7984553

10. Radzi, S. A., Alif, M. K. M., Athirah, Y. N., Jaafar, A. S., Norihan, A. H., & Saleha, M. S. (2020). IOT based facial recognition door access control home security system using Raspberry Pi. *International Journal of Power Electronics and Drive Systems (IJPEDS), 11*(1), 417. https://doi.org/10.11591/ijpeds.v11.i1.pp417-424

11. Kumar, P. M., Gandhi, U., Varatharajan, R., Manogaran, G. Jidhesh, R., & Vadivel, T. (2017). Intelligent face recognition and navigation system using neural learning for smart security in internet of things. *Cluster Computing, 22*(S4), 7733–7744. https://doi.org/10.1007/s10586-017-1323-4

12. Welsh, B. C., & Farrington, D. P. (n.d.). Closed-circuit television surveillance. *Preventing Crime*, 193–208. https://doi.org/10.1007/1-4020-4244-2_13

13. Albalooshi, F. A., Smith-Creasey, M., Albastaki, Y., & Rajarajan, M. (2018). Facial recognition system for secured mobile banking. *KnE Engineering*, 92–101.

14. Sharma, S., & Kumar, V. (2020). 3D landmark based face restoration for recognition using variational autoencoder and triplet loss. *IET Biometrics, 10*(1), 87–98. https://doi.org/10.1049/bme2.12005

15. Sharma, S., Bhatt, M., & Sharma, P. (2020). Face recognition system using machine learning algorithm. In *2020 5th International Conference on Communication and Electronics Systems (ICCES)*. https://doi.org/10.1109/icces48766.2020.9137850

16. VenkateswarLal, P., Nitta, G. R., & Prasad, A. (2019). Ensemble of texture and shape descriptors using support vector machine classification for face recognition. *Journal of Ambient Intelligence and Humanized Computing*. https://doi.org/10.1007/s12652-019-01192-7

17. Apoorva, P., Impana, H. C., Siri, S. L., Varshitha, M. R., & Ramesh, B. (2019). Automated criminal identification by face recognition using Open Computer Vision classifiers. In *2019 3rd International Conference on Computing Methodologies and Communication (ICCMC)*. https://doi.org/10.1109/iccmc.2019.8819850

18. Yang, C.-W., Phung, T. H., Shuai, H.-H., & Cheng, W.-H. (2022). Mask or non-Mask? robust face mask detector via triplet-consistency representation learning. *ACM Transactions on Multimedia Computing, Communications, and Applications, 18*(1s), 1–20. https://doi.org/10.1145/3472623

19. Chan, C. H., Zou, X., Poh, N., & Kittler, J. (2018). Illumination invariant face recognition. *Computer Vision*, 58–79 https://doi.org/10.4018/978-1-5225-5204-8.ch003

20. Anwarul, S., & Dahiya, S. (2019). A comprehensive review on face recognition methods and factors affecting facial recognition accuracy. In *Lecture notes in electrical engineering* (pp. 495–514). https://doi.org/10.1007/978-3-030-29407-6_36

Personality as a Predictor of Computer Science Students' Learning Motivation

Amanpreet Kaur and Kuljit Kaur Chahal

Abstract Learning motivation is a crucial factor to improve students' academic performance. However, the role of personality traits in developing learning motivation in computer-related studies has been poorly defined. Understanding of inter-relationships between personality and learning motivation of students pursuing graduation in computer science and engineering (CSE) can help increase the chances of success in computer education. The study aimed to investigate the impact of personality traits on the learning motivation of second-year CSE students with the help of structural equation modeling (SEM) and to determine to what extent personality factors explain learning motivation. The study included 92 students pursuing undergraduate degrees in CSE. Based on the study results, it was concluded that personality significantly affects the learning motivation of the students. Personality explains motivation by 55%. In the proposed model, the most influential predictor of motivation is consciousness of the students. It was found that self-efficacy is the most significant among the six learning motivation factors.

Keywords Introductory programming · Computer education · Personality traits · Learning motivation · Human–computer interaction

1 Introduction

Recent years have shown a growing interest in intellectual and non-intellectual factors which can predict students' academic performance. Factors impacting computer students' performance have become an active research topic. Rountree et al. [1] have presented the results of a survey of introductory programming course students, taken to find out the factors that influenced students' success in the course.

A. Kaur (✉) · K. K. Chahal
Department of Computer Science, Guru Nanak Dev University, Amritsar 143005, India
e-mail: amanpreetcs.rsh@gndu.ac.in

K. K. Chahal
e-mail: kuljitchahal.cse@gndu.ac.in

© The Author(s), under exclusive license to Springer Nature Singapore Pte Ltd. 2023   651
S. Shakya et al. (eds.), *Sentiment Analysis and Deep Learning*,
Advances in Intelligent Systems and Computing 1432,
https://doi.org/10.1007/978-981-19-5443-6_50

Research on non-intellectual constructs has established that personality and motivation are the important determinants of academic performance [2, 3]. To predict students' success in the first year of study, Gray et al. [4] took four psychometric indicators, personality, and motivation along with aptitude and learning strategies. Research in computing education has also widely taken computational thinking (CT) into consideration along with personality and motivation. Román-González et al. [5] have found the correlations between CT, self-efficacy, and personality to better understand and assess students' computing efficiency. In particular, the learning motivation constructs of intrinsic goals, extrinsic goals, task value, and self-efficacy have been referred to as (a) learning goal orientation, (b) performance approach goal, (c) academic intrinsic motivation, and (d) academic self-efficacy, respectively, by some related research works [2].

Given the recent interest in understanding the different predictors of computation interest and performance, this paper explores the degree to which personality factors predict motivation of the second-year undergraduate CSE students using SEM. In doing so, it also attempts to assess the confirmatory factor analysis of the learning motivation factors.

2 Literature Review

This section reviews research works related to the prediction of academic performance based on behavioral factors, personality, and learning motivation of the students. First, it describes the research related to personality and motivation taken together and their association with academic performance. Second, it considers the studies focused on personality and motivation individually.

Many empirical studies have investigated how personality traits and learning motivation affect academic success. Busato et al. [6] integrated intellectual ability, learning style, personality, and achievement motivation, in order to examine how these variables correlate with academic success in higher education. High levels of conscientiousness, intrinsic motivation, and low levels of extrinsic motivation have been found to be related to first-quarter school success [7, 16]. Cheng and Ickes [8] found that conscientiousness and self-motivation are "mutually compensatory predictors" of university-level grades. Students who are low in both these factors were unable to perform as well as students who had either a high level of consciousness or a high level of self-motivation. Phillips et al. [9] also concluded that personality can affect achievement through motivation at university-level academic performance.

A number of these studies examined the relationship between the Big Five model of personality dimensions and academic achievement. Richardson and Abraham [10] found the pathways by which Big Five personality dimensions impact academic performance. Poropat [11] also concluded that conscientiousness added slightly more to graded point average (GPA) prediction controlling for the effects of intelligence. Conscientiousness, the most influential factor of personality, is a reasonably constant predictor of academic achievement [6, 10]. So, students with high conscientiousness are more likely to obtain strong academic results because they can create an orderly study plan, acquire the resources needed to make it work, and follow it out responsibly. Schrempft et al. [12] stated that both consciousness and perceived education environment predict medical performance taken as both independent and interactive. However, the remaining four factors extraversion, agreeableness, neuroticism, and openness have mixed outcomes [5, 11, 13].

Learning motivation, the main factor of the study, is considered to be an important predictor of students' academic performance. Motivation, self-regulatory learning strategies, and learning styles have also been concluded to predict academic performance [14, 15]. According to Richardson et al. [2], among the 6 factors of motivation, intrinsic motivation positively correlated with GPA, whereas extrinsic motivation was not significantly correlated with GPA. Also, the work stated that self-efficacy was found to have a large positive correlation with grade goal. Besides, students' motivation and their performance largely get influenced by the type of goals students pursue during their course [22].

No studies, to the best of our knowledge, have found the impact of personality traits on the learning motivation of students in the stream of CSE. There is a gap to explore the impact of personality on students' motivation to perform well on programming assignments and to get expertise in computer education.

3 Methodology

3.1 Research Model and Hypothesis

H1: Personality of second-year CSE students impacts their learning motivation level significantly.

H2: Personality is the major factor that influences second-year CSE students' learning motivation.

3.2 Study Group and Its Characteristics

The study group consists of 92 undergraduate students of CSE, who voluntarily took part in the research. Students pursuing undergraduate degrees in CSE are selected for the study to find the role of personality traits in enhancing learning motivation of the CSE students specifically. The students of the study have already completed their first year of the study. So, it can be assumed that their motivational aspects have some influence on their programming course. The distribution of students by CSE courses and degrees is given in Table 1, of the participants, 47% were female and 53% were male.

Table 1 Sample distribution by course and gender

Course	Total	Male	Female
Bachelors in Computer Application	40	22	18
B. Tech (CSE)	52	28	24

3.3 Data Collection Instruments

Validated psychological tests for personality and learning motivation were used. Figures 1 and 2 briefly define the underlying factors of personality and learning motivation, respectively.

The first data collection tool used to measure personality traits is the 5-point Likert scale "NEO-FFI" developed by Costa and McCrae [17]. 60 items of the scale were divided into 5 factors. Cronbach alpha consistency coefficient calculated for the questionnaire is at an acceptable level (0.645) considering three out of five factors (extraversion, agreeableness, conscientiousness).

The second data collection tool to measure learning motivation, the 7-point Likert scale "Motivated Strategies for Learning Questionnaire (MSLQ)", developed by Pintrich et al. [18], is used. The scale includes 31 items of 6 factors. Cronbach alpha consistency coefficient calculated for the questionnaire is at an acceptable level (0.787).

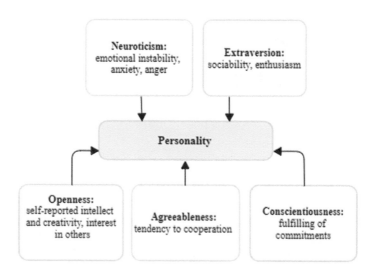

Fig. 1 Factors of personality measured with NEO-FFI

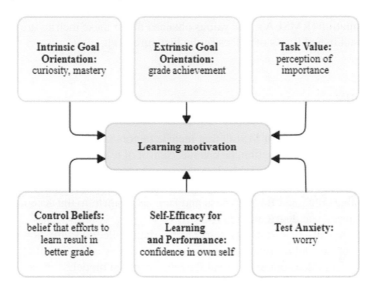

Fig. 2 Factors of learning motivation measured with MSLQ

3.4 Data Collection

For data collection, the validated questionnaires, NEO-FFI and MSLQ, were converted into online forms. Students answered the online-created questionnaires for data collection. Before answering the data collection instrument, students were provided with information about the research's goal. Participants who volunteered had not given any personal information. The participants were given 30 min to mark their choices.

3.5 Data Analysis

For data analysis, Python was used for performing data preprocessing. After that, SEM was implemented with the AMOS 26 software of SPSS. To find the existence of relations between students' personality and motivation factors, AMOS 26 software calculates SEM coefficient using maximum likelihood technique. After running SEM analysis, AMOS provides output both in text and graphic form including un-standard and standard regression weights to estimate the relation, model fit indices, etc. For training, regression weight for error to dependent variable was set to 1 and for one factor from each latent variable was also set to 1. For finding the compatibility level of relation patterns in the SEM and to validate the proposed model, the following fit index tests were used: chi-square fit index test (χ^2), goodness of fit index (GFI), normed fit index (NFI), comparative factor index (CFI), the root mean square error of approximation (RMSEA). The values obtained under these metrics are compared with the established limits of values in the literature [21].

4 Results

The descriptive statistics of MSLQ that include factors such as intrinsic goal orientation, extrinsic goal orientation, task value, control of learning beliefs, self-efficacy, and test anxiety are given in Table 2. Analyzing the results, it can be stated that: (a) all factors display negative skewness, (b) kurtosis values range between -0.75 and $+2$. The values obtained for skewness and kurtosis confirm to the acceptable range of normal univariate distribution [19, 20].

The coefficients of constructs selected for data collection are shown as SEM in Fig. 3. Before finding relations between constructs, it is important to verify the goodness of fit indices obtained to validate the proposed model.

Table 3 represents the range and the value obtained for each goodness of fit index and confirms a good fit as compared to good fit values from Hu and Bentler [21]. It is found as $\chi^2 = 72.50$, $p = 0.002$, CMIN/DF $= 1.77$, RMSEA $= 0.09$, GFI $= 0.88$, CFI $= 0.90$, NFI $= 0.81$, and IFI $= 0.91$. According to these values, it can be seen that the SEM developed for the study validates the condition of good agreement of data with the model.

The SEM as in Fig. 3 formed learning motivation as latent variable with chi-square $= 72.50$ and $p = 0.002$. When the model was analyzed, it was found that personality estimated motivation positively and significantly ($\gamma = 0.55$).

Among the 6 factors of motivational strategies for learning latent variable, task value has the highest factor loading. Personality has consciousness as the highest factor loading among the 5 factors.

Table 2 Statistics for MSLQ

		Intrinsic goal orientation	Extrinsic goal orientation	Task value	Control of learning beliefs	Self-efficacy	Test anxiety
N	Valid	91	91	91	91	91	91
	Missing	00	0	0	0	0	0
Mean		5.50	5.41	5.64	5.31	5.38	3.45
Std. error of mean		0.11	0.11	0.09	0.10	0.10	0.12
Median		5.7	5.50	5.83	5.25	5.62	3.50
Mode		5.00	6.25	6.00	5.25	6.00	3.83
Std. deviation		1.05	1.11	0.92	0.97	1.01	1.19
Skewness		-0.88	-0.48	-1.10	-0.70	-0.82	-0.18
Kurtosis		0.73	-0.56	2.00	0.43	0.44	-0.75

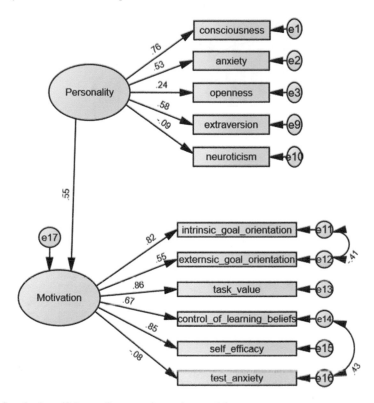

Fig. 3 Standard coefficients of structural equation model

Table 3 Values of SEM goodness of fit indices

	χ^2/SD	RMSEA	GFI	CFI	NFI	IFI
Acceptable values	≤ 3	<0.09	≥ 0.80	≥ 0.90	≥ 0.80	≥ 0.80
Structural model	1.77	0.09	0.88	0.90	0.81	0.91

Figure 4 elaborates the confirmatory factor analysis results of the MSLQ scale. The results of the analysis validated the scale in the context of this study. The fit indices of the scale were acceptable ($\chi^2 = 963.35$, $p = 0.000$, CMIN/DF $= 2.31$, RMSEA $= 0.09$, GFI $= 0.72$, CFI $= 0.83$, NFI $= 0.74$, and IFI $= 0.84$) [21].

The variance values of factors indicating the most influential factor are given in Table 4; it can be observed that the variance value of extrinsic goal orientation was the highest, and the variance value of control beliefs was found to be the lowest. Further as in the review of covariance values between sub-factors, as given in Table 5, intrinsic goal orientation and control beliefs were found to be highly correlated with each other, whereas self-efficacy and test anxiety were correlated at the lowest value.

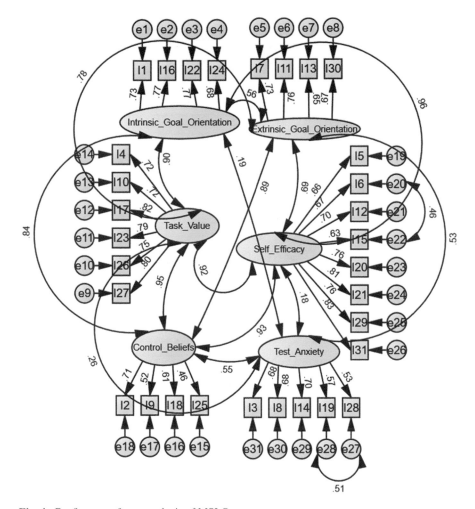

Fig. 4 Confirmatory factor analysis of MSLQ

Table 4 Variance matrix for motivation

	Intrinsic goal orientation	Extrinsic goal orientation	Task value	Control beliefs	Self-efficacy	Test anxiety
Variance	1.14	**1.43**	1.07	0.59	0.88	0.94

According to results, H1: Personality of second-year CSE students impacts their learning motivation level significantly is accepted as obtained *p*-value < 0.01 for the path coefficient 0.55. As the results show personality significantly influences students' motivation to achieve success in CSE.

Table 5 Covariance matrix for motivation

		1	2	3	4	5
1	Intrinsic goal orientation					
2	Extrinsic goal orientation	0.71				
3	Task value	**0.99**	0.96			
4	Control beliefs	0.69	0.82	0.76		
5	Self-efficacy	0.96	0.77	0.90	0.67	
6	Test anxiety	0.19	0.62	0.26	0.41	0.17

H2: Personality is the major factor that influences second-year CSE students' learning motivation is rejected. There are other factors that influence motivation as personality defines only by the variance of 0.36.

5 Discussion and Future Work

The aim of this study was to find the impact of personality on learning motivation of second-year undergraduate students pursuing programming courses. A structural equation model was developed to predict the relationships between two key factors personality traits and learning motivation, which influence students' programming learning. The proposed hypotheses were supported by the results.

The results provide empirical evidence that second-year computer CSE students' learning motivation is majorly influenced by students' personality traits. The findings are in parallel with studies that elaborated on the significance of the relationship between students' motivation and their personality in university-level academic performance, where non-CSE students from other disciplines were taken [6–10].

The results also indicated that the task value factor among other factors of motivation plays a major role in CSE students. It can be concluded that students' motivation is driven by the students' perceptions of the course in terms of interest and importance. The model obtained self-efficacy with the second-highest factor loading. Students' self-perception of the skills to learn may have an influence on their programming performance [23]. This finding overlaps with the previous research works which argued that self-efficacy and intrinsic motivation have major factor loadings [2, 5].

In the case of personality traits, consciousness is the most influencing sub-factor in this research. The result is in parallel with studies that also emphasized that consciousness is more strongly associated with students' academic performance among the five-factor personality traits [2, 11, 12]. Chamorro-Premuzic and Furnham [14] showed consciousness as the main personality factor positively correlated with the academic performance of non-CSE students. Higgins et al. [16] demonstrated that students high in consciousness are more motivated to perform good in the course. Pintrich [15] also related consciousness with effort regulation and academic self-efficacy of college students of various streams.

The study can further explore the individual differences of CSE students. Further analysis can be done at the end of the course that compares students' final performance who expressed high motivation with those who expressed low motivation. The analysis can reveal interesting patterns related to students' motivation and success.

Acknowledgements The authors pay homage to the Department of Computer Science and all students of GNDU for their support in conducting this study. This work is financially supported by the Maulana Azad National Fellowship of India.

References

1. Rountree, N., Rountree, J., & Robins, A. (2002). Predictors of success and failure in a CS1 course. *ACM SIGCSE Bulletin, 34*(4), 121–124.
2. Richardson, M., Abraham, C., & Bond, R. (2012). Psychological correlates of university students' academic performance: A systematic review and meta-analysis. *Psychological Bulletin, 138*(2), 353–387.
3. Farsides, T., & Woodfield, R. (2003). Individual differences and undergraduate academic success: The roles of personality, intelligence, and application. *Personality and Individual Differences, 34*(7), 1225–1234.
4. Gray, G., Mcguinness, C., & Owende, P. (2014). An application of classification models to predict learner progression in tertiary education. In *2014 IEEE International Advance Computing Conference (IACC)*. IEEE (pp. 549–54)
5. Román-González, M., Moreno-león, J., Robles, G., & Pérez-González, J.-C. (2018). Extending the nomological network of computational thinking with non-cognitive factors. *Computers in Human Behavior [Internet], 80*, 441–459.
6. Busato, V. V., Prins, F. J., Elshout, J. J., & Hamaker, C. (2000). Intellectual ability, learning style, personality, achievement motivation and academic success of psychology students in higher education. *Personality and Individual Differences, 29*, 1057–1068.
7. Kaufman, J. C., Agars, M. D., & Lopez-wagner, M. C. (2008). The role of personality and motivation in predicting early college academic success in non-traditional students at a Hispanic-serving institution. *Learning and Individual Differences [Internet], 18*(4), 492–496.
8. Cheng, W., & Ickes, W. (2009). Conscientiousness and self-motivation as mutually compensatory predictors of university-level GPA. *Personality and Individual Differences [Internet], 47*(8), 817–822.
9. Phillips, P. R. U., Abraham, C., & Bond, R. O. D. (2003). Personality, cognition, and university students' examination performance. *European Journal of Personality, 17*(6), 435–48.
10. Richardson, M., & Abraham, C. (2009). Conscientiousness and achievement motivation predict performance. *European Journal of Personality, 23*(7), 589–605.
11. Poropat, A. E. (2009). A meta-analysis of the five-factor model of personality and academic performance. *Psychological Bulletin, 135*(2), 322–338.
12. Schrempft, S., Piumatti, G., & Baroffio, M. W. G. A. (2021). Pathways to performance in undergraduate medical students: Role of conscientiousness and the perceived educational environment. *Advances in Health Sciences Education [Internet], 26*, 1537–1554.
13. Zuf, A., Alessandri, G., Gerbino, M., Paula, B., Kanacri, L., Di, L., et al. (2013). Academic achievement: The unique contribution of self-efficacy beliefs in self-regulated learning beyond intelligence, personality traits, and self-esteem. *Learning and Individual Differences, 23*, 158–162.
14. Chamorro-Premuzic, T., & Furnham, A. (2008). Personality, intelligence and approaches to learning as predictors of academic performance. *Personality and Individual Differences, 44*, 1596–1603.

15. Pintrich, P. R. (2004). A conceptual framework for assessing motivation and self-regulated learning in college students. *Educational Psychology Review, 16*(4), 385–407.
16. Higgins, D. M., Peterson, J. B., Pihl, R. O., & Lee, A. G. M. (2007). Prefrontal cognitive ability, intelligence, big five personality, and the prediction of advanced academic and workplace performance. *Journal of Personality and Social Psychology, 93*(2), 298–319.
17. Costa, Jr, P. T., McCrae, R. R. (2008). The revised NEO personality inventory (NEO-PI-R). In G. J. Boyle, G. Matthews, & D. H. Saklofske (Eds.), *The SAGE handbook of personality theory and assessment* (Vol. 2, pp. 79–198). Sage Publications.
18. Pintrich, P. R., Smith, D. A., Garcia, T., & McKeachie, W. J. (1991). A manual for the use of the motivated strategies for learning questionnaire (MSLQ). Office of Educational Research and Improvement (ED).
19. George, D. (2011). *SPSS for windows step by step: A simple study guide and reference, 17.0 update, 10/e.* Pearson Education India (2011).
20. Gravetter, F. J., Wallnau, L. B., Forzano, L. A., & Witnauer, J. E. (2020). *Essentials of statistics for the behavioral sciences.* Cengage Learning.
21. Hu, L. T., & Bentler, P. M. (1999). Cutoff criteria for fit indexes in covariance structure analysis: Conventional criteria versus new alternatives. *Structural Equation Modeling: A Multidisciplinary Journal, 6*(1), 1–55.
22. Coffin, R. J., & MacIntyre, P. D. (1999). Motivational influences on computer-related affective states. *Computers in Human Behavior, 15*(5), 549–569.
23. Durak, H. Y., Saritepeci, M., & Durak, A. (2021). Modeling of relationship of personal and affective variables with computational thinking and programming. *Technology, Knowledge and Learning,* 1–20.

Hybrid Sign Language Learning Approach Using Multi-scale Hierarchical Deep Convolutional Neural Network (MDCnn)

Swati Sharma and Varun Prakash Saxena

Abstract Sign language is now the only means of communication for deaf or hard-of-hearing people. SLR has been the trendiest research area in recent years. Determining an accurate result from many photo data sets is crucial for an automatic sign language recognition system. Regardless, convolutional neural networks (CNNs) outperform traditional neural networks in a wide range of visual tasks. Improved results in numerous measures, including as accuracy, recall, and FI score, are required for many resources and methodologies. In this context, we present convolutional neural network model approach with multi-scale layers and various filters that serve as an additional layer in the design. This model could be used on a variety of data sets and is also useful for real-time picture streaming. Two self-created data sets were utilized to test the performance of the proposed model: ASL and ISL data sets. Transfer learning models like as VGG16, MobileNet, VGG19, MobileNetV2, ResNet50, and InceptionV3 are also used to evaluate these data sets. The experimental findings reveal that our suggested model performs better than the basic CNN model, with an accuracy of 98% using ISL data set and 95% using ASL data set.

Keywords American sign language (ASL) · Convolutions neural network (CNN) · Indian sign language (ISL) · Sign language recognition (SLR)

S. Sharma
Government Women Engineering College, Ajmer, Rajasthan, India
e-mail: swati.s2207@gmail.com

V. P. Saxena (✉)
Department of CSE, Government Women Engineering College, Ajmer, Rajasthan, India
e-mail: varunsaxena82@gmail.com

1 Introduction

1.1 Sign Language

The origins of sign language may be traced back to the seventeenth century. In the recent years, more than 300 sign languages have been employed to bridge the gap between normal and hearing-impaired people. Hand gestures, body motions, and facial expressions are used in sign language. Signs can be done with either one or both hands. Three strategies are primarily utilized to recognize signs. Sensor-based technique, vision-based technique, and hybrid approach are all part of it. Sensor-based techniques make use of gloves; hence, they are also known as glove-based techniques. Sensors such as accelerometers, proximity sensors, and flexion sensors are commonly employed for this purpose. The biggest downside of this approach is its low cost. Not everyone can afford a cyber-glove for gesture recognition. To address the issue of sensor-based procedures, vision-based approaches were implemented. The vision-based method often employs machine learning approaches with an in-built camera and digital image processing techniques [1] to extract the attributes required to detect the sign. Raw gesture data is captured and merged with glove-camera-based devices in the hybrid approach. This approach involves error elimination to increase overall accuracy and precision. Little effort has been done on this due to the price and computational overheads of the entire system. Enhanced systems, on the other hand, produce promising results when applying hybrid tracking approaches.

1.2 Convolutional Neural Network (CNN)

CNN is just a feed-forward neural network that is widely used to analyse text and visual pictures using a grid-like structure to process input. CNN is often referred to as "ConvNet". CNN is made up of the following layers as shown in Fig. 1.

1. **Input Layer**—Input layer receives an array of data as input.
2. **Hidden Layer**—Hidden layers extract features by conducting various computations and manipulations. There are several levels concealed beneath the surface, as follows:

   - **Convolution Layer**—This layer has a number of filters. To find patterns in the data, this layer employs the matrix filter and conducts a convolution operation (image). Every CNN starts with a convolution process. Convolution processes the incoming pictures through a series of convolutional filters, each of which activates different aspects of the images.
   - **Rectified Linear Unit (ReLU Layer)**—A rectified feature map of the data is obtained by applying the ReLU activation function to the convolution layer (image). It executes an element-by-element action. All negative pixels are set to zero . Nonlinearity is introduced to the network. This layer produces

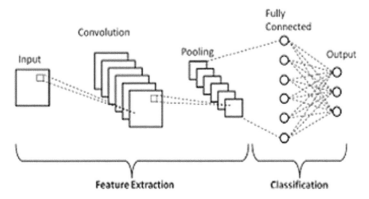

Fig. 1 Layers of CNN

a corrected feature map as its output. By mapping negative values to zero and preserving positive values, the rectified linear unit (ReLU) enables for faster and more efficient training. A feature map is inherently dependent on the learning model class utilized and the data's "input space" χ. A feature map for $\mathcal{F}$ on an input space χ is a (just) function: $\chi \rightarrow \mathrm{R}^d$ if $\mathcal{F}$ is a class of models ranging from R^d to R^D

$$\phi : \chi \rightarrow R^d$$

- **Pooling Layer**—To detect distinct [2] sections of the data, the pooling layer employs numerous filters (image). It collects data by pooling information. A pooling layer is now applied to the corrected feature map. Pooling is a down sampling process that decreases the feature map's dimensionality. The resultant 2D arrays formed from pooled feature maps are converted to a 1D continuous linear vector during the flattening process. Pooling reduces the number of parameters that the network has to learn by conducting nonlinear down-sampling on the output.

  Over tens or hundreds of layers, these three steps are repeated, with each layer learning to recognize distinct characteristics.

3. **Output (Fully Connected) Layer**—At last, there is a fully connected layer which identifies the data items (image). The fully connected layer uses the flattened matrix formed by the pooling layer to identify or classify the data (image).

1.3   Transfer Learning Models

Thousands of data sets have already been used to build transfer learning models. Transfer learning is basically the reuse of previously learned models. It can do picture classification, image prediction, and natural language processing utilizing the information obtained from prior tasks. For the vast data set, deep learning fared much better than machine learning in the survey. For picture prediction and classification, we employed pre-trained models such as VGG16, ResNet50, MobileNetV2, MobileNet, VGG19, and InceptionV3 in this study.

2   Literature Survey

In 2021, Sharma and Singh [3] created vision-based hand motion detection with deep learning for sign language interpretation. They provide a deep learning-based CNN model for sign language recognition. They compared the results to VGG16 and VGG11 while evaluating the efficacy of their proposed model. The model was tested and trained on two data sets: Indian sign language (ISL) gestures [4, 5], which consist of 2150 images recorded with an RGB camera, and an openly available American sign language (ASL) data set, which was used in this study. The accuracy of the ISL [6] and ASL databases was 99.96 and 100%, respectively. It is also tested with data augmentation, and it is shown to be rotation and scaling invariant.

Pre-trained models like MobileNetV2, MobileNet, VGG19, VGG16, and ResNet50 are used in this research by Ja et al. [7]. The findings show that diverse models work effectively in different scenarios. Huang et al. [8] proposed a sensor-based system that uses deep neural networks in the past. It uses finger joint coordinates as input and two data sets to compare the performance of DNN and SVM. Badhe and Kulkarni [9] has created an algorithm [10, 11] for ISL to English translation [12] that has a 97.5% accuracy.

Hafiz et al. [13] presented an innovative MFIF technique in the past. To extract the most essential shared properties from a pair of multi-focus photos, they used multi-scale convolutions and skip connections as part of the recommended network. Furthermore, the combined images are reconstructed in a multi-scale manner to enable a more exact restoration. For testing and assessing the recommended model, images of various sizes might be used. According to the results of testing on a number of test photographs, this strategy produces higher-quality fused images than state-of-the-art image fusion approaches.

In 2021, Shruti [14, 15] proposed a framework to identify sign language using American sign language data sets. It may forecast indications ranging from 0 to 9 on a scale of 0–9. Through image processing, RGB images are transformed to grayscale. Its purpose is to search a way to combine deep learning architecture with image processing in mobile applications or embedded single board computers that is easy to execute. With LeNet-5 and AlexNet, deeper networks such as VGG16 and

MobileNetV2 are used with this. Only 10 layers make up the final design, which includes a dropout layer that boosted training and testing accuracy to 91.37 and 87.5%, respectively. Aditya and Rajesh [16] created a CNN [17] and tested it on different data sets using the stochastic gradient descent with momentum optimization function, which also has a momentum value of 0.90 and a learning rate of 0.01 as well as an accuracy of 94.7% for the NUS data set and 99.96% for the ASL data set.

Manasi et al. [18] proposes a four-step approach for recognizing hand gestures: (1) Using a webcam, create a live stream of hand movements. (2) Using video frames, create visuals from the video. (3) Prepare these photographs by pre-processing them. (4) Recognize hand motions in sign language and transform them to text or audio output. The image processing and neural network ideas are used to construct the system. Kaggle data set is used with their data set, and a data set created by merging the two to evaluate the suggested models. Background variation is introduced to reduce the uncertainty in the results. The majority of models produce identical test results for both basic and complex backgrounds.

Owing to inter-class similarities and intra-class heterogeneity, several conventional ways to understanding sign language are increasingly being applied, although achieving high accuracy remains a challenging problem, and vision-based finger-spelling recognition remains difficult. The RGB and depth modalities are taken into account by the Rajan's model [19]. Depth information can properly identify and manage finger occlusion and hand forms. In comparison with current algorithms that analyse RGB-D pictures individually, a fine-tuned dual-path network is proposed, which recognizes finger-spelling representations in distinct RGB and depth channels and eventually combines the characteristics gathered from both tracks.

3 Methodology

We built a simple CNN model with some advanced hybrid features in this article. In this technique, we obtain some exceptional performance of convolutional neural networks (CNNs) for analysing visual images and image categorization during sign language recognition. Furthermore, we discovered some improved outcomes in a variety of measures including as accuracy and precision. The model works with a variety of resource data sets which are self-created. TensorFlow version 2.2.0 and several Keras libraries with initial image size and batch size of 64 are the platforms for our hybrid CNN model. For the recognition of image using CNN, the following steps are used:

1. Images are taken from web camera as input.
2. Pre-processing of images, removal of background images, and converting to grayscale.
3. Feature extraction and classification.
4. Output in the form of text and speech.

The data set for pattern recognition is split into two parts: training and testing. The photos that the system is trained on make up the training set. The system is put to the test using available data. It is a set of data that is used to check if a system's output is correct once it is been trained. In most situations, just 20% of the data in the data set is used for testing. Data from the tests is used to evaluate the system's accuracy. Classic CNNs learn by collecting local information from pictures and using convolutional layers to train them. In this feature extraction, the size of the convolutional kernel is a crucial issue. Small kernels can extract short edges or low-frequency material, but not high-frequency picture features. Large kernels, on the other hand, extract more specific information from pictures while avoiding the use of low-frequency material. As the filter size or even alternative sizes are utilized, the process of a convolutional layer becomes more complex, and calculations grow more complicated. As a result, training progresses slowly. To overcome this challenge, we opted to collect low- and high-frequency characteristics in each layer using multi-scale convolutions on the same layers of the CNN (Table 1).

The multi-scale block in the first layer, which is made up of four unique filters or 4 distinct sets of convolutional filters with sizes of $1 \times 1, 3 \times 3, 5 \times 5$, and 7×7, introduces multi-scale feature extraction in our recommended model. The output in feature maps is blended with the spectral dimension and transmitted into the Max Pooling layer through a concatenation operation at the conclusion of each convolution phase. As mentioned in the algorithm, we input it into the dropout layer, which is accompanied by two convolutional layers, each of which includes another Max Pooling and dropout layer. After converting the input picture, we flatten the images into the column vector. The flattened output is given to a feed-forward neural network with back propagation in the training set. Over a number of epochs, the model will be able to discriminate between dominating and low-level characteristics in pictures and categorize them using the Softmax classification method as illustrated in Fig. 2.

Figure 3 shows the summary of flow chart shown as first the input layer size is same as that image is the same size (64, 64, 1). Keras then adds a new dimension for processing multiple batches, or training numerous pictures in each epoch step. Because batch size varies, none is used to represent it. As a result, the input shape changes (none, 64, 64, 1). Four convolutional layers each of filter size $1 \times 1, 3 \times 3, 5 \times 5$, and 7×7 are merged or concatenated and fed to Max Pooling layer and dropout layer. The series steps are repeated two times for convolutional, Max Pooling, and dropout. Then, the flatten layer generates a 1D vector from all pixels in all channels. After that dense layer (none, 256) and (none, 512) are followed by batch normalization, ReLU activation, and dropout layer. Lastly, dense layer with 27 classes and Softmax activation function is applied with a learning rate of 0.0005.

To create the model, three convolutional layers are necessary, which are related to the two dense layers or completely connected layers among the neurons. The suggested technique was evaluated using two self-created data sets. Both the double handed sign language gestures (ISL) and the single handed sign language (ASL) gestures have a black and white backgrounds with 1–26 hand gesture classes A–Z alphabets, with the 27th blank. A data set of Indian and American sign languages

Table 1 Precision for all models

Model	Proposed CNN		CNN		MobileNet		MobileNetV2		VGG16		VGG19		ResNet50		InceptionV3	
Data set	ISL	ASL	ISL	ASL	ISL	ASL	ISL	ASL	ISL	ASL	ISL	ASL	ISL	ASL	ISL	ASL
A	1	1	1	1	1	0.98	0.98	0.94	0.94	0.96	0.99	0.96	0.95	0.90	0.93	0.94
B	0.95	0.92	0.85	0.97	1	1	0.99	0.98	0.99	0.91	0.98	0.98	0.93	0.98	0.97	0.96
C	1	0.95	0.99	1	1	1	1	1	0.99	1	0.99	1	1	0.99	0.92	0.98
D	1	1	0.98	1	0.99	0.91	1	0.97	0.97	0.98	0.99	0.97	0.99	0.92	0.94	0.80
E	0.89	0.95	0.98	0.96	0.98	0.93	0.90	1	0.96	0.96	0.96	0.98	0.97	0.96	0.82	0.87
F	0.96	0.98	0.93	1	0.98	0.92	0.95	1	0.96	0.96	0.95	0.99	0.97	0.99	0.84	0.83
G	1	0.96	0.93	0.99	0.98	1	0.99	0.88	0.97	0.99	1	1	0.98	0.98	0.92	0.96
H	0.98	1	0.97	1	1	0.98	0.971	1	0.92	0.98	0.99	0.99	0.97	0.97	0.97	0.86
I	1	0.95	0.83	0.78	0.99	1	0.97	0.97	0.97	0.93	0.94	0.92	0.95	0.90	0.97	0.82
J	0.99	0.91	1	0.99	0.99	0.98	0.97	0.99	0.99	0.93	0.99	0.99	0.99	0.99	0.97	0.92
K	0.92	1	1	0.46	1	0.97	1	0.99	1	0.97	0.99	0.91	0.99	0.81	0.91	0.85
L	0.92	1	0.601	1	0.98	0.99	0.99	0.99	0.98	0.99	0.96	1	1	0.97	0.87	0.94
M	0.95	0.90	0.93	0.71	0.93	0.93	0.78	0.82	0.93	0.95	0.96	0.95	0.94	0.88	0.69	0.93
N	1	0.87	1	0.97	0.96	0.85	0.67	0.87	0.89	0.90	0.87	0.66	0.78	0.91	0.67	0.84
O	1	1	1	0.96	0.98	1	0.99	1	0.97	1	0.98	1	0.94	1	0.99	0.94
P	1	1	1	1	0.99	0.99	0.91	0.98	1	0.99	0.99	0.91	0.98	0.98	0.94	0.88
Q	0.99	1	1	1	0.971	1	0.97	0.99	0.97	0.99	0.99	1	0.89	0.95	0.89	0.84
R	1	0.95	0.95	1	0.94	0.95	1	0.80	0.95	0.85	0.96	0.85	0.91	0.86	0.92	0.83
S	0.99	0.87	0.99	0.84	0.98	0.92	0.98	0.31	0.97	0.93	0.97	0.98	0.97	0.79	0.78	0.56
T	1	0.94	0.98	0.95	0.99	0.76	1	0.74	0.98	0.66	0.99	0.59	0.97	0.83	0.89	0.86

(continued)

Table 1 (continued)

Model	Proposed CNN		CNN		MobileNet		MobileNetV2		VGG16		VGG19		ResNet50		InceptionV3	
Data set	ISL	ASL	ISL	ASL	ISL	ASL	ISL	ASL	ISL	ASL	ISL	ASL	ISL	ASL	ISL	ASL
U	0.99	1	0.96	1	0.98	1	0.97	0.93	0.90	0.86	0.96	0.86	0.95	1	0.90	0.72
V	0.98	0.99	0.86	0.99	0.97	0.99	0.96	0.78	0.90	1	0.97	0.95	0.83	0.90	0.90	1
W	1	0.87	0.99	1	1	0.88	0.98	1	1	0.81	1	1	0.99	1	0.86	0.87
X	0.88	0.90	1	0.72	10.98	0.98	0.95	0.97	0.91	0.97	0.93	0.94	0.96	0.97	0.73	0.81
Y	1	0.63	1	1	0.99	1	0.90	1	0.98	0.99	0.96	0.98	0.92	1	0.84	0.86
Z	1	0.96	0.86	0.72	1	0.99	1	0.97	1	0.91	0.98	0.80	1	0.95	0.98	0.94
BLANK	1	1	0.95	1	1	1	1	1	1	1	1	1	1	1	1	1

Fig. 2 Flow chart of our proposed model

Layer (type)	Output Shape	Param #	Connected to
input_1 (InputLayer)	[(None, 64, 64, 1)]	0	
conv2d (Conv2D)	(None, 64, 64, 16)	32	input_1[0][0]
conv2d_1 (Conv2D)	(None, 64, 64, 16)	160	input_1[0][0]
conv2d_2 (Conv2D)	(None, 64, 64, 16)	416	input_1[0][0]
conv2d_3 (Conv2D)	(None, 64, 64, 16)	800	input_1[0][0]
concatenate (Concatenate)	(None, 64, 64, 64)	0	conv2d[0][0] conv2d_1[0][0] conv2d_2[0][0] conv2d_3[0][0]
max_pooling2d (MaxPooling2D)	(None, 32, 32, 64)	0	concatenate[0][0]
dropout (Dropout)	(None, 32, 32, 64)	0	max_pooling2d[0][0]
conv2d_4 (Conv2D)	(None, 32, 32, 128)	73856	dropout[0][0]
max_pooling2d_1 (MaxPooling2D)	(None, 16, 16, 128)	0	conv2d_4[0][0]
dropout_1 (Dropout)	(None, 16, 16, 128)	0	max_pooling2d_1[0][0]
conv2d_5 (Conv2D)	(None, 16, 16, 256)	295168	dropout_1[0][0]
max_pooling2d_2 (MaxPooling2D)	(None, 8, 8, 256)	0	conv2d_5[0][0]
dropout_2 (Dropout)	(None, 8, 8, 256)	0	max_pooling2d_2[0][0]
flatten (Flatten)	(None, 16384)	0	dropout_2[0][0]
dense (Dense)	(None, 256)	4194560	flatten[0][0]
batch_normalization (BatchNorma	(None, 256)	1024	dense[0][0]
activation (Activation)	(None, 256)	0	batch_normalization[0][0]
dropout_3 (Dropout)	(None, 256)	0	activation[0][0]
dense_1 (Dense)	(None, 512)	131584	dropout_3[0][0]
batch_normalization_1 (BatchNor	(None, 512)	2048	dense_1[0][0]
activation_1 (Activation)	(None, 512)	0	batch_normalization_1[0][0]
dropout_4 (Dropout)	(None, 512)	0	activation_1[0][0]
dense_2 (Dense)	(None, 27)	13851	dropout_4[0][0]

Total params: 4,713,499
Trainable params: 4,711,963
Non-trainable params: 1,536

Fig. 3 Structure of multi-scale CNN

which is self-created was used to train the model. In ISL training set with 20,035 images belonging to 27 classes and 6750 testing images belonging to 27 classes. In ASL training set with 17,649 images belonging to 27 classes and 6230 testing images belonging to 27 classes. During the training phase, the images are fed into a three-with four filters of sizes $1 \times 1, 3 \times 3, 5 \times 5$, and 7×7 to extract the features, which is then fed into a Max Pooling layer with filter size $(2, 2)$ and dropout layer. It is then flattened and sent through two completely connected layers after going through two convolution layers, Max Pooling and dropout. The Softmax activation with a learning rate of 0.005 was employed in this study. In double handed gestures (ISL data set), the training procedure took 5 epochs with a 97.79%, and 15 epochs with a 95.21% in single handed gestures (ASL data set). The experiment is also carried out with the data sets for the transfer learning models VGG16, VGG19, MobileNet, MobileNetV2, ResNet50, and Inception V3. The comparison Table 2 illustrates the results of our suggested model when tested with both our self-created data sets and indicates that it outperforms the standard CNN model. Both of our self-created data sets are put through their paces with transfer learning models.

4 Result and Conclusion

Multi-scale is used to modify CNN for SLR recognition, in which four sets of convolutional filters of Size $1 \times 1, 3 \times 3, 5 \times 5$, and 7×7 are utilized with the input in the first layer, and the outputs of these filters are concatenated and fed into further convolutional layers. Layered convolutional operation refers to the step by step convolutional process as shown in the flow chart (Fig. 2) which manifest that adding a first layer as multi-scale layer unveil more accurate result. Both of our self-created data sets have been evaluated using CNN and transfer learning models like as VGG16, VGG19, MobileNet, MobileNet V2, ResNet50, and Inception V3 and have shown to be accurate in all of them as shown in Fig. 4. Statistical measurements including as accuracy and precision are used to assess each model's performance, revealing that the proposed model has a greater recognition capacity.

For both the ASL and ISL data sets, the epoch values achieved for train accuracy and validation accuracy are shown in the graph above as in Fig. 5. Our model shows better performance and accuracy when compared to traditional CNN architecture. In future, proposed model can further be improved for detecting sentences in sign language recognition.

Table 2 Comparison table for all models

Model	Data set	Epochs	Step-loss	Accuracy	Val-loss	Val-accuracy
Proposed model	ISL	5	0.0225	0.9930	0.0939	0.9779
CNN	ISL	5	0.0080	0.9987	0.2388	0.9418
VGG16 with data augmentation	ISL	5	0.1112	0.9882	0.4452	0.9604
VGG19 with data augmentation	ISL	5	0.1172	0.9882	0.5262	0.9601
MobileNet with data augmentation	ISL	5	0.0403	0.9952	0.0834	0.9888
MobileNetV2 with data augmentation	ISL	5	0.1867	0.9830	0.7425	0.9501
Resnet50 with data augmentation	ISL	5	0.3867	0.9793	1.0454	0.9510
InceptionV3 with data augmentation	ISL	5	0.1970	0.9395	0.3713	0.8951
Proposed CNN model	ASL	5	0.0143	0.9955	0.3347	0.9521
CNN	ASL	5	0.0060	0.9983	0.3733	0.9311
VGG16 with data augmentation	ASL	5	0.2008	0.9783	0.9097	0.9293
VGG19 with data augmentation	ASL	5	0.2169	0.9774	1.6134	0.9257
MobileNet with data augmentation	ASL	5	0.08434	0.9860	0.3391	0.9618
MobileNetV2 with data augmentation	ASL	5	0.2811	0.9706	1.0739	0.9280
Resnet50 with data augmentation	ASL	5	0.7250	0.9592	0.9364	0.9514
InceptionV3 with data augmentation	ASL	5	0.2843	0.9124	0.4979	0.8684

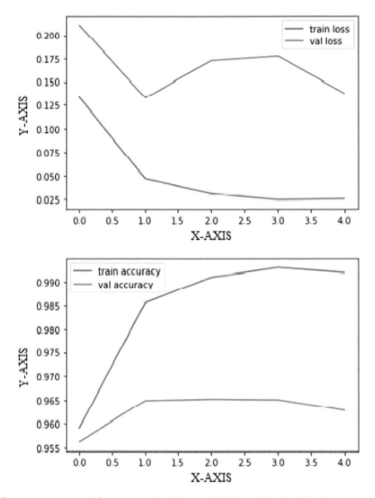

Fig. 4 Loss accuracy graph for proposed multi-scale CNN model using ISL data set

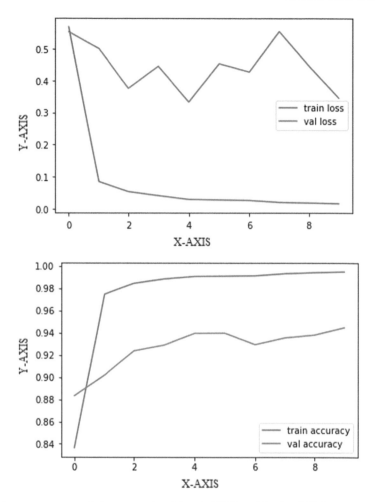

Fig. 5 Loss accuracy graph for proposed multi-scale CNN model using ASL data set

References

1. Mathur, S., & Saxena, V. P. (2014). Hybrid appraoch to English-Hindi name entity transliteration. In *2014 IEEE Students' Conference on Electrical, Electronics and Computer Science* (pp. 1–5). https://doi.org/10.1109/SCEECS.2014.6804467
2. Sharma, S., Saxena, V. P., & Satish, K. (2019). Comparative analysis on sign language recognition system. *International Journal of Scientific Technology Research, 8*, 981–990.
3. Sharma, S., & Singh, S. (2021). Vision-based hand gesture recognition using deep learning for the interpretation of sign language. *Expert Systems With Applications, 182*, 115657.
4. Abraham, E., Nayak, A., & Iqbal, A. (2019). Real-time translation of Indian sign language using LSTM. In *2019 Global Conference for Advancement in Technology (GCAT)* (pp. 1–5). Bangaluru, India. https://doi.org/10.1109/GCAT47503.2019.8978343
5. Muthu Mariappan, H., & Gomathi, V. (2019). Real-time recognition of Indian sign language. In *2019 International Conference on Computational Intelligence in Data Science (ICCIDS)*

(pp. 1–6). Chennai, India. https://doi.org/10.1109/ICCIDS.2019.8862125

6. Intwala, N., Banerjee, A. M., & Gala, N. (2019). Indian sign language converter using convolutional neural networks. In *2019 IEEE 5th International Conference for Convergence in Technology (I2CT)* (pp. 1–5). Bombay, India. https://doi.org/10.1109/I2CT45611.2019.9033667

7. Ja, P. G., Rb, P. K. H., & Chiplunkar, N. N. (2021). Image classification and prediction using transfer learning in colab notebook. In *Global Transitions Proceedings*. https://doi.org/10.1016/j.gltp.2021.08.068

8. Huang, J., Zhou, W., Li, H., & Li, W. (2018). Sign language using real-sense. In *IEEE University Of Science and Technology of China, Hefei* (Vol. 230027). China; Shenoy, T. D., Rao, V., & Vyavaharkar, D. (2018). Real-time Indian Sign Language (ISL) Recognition 9th ICCCNT 2018. IISC, Bengaluru, India.

9. Badhe, P. C., & Kulkarni, V. (2015). Indian sign language translator using gesture recognition algorithm. In *2015 IEEE conference on Computer Graphics, Vision and Information Security (CGVIS)*.

10. Rethna Virgil Jeny, J., Anjana, A., Monica, K., Sumanth, T., & Mamatha, A. Hand gesture recognition for sign language using convolutional neural network. In *Proceedings of the Fifth International Conference on Trends in Electronics and Informatics (ICOEI)*. IEEE Xplore Part Number: CFP21J32-ART; ISBN:978-1-6654-1571-2.

11. Wang, F., Hu, R., & Jin, Y. (2021). Research on gesture image recognition method based on transfer learning. *Procedia Computer Science, 187*, 140–145. ISSN 1877-0509, https://doi.org/10.1016/j.procs.2021.04.044

12. Mathur, S., & Saxena, V. P. (2014). Hybrid approach to English-Hindi name entity transliteration. In *2014 IEEE Students' Conference on Electrical, Electronics and Computer Science*.

13. Mustafa, H. T., Yang, J., & Zareapoor, M. (2019). Multi-scale convolutional neural network for multi-focus image fusion. *Image and Vision Computing, 85*, pp. 26–35. ISSN: 0262-8856.

14. Chavan, S., Yu, X., & Saniie, J. (2021). Convolutional neural network hand gesture recognition for American sign language. In *2021 IEEE International Conference on Electro Information Technology (EIT)*. 978-1-6654-1846-1/21/31.00 ©2021 IEEE. https://doi.org/10.1109/EIT51626.2021.9491897

15. Hamdan, Y. B. (2021). Construction of statistical SVM based recognition model for handwritten character recognition. *Journal of Information Technology, 3*(02), 92–107.

16. Adithya, V, & Rajesh, R. A deep convolutional neural network approach for static hand gestures recognition. In *Third International Conference on Computing and Network Communications (CoCoNet'19)*.

17. Manoharan, J. S. (2021). Capsule network algorithm for performance optimization of text classification. *Journal of Soft Computing Paradigm (JSCP), 3*(01), 1–9.

18. Agrawal, M., Ainapure, R., Agrawal, S., Bhosale, S., & Desai, S. (2020). Models for hand gesture recognition using deep learning. In *2020 IEEE 5th International Conference on Computing Communication and Automation (ICCCA)*. Galgotias University, Greater Noida, UP, India.

19. Rajan, R. G., & Selvi Rajendran, P. (2021). Gesture recognition of RGB-D and RGB static images using ensemble-based CNN architecture. In *Proceedings of the Fifth International Conference on Intelligent Computing and Control Systems (ICICCS 2021)*. IEEE Xplore Part Number: CFP21K74-ART; ISBN: 978-0-7381-1327-2.

Prediction and Analysis of Liver Disease Using Extreme Learning Machine

Geetika Singh and Charu Agarwal

Abstract Hepar, also known as the liver, is a vital organ of the digestive system which performs hepatic blood flow, helps to clot, etc. Poor eating habits and a sedentary lifestyle, unfortunately, have an impact on the liver. Developing a fully automatic computer-assisted disease prediction model would be extremely beneficial to medical professionals as manual analysis is a tedious and time-taking task. To accomplish this, we provide an automated liver disease prediction model based on the extreme learning machine classifier based on his/her biological parameters. ELM is a fast single-layer feed-forward neural network with high generalization capabilities that have previously been used to generate a variety of classification models. The proposed model is evaluated using the ILPD dataset. The performance of the proposed model is assessed using a range of activation functions and hidden neuron counts. According to the calculations, the proposed model beats other models currently in use.

Keywords ELM · Activation function · Hidden neurons · Liver function test

1 Introduction

The liver, also known as the 'hepar' [1], is an important organ of the digestive system which is located in the upper right part of the stomach and the ribcage protects it from external injuries. The wedge-shaped organ contains two lobes [2] separated by the falciform ligament, and the right lobe is larger than the left lobe. The color of the organ is reddish-brown. Its major functions involve hepatic blood flow, production of bile, and proteins, excess glucose is converted to glycogen for storage, etc. The hepatic flow of blood [3] is the important function of the liver where it receives the blood from the hepatic artery and portal vein, and after processing, the blood is supplied to various organs through the hepatic vein. As a result, a healthy liver is critical for the body's proper functioning.

G. Singh (✉) · C. Agarwal
Ajay Kumar Garg Engineering College, Dr. A.P.J. Abdul Kalam Technical University Uttar Pradesh, Lucknow, India
e-mail: singh.geetika93@gmail.com

S. Shakya et al. (eds.), *Sentiment Analysis and Deep Learning*,
Advances in Intelligent Systems and Computing 1432,
https://doi.org/10.1007/978-981-19-5443-6_52

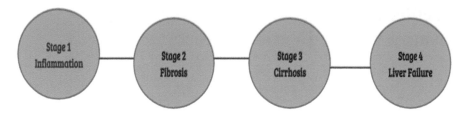

Fig. 1 Stages of liver disease

However, we tend to damage the condition of the liver as a result of poor eating habits and a sedentary lifestyle, which has an impact on a person's health. Liver disease can be caused due to various factors, involving the consumption of excess alcohol, liver infection due to virus which causes inflammation of the liver, immune system abnormalities like autoimmune hepatitis and genetic liver diseases like hemochromatosis, etc. The stages of the liver disease [4] are shown in Fig. 1.

The preliminary stage is inflammation of the liver, often known as hepatitis, caused by a virus or parasite. Fibrosis is the next stage. The liver repairs itself whenever it is injured by producing new liver cells. Repeated injury, on the other hand, stimulates the liver to repair the liver cells, resulting in the creation of scar tissue known as fibrosis. Cirrhosis is the third stage which involves severe and permanent scarring of liver tissues. Liver failure is the final stage and the only treatment available is liver transplantation, which is very expensive.

A person's ability to live a healthy life is improved by early detection of liver disease. The diagnosis of illness can be done with a liver function test (LFT) and imaging. After collecting a person's blood sample and analyzing it, a report is generated that includes metrics such as total albumin, total bilirubin, and many more. The hepatologist prescribes medicine and preventative measures based on these factors to treat the patient.

Liver disease is adversely affecting the healthy living. It is the tenth most common cause of death in India, the second most prevalent in the United States, and the fifth most common in the United Kingdom. Liver disease claims the lives of over two million individuals each year.

We can deduce from the foregoing that manual analysis by a hepatologist would be a time-consuming and difficult task. Analyzing manually is also prone to errors. A fully automated analysis system can be constructed employing numerous latest technologies to assist our medical fraternity and provide efficient and accurate results. For the building of such models, a variety of machine learning algorithms can be applied. Many researchers have sought to construct the following model.

Thirunavukkarasu et al. [5] presented a study for predicting liver illness. The authors compared the accuracy of three classification algorithms: support vector machine (SVM), k-nearest neighbor (KNN), and logistic regression (LR), and found that while LR and KNN have equal accuracy (73.97%), LR is the best due to its high specificity. The authors conducted their study using the ILPD dataset.

Kumar et al. [6] proposed a study to diagnose liver sickness using KNN, k-means, Naive Bayes, C5.0 decision tree, and random forest, and found that random forest outperforms with an accuracy of 72.18%. The accuracy of C5.0 with adaptive boosting was 75.19%, according to the authors.

Sumedh et al. [7] used the ILPD dataset to analyze the performance of support vector machine (SVM) and back propagation. Back propagation's accuracy is 73.2%, which is greater than SVM's 71%, according to the authors.

In this article, Sa'diyah et al. [8] used the decision tree (DT), Naive Bayes (NB), and NBTree algorithms. The authors determined that NBTree produced the best results, with a disease prediction accuracy of 67.01%. The ILPD dataset was used for the study by the authors.

For the prediction of liver disease, Nazmun et al. [9] assessed the performance of seven decision tree algorithms: J48, LMT, random forest, random tree, REPTree, decision stump, and Hoeffding tree using the ILPD dataset. Their research found that decision stump was the most accurate decision tree algorithm for predicting liver illness, with a 70.67% accuracy rate.

Particle swarm optimization (PSO), genetic algorithm (GA), alternating decision tree (ADT), and multi-linear regression (MReg) techniques were employed by Somaya et al. [10] in their research. The Egyptian National Committee for Control of Viral Hepatitis database was used for the study. Following the experiment, the genetic algorithm produced the best result, with a precision of 69.6%.

Based on the foregoing literature review, it is clear that a computer-based model that can more correctly predict liver disease with the less computational cost is needed.

In this research, we use an extreme learning machine (ELM) as a classifier to propose a more computationally efficient and accurate machine learning-based model for the prediction of liver disorder. ELM is a rapid single-layer feed-forward neural network with excellent generalization. ELM has been effectively employed in a variety of different classification tasks, including ECG classification [11], fingerprint detection [12], and so on. On the ILPD dataset, the suggested model is trained and tested. We also investigated the performance of ELM using various activation functions because it aids the network in comprehending complex data and translates input from previous layers into a form that is then passed on to the next layer. Nonlinear activation functions are used in ELM. ReLU, leaky ReLU, TanH, and TanHRe are the activation functions employed in our model, each having a different set of hidden neurons (8, 16, 32, 64). The key contribution of this research is to determine whether or not the ELM algorithm can be used to diagnose liver disease. The following is a breakdown of the paper's structure. The mathematical description of ELM and its activation functions are presented in Sect. 2. The ILPD dataset is described in depth in Sect. 3. The proposed methodology is presented in Sect. 4. The results and comments are presented in Sect. 5. Finally, Sect. 6 brings the paper to a close.

2   Extreme Learning Machine (ELM)

Guang-Bin and Qin-Yu [13] proposed ELM, a feed-forward neural network with a single hidden layer [14]. The approach is based on the universal approximation theorem, which states that a single hidden layer feed-forward network with a finite number of neurons may approximate continuous functions on a compact subset under modest activation function requirements. It was designed to train a single hidden layer feed-forward neural network, which has three basic layers: input layer, hidden layer, and output layer, as illustrated in Fig. 2.

ELM assigns random values to the weights between the input and hidden layers, as well as the biases in the hidden layer. The hidden layer's nonlinear activation functions supply the system's nonlinearity. As a result, it can be characterized as a linear system. The only parameter that has to be learned is the weight between the hidden and output layers. ELM converges much faster than typical algorithms since

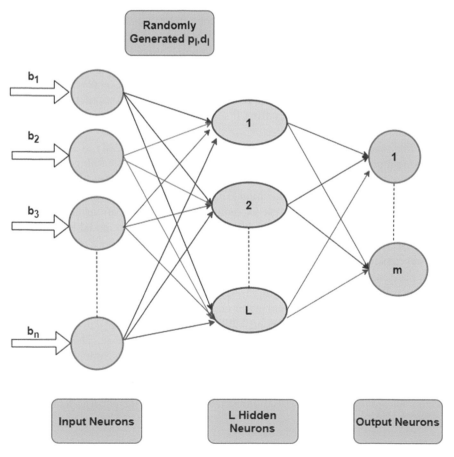

Fig. 2 Architecture of ELM

Notations	Meaning
b_1	Input vector of lth value
k_1	Target vector of lth value
p_1	Weight vector from the input layer to lth hidden node
r_1	The bias of lth node which is hidden
L	Count of hidden nodes
β_l	Weight vector from lth hidden node to output layer

Table 1 Meaning of notations

it learns without iteration. It learns more quickly than gradient boosting algorithms [15].

Let us look at the training of a single feed-forward neural network in detail. Assume we have been provided a collection of training set $R = \{(b_1, k_1)\}| b_1 = (b_{11}, b_{12}, b_{13},\dots, a_{kx})^Y \in T^x, k_1 = (k_{11}, k_{12}, k_{13},\dots,k_{1w})^Y \in T^w$ and $1 = 1, 2, 3\dots N$ where input values are shown by b_1 and k_1 showcases target values. Table 1 gives the definitions of notations. Equation (1) can be used to express the output of the ELM with L hidden neurons:

$$\sum_{l=1}^{L} \beta_l g_l\left((p_l, d_j) + r_l\right) = o_j \tag{1}$$

The integer N represents the total number of training samples, while l and j are the training sample and output layer indexes, respectively. It can be proved that a single hidden layer neural network can approximate all of the training data when the number of hidden nodes reaches infinity, which is known as universal approximation ability.

$$\sum_{j=1}^{N} \|o_j - z_j\| = 0 \tag{2}$$

Thus, there must be a set of p_1, r_1, and β_l which satisfies Eq. (3).

$$\sum_{l=1}^{k} \beta_l g\left((p_l, d_j) + r_l\right) = z_j \text{ where } j = 1\dots, k \tag{3}$$

The formula given above can be abbreviated as shown below in Eq. (4)

$$O\beta = A \tag{4}$$

$$O(p_1, \ldots, p_L, r_1 \ldots r_L, d, \ldots d_N)$$

$$= \begin{bmatrix} g(p_1.d_1 + r_1) & \cdots & g(p_L.d_1 + r_L) \\ \cdots & \cdots \cdots \\ \cdots & \cdots \cdots \\ \cdots & \cdots \cdots \\ g(p_1.d_N + r_1) & \cdots & g(p_P.d_N + r_L) \end{bmatrix}$$

$$\beta = \begin{bmatrix} \beta_1^A \\ \beta_2^A \\ \beta_3^A \\ \vdots \\ \beta_L^A \end{bmatrix}_{L \times m} \qquad A = \begin{bmatrix} z_1^A \\ z_2^A \\ z_3^A \\ \vdots \\ z_L^A \end{bmatrix}_{N \times m}$$

ReLU, leaky ReLU, TanH, and TanHRe are the activation functions employed in our study for the examination of liver disease.

Rectified Linear Unit: ReLU is the most commonly utilized activation function for many neural networks. The method returns zero for negative input and the value itself for positive input. Equation (5) states the mathematical formula for ReLU:

$$f(x) = \max(0, x) \tag{5}$$

Leaky Rectified Linear Unit: The improvised form of ReLU is the leaky ReLU. The neurons in the area where the value is less than 0 are inactive since there was a problem with dying ReLU. For any positive input value, the function $f(x)$ returns the value itself, but for a negative input value, it returns 0.01 times the value of x. Equation 6 gives us a leaky ReLU:

$$f(x) = \max(0.01 * x, x) \tag{6}$$

Hyperbolic Tangent (TanH): This method returns a number ranging from -1 to 1. The graph's slope is similar to the sigmoid function; however, it is steeper. Given by Eq. (7):

$$f(x) = \frac{e^x - e^{-x}}{e^x + e^{-x}} \tag{7}$$

TanHRe: This activation function is a mix of TanH and ReLU, both of which have been discussed previously. The function can be written as shown in Eq. (8):

$$f(x) = \begin{cases} x, & \text{if} x > 0 \\ \tanh(x), & \text{if} x \leq 0 \end{cases} \tag{8}$$

3 ILPD Dataset

To create a machine learning model, you will need a dataset to train and test it on. We chose the Indian Liver Patient Dataset (ILPD) [16] from the Kaggle repository for our research because it is the only dataset that is openly available to the research community. There are 583 records in this dataset, with 167 records of healthy patients and 416 records of patients with liver disease. There are 11 attributes in this dataset. Except for age, gender, and is patient, all columns are biological parameters connected to the liver that are present in the liver function test (LFT) report and used to determine whether or not a person has liver disease.

4 Proposed Methodology

The technology used in this study employs an extreme learning machine as a classifier to determine whether or not a patient has liver illness based on his data. The framework of the suggested methodology for identifying liver disease is depicted in Fig. 3.

The steps followed for building a model are mentioned below:

Step 1. Collection of data

The ILPD dataset, as described in Sect. 3, is gathered from the data repository and sent into the system as an input.

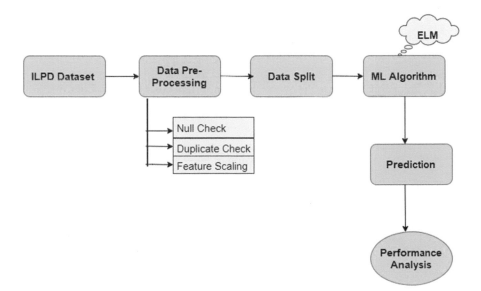

Fig. 3 Framework of the proposed methodology

Step 2: Dataset Pre-processing

(a) Null Check: Attributes with no value are dealt with. This can be accomplished by either deleting such entries or substituting a mean, average value for them. In the ILPD dataset, four null values were discovered and substituted with the mean value of 0.94.

(b) Duplicate Check: Duplicate values are dealt with here. The ILPD dataset had 13 duplicate values, which were removed in order to retrieve the clean dataset.

(c) Feature Scaling: Feature scaling deals with categorical values. The 'gender' attribute has two category values: 'female' and 'male.' These category values were replaced with a value of 0,1 in each case.

Step 3: Splitting the Dataset: During this phase, the dataset is split into training and test data in a specific ratio. We divided the dataset into 80:20, where 80% of the data is used to train ELM and 20% is used to test the model.

Step 4: ELM Architecture Training and Testing: The extreme learning machine (ELM) technique was used to construct the suggested model, with details provided in Sect. 2. The model is first trained on a training dataset before being evaluated to see if it produces accurate results.

Step 5: Performance Analysis: The model is then evaluated based on several characteristics such as accuracy, precision, and so on. We evaluated the algorithm's performance using the accuracy score [17], precision [18], recall [18], and F1-score [18]. The following are the performance parameters in detail:

(a) Accuracy: Accuracy represents the number of records that has been correctly categorized out of a total number of records. Equation (9). shows the mathematical formulation.

$$\text{Accuracy} = \frac{\text{TN} + \text{TP}}{\text{TN} + \text{FP} + \text{TP} + \text{FN}} \qquad (9)$$

(b) Precision: Refers to the overall number of right positive predictions made out of all positive predictions made. Equation (10) represents the mathematical formulation of precision.

$$\text{Precision} = \frac{\text{TP}}{\text{FP} + \text{TP}} \qquad (10)$$

(c) Recall: This shows how many right positive values were obtained out of the total number of positive values received. The recall mathematical formula is represented in Eq. (11).

$$\text{Recall} = \frac{\text{TP}}{\text{TP} + \text{FN}} \qquad (11)$$

(d) $F1$-Score: The weighted average of precision and recall can be called $F1$-score. Equation (12) shows the mathematical representation of the F1-score.

$$F\ 1\text{-Score} = 2*\frac{\text{Precision *Recall}}{\text{Precision + Recall}} \tag{12}$$

5 Result and Discussion

We used the following setup of the system in this study: Intel i5 processor (8 GB RAM) with Python 3.7 and Google Collaboratory software. To carry out this investigation, we employed the extreme learning machine as a classifier. As a result, the performance parameter values are generated using different numbers of hidden neurons to determine the optimal value of classification accuracy. Different activation functions in neurons are employed to further examine the performance. As a result, we generate performance parameter values based on the number of hidden neurons (8, 16, 32, 64) and different activation functions (ReLU, leaky ReLU, TanH, sin, TanHRe).

5.1 ELM Performance Analysis with a Training-to-Testing Ratio of 80:20

In the experiment, we used 80% of the data to train the ELM and 20% of the data to evaluate the ELM. Figure 4 shows the accuracy of all activation functions (ReLU, leaky ReLU, TanH, sin, TanHRe) applied to the trained ELM model with varying numbers of hidden neurons. The ReLU activation function with 32 neurons achieve the highest accuracy of 77.77%.

Table 2 gives the precision score for all activation functions for various hidden neurons for the 80:20 data split.

Table 3 gives the recall for all activation functions for distinct hidden neurons for the 80:20 data split.

The $F1$-scores for the 80:20 data split is given in Table 4.

To further demonstrate the efficacy of the suggested model, we compared the accuracy score obtained by the proposed model to that of other published works in the same domain in the section below.

5.2 Evaluation of Other Published Works in the Same Field

Table 5. gives a comparison of our experimental results with the results of other authors' work in terms of accuracy.

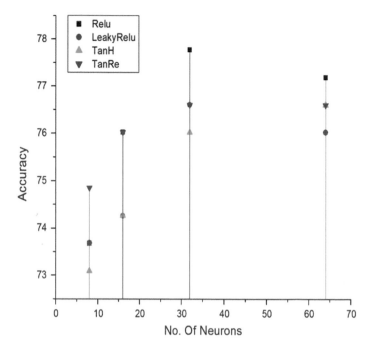

Fig. 4 Depicts the accuracy of 80:20 data split

Table 2 Precision for 80:20 data split

	ReLU	Leaky ReLU	TanH	TanHRe
8	0.57	0.6	0.5	0.57
16	0.63	0.55	0.54	0.66
32	0.7	0.71	0.72	0.66
64	0.66	0.64	0.63	0.68

Table 3 Recall for the 80:20 data split

	ReLU	Leaky ReLU	TanH	TanHRe
8	0.08	0.06	0.45	0.23
16	0.26	0.21	0.26	0.21
32	0.3	0.21	0.17	0.26
64	0.30	0.23	0.30	0.23

According to the Table 5, the proposed methodology based on the ELM classifier achieves 77.77% accuracy, which is the highest accuracy among previous published research in the same domain. As a result, we can infer that the liver disease detection model created with ELM as a classifier performed the best in terms of prediction and

Table 4 $F1$-Score for 80:20 data split

	ReLU	Leaky ReLU	TanH	TanHRe
8	0.15	0.11	0.47	0.33
16	0.36	0.31	0.35	0.32
32	0.42	0.33	0.28	0.37
64	0.41	0.34	0.41	0.35

Table 5 Comparison of the proposed method with work done by other authors based on the accuracy

Author	Dataset	ML technique	Accuracy (%)
Thirunavukkarasu et al. [4]	ILPD	Logistic regression SVM K-nearest neighbor	73.97
Nazmun et al. [5]	ILPD	J48, LMT, RF, RT, REPTree, DS, Hoeffding Tree	70.67
Sanjay et al. [6]	ILPD	K-nearest neighbor K-means Naive Bayes C5.0 Random forest	75.19
Somaya et al. [7]	Egyptian National Committee for Control of Viral Hepatitis database	Particle swarm optimization, decision tree, multi-linear regression, and genetic algorithm models	69.6
Sumedh et al. [8]	ILPD	SVM, back propagation	73.2
Sa'diyah et al. [9]	ILPD	Decision tree C4.5, Naïve Bayes, and NBTree	67.01
Proposed methodology	ILPD	Extreme learning machine (ELM)	**77.77**

can be employed in the healthcare domain. The current work can be expanded by running the proposed model on different datasets.

6   Conclusion

As we all know, detecting liver illness early on can help a person live a long and healthy life. We can assist the medical department with a model that predicts liver illness with high accuracy and efficiency, as manual analysis is a time-consuming task. We used the extreme learning machine (ELM) classifier to create a liver disease prediction model in this paper. Because it uses a feed-forward method rather than backpropagation, it is a fast-learning algorithm when compared to other neural networks. Due to its speedy speed, strong generalization, and ease of implementation, ELM is typically preferred over alternative approaches for AI issues.

References

1. Riva, M. A., Riva, E., Spicci, M., Strazzabosco, M., Giovannini, M., & Cesana, G. (2011). The city of Hepar: Rituals, gastronomy, and politics at the origins of the modern names for the liver. *Journal of hepatology, 55*(5), 1132–1136.
2. Abdel-Misih, S. R., & Bloomston, M. (2010). Liver anatomy. *The Surgical Clinics of North America, 90*(4), 643–653.
3. Crawford, J., Bioulac-Sage, P., & Hytiroglou, P. (2018). Structure, function, and responses to injury.
4. https://www.healthline.com/health/liver-failure-stages
5. Thirunavukkarasu, K., Singh, A. S., Irfan, M., & Chowdhury, A. (2018). Prediction of liver disease using classification algorithms. In *4th International Conference on Computing Communication and Automation (ICCCA)* (pp. 1–3).
6. Kumar, S., & Katyal, S. (2018). Effective analysis and diagnosis of liver disorder by data mining. In *International Conference on Inventive Research in Computing Applications (ICIRCA)* (pp. 1047–1051).
7. Sontakke, S., Lohokare, J, & Dani, R. (2017). Diagnosis of liver diseases using machine learning. In *International Conference on Emerging Trends & Innovation in ICT (ICEI)* (pp. 129–133).
8. Alfisahrin, S. N. N. & Mantoro, T. (2013). Data mining techniques for optimization of liver disease classification. In *International Conference on Advanced Computer Science Applications and Technologies* (pp. 379–384).
9. Nahar, N., & Ara, F. (2018). Liver disease prediction by using different decision tree techniques. *International Journal of Data Mining & Knowledge Management Process, 8*, 01–09.
10. Hashem, S., et al. (2018). Comparison of machine learning approaches for prediction of advanced liver fibrosis in chronic hepatitis C patients. *IEEE/ACM Transactions on Computational Biology and Bioinformatics, 15*(3), 861–868.
11. Kim, J., Shin, H., Lee, Y., & Lee, M. (2007). Algorithm for classifying arrhythmia using extreme learning machine and principal component analysis. In *Conference Proceeding IEEE Engineering Medical Biological Social 2007* (pp. 3257–3260).
12. Yang, J., Xie, S., Yoon, S., Park, D., Fang, Z., & Yang, S. (2013). Fingerprint matching based on extreme learning machine. *Neural Computing and Applications* (pp. 435–445).
13. Guang-Bin, H., Qin-Yu, Z., & Chee-Kheong, S. (2006). Extreme learning machine: Theory and applications. *Neurocomputing*, 489–501.
14. Huynh, H., & Won, Y. (2009) Online training for single hidden-layer feedforward neural networks using RLS-ELM. In *Proceedings of IEEE International Symposium on Computational Intelligence in Robotics and Automation, CIRA* (pp. 469–473).
15. Wang, J., Lu, S., Wang, S. H. et al. (2021) A review on extreme learning machine. *Multimedia Tools Application.*
16. UCI data set, http://archive.ics.uci.edu/ml/machine-learning-databases/00225/
17. Ravuri, S.V., & Vinyals, O. (2019). Classification accuracy score for conditional generative models. *NeurIPS.*
18. Powers, D. M. W. (2011). Evaluation: From precision, recall and f-measure to roc., informedness, markedness & correlation. *J Mach Learn Technol 2*, 37–63.

Identifying the Predictors from Lung Cancer Data Using Machine Learning

Anu Maria Sebastian and David Peter

Abstract Lung cancer is one of the most common deaths causing cancers in the world. The late diagnosis would lead to the rapid spread of cancer to other organs in the body and becomes life threatening. Studies say that the five-year survival rate of late diagnosed lung cancer cases are only 4%, and only about 16% of the cases are diagnosed at an early stage. The major cause that paves for the late diagnosis of lung cancer is that the initial clinical symptoms are often neglected as inconsequential. The widely used method for lung cancer diagnosis is radiomics. However, radiomics is expensive and the recurrent subjection to it itself can be risky. Also, in most of the cases, it is found to be performed in a late stage. Use of clinical data is a possible alternative for radiomics in early diagnosis of lung cancer. In this work, we tried to identify the predictors from clinical lung cancer data using machine learning (ML) techniques. Through the secondary data analysis, we identified the most influential predictors of lung cancer from the clinical data. These prognostic markers can aid in the early detection of lung cancer. The patients that hit the specified marker values can be classified as potential risky group to develop lung cancer. Initially, in our approach, the data is pre-processed and feature selection is applied using data filtering method. Then, we applied ML models such as support vector machine (SVM) and random forest (RF) to classify the patients.

Keywords Prognostic markers · Attribute selection · Machine learning · Lung cancer diagnosis · Feature reduction · Feature significance

1 Introduction

Carcinogenic agents often stimulate uninhibited cell division in pulmonary organs which lead to the development of life-threatening tumours in lungs, leading to lung cancer. Cancer can be caused by genetic disorders, sedentary lifestyles and eating

A. M. Sebastian (✉) · D. Peter
Department of Computer Science, Cochin University of Science and Technology, Kochi, Kerala, India
e-mail: anumseb@gmail.com

© The Author(s), under exclusive license to Springer Nature Singapore Pte Ltd. 2023
S. Shakya et al. (eds.), *Sentiment Analysis and Deep Learning*,
Advances in Intelligent Systems and Computing 1432,
https://doi.org/10.1007/978-981-19-5443-6_53

habits, or by the habit of people consuming toxic and unhealthy food products. The chances of survival through treatment are directly related to the stage of detection of this disease, where earlier the better. Once it is rightly detected, the patient has to undergo a proper course of treatment to increase the chances of survival.

In the year 2012, there were 14.1 million new cases of cancer and 8.2 million deaths globally recorded by the "GLOBOCAN series of the International Agency for Research on Cancer". According to their statistics, the highest diagnosed cancer was lung cancer with 1.82 million cases, breast with 1.67 million cases, and colorectal with 1.36 million cases, respectively [1]. Among the cancers, lung cancer had an alarming rates of increased deaths (1.6 million deaths) compared to subsequent death causing cancers such as liver cancer (745,000 deaths), and stomach cancer (723,000 deaths). According to the "American Cancer Society's 2022 report", in the United States, 117,910 and 118,830 new cases of lung cancer were found in men and women, respectively. The deaths caused from lung cancer were about 68,820 in men and 61,360 in women [2]. The five-year survival rate for lung cancer was 55 percent for cases detected when the disease was still localized (within the lungs). The lung cancer death rates were found to be the highest among people of 65 years and older. The percentage of people diagnosed with lung cancer before age 45 was found to be very low. The average age of lung cancer detection was observed as 70 years [3]. Percentage of early detection of lung cancer is reported to be only about sixteen percentage. In case of metastasis, the five-year survival rate is said to be only four percent [1].

In developed countries, the estimated percentage of lung cancer caused by smoking in men is over ninety percentage, and about seventy percentage among women. The early-stage symptoms of lung cancer are often regarded as common ailments and discarded as inconsequential. Oncologist from "University College London's Cancer Institute" says, "People ignore cough, and often come to my clinic with metastatic disease, but by that stage the effective treatment might already be out of reach [4]". The initial symptoms of lung cancer are mostly mistaken as signs of other ailments, for example, severe cough or tiredness are often regarded as inconsequential. The clinical staging plays a decisive role in determining the treatment and survival probability. Therefore, the guidelines are regularly revised as per the data available to identify the exact predictors in order to support early diagnosis [5].

All the clinical doctors might not be equally skilled as the domain experts to diagnose the early symptoms of lung cancer. Radiation screening is a widely used non-invasive lung cancer diagnosis method, which is mostly performed at a later stage only as the initial symptoms will be neglected as some other health issue [6–9]. The major limitations of radiomics include lack of standardization of acquisition parameters, inconsistency among the radiomic methods, and lack of reproducibility. Researchers are working on overcoming these limitations [10]. In particular, the overall ionizing radiations exposure associated with the low dose computed tomography (LDCT) screening procedure may imply an increased incidence of solid cancers and leukaemia [11]. Computed tomography (CT) imaging involves radiation to find lung cancer but with a chance to actually induce it [12]. Another noted drawback of LDCT screening is that it has an increased false positive rate (FPR)

[13]. Given concerns over the high cost, cumulative radiation exposure, and high rate of false positives from LDCT screening, the recent researches are focused more on identifying alternative approaches to detect lung cancer at an early stage.

ML has huge relevance in the field of health care due to its ability to process huge data sets and convert them into clinical insights. It uses algorithms to learn the pattern from the healthcare data to perform detection, prediction, and classification of diseases [14]. Syed et al. [15] developed a system for the early detection of breast cancer using ML techniques. In their experiment results, multi-layer perceptron was the optimal classifier, followed by RF and KNN. Hornbrook et al. [16] developed an ML-based algorithm for colorectal cancer detection and its classification. The model has excelled in detection of right-sided (than left-sided) colorectal cancers. Zhang et al. [17] discovered a set of biomarkers for detecting lung cancer from urine samples. Pearson correlation was used for selecting the biomarkers. The RF algorithm used five biomarkers which could identify the lung cancer patients from the control group and also from other cancer patients. From the literature, we can see the vital role of ML in detection and classification of cancers.

2 Materials and Methods

2.1 Data Description

The plausible risk factors considered by [18] for increased lung cancer risk are "*age, alcohol consumption status, risky chemicals, chronic lung inflammation, diet, hormones, immunosuppression, infectious agents, obesity, radiation*, and *tobacco*". The data used in this research work is secondary lung cancer clinical data from the Dataworld repository [19]. The data has 1000 records with 24 input attributes, namely "'*Patient-Id*', '*Age*', '*Gender*', '*Air Pollution*', '*Alcohol use*', '*Dust Allergy*', '*Occupational Hazards*', '*Genetic Risk*', '*chronic Lung Disease*', '*Balanced Diet*', '*Obesity*', '*Smoking*', '*Passive Smoker*', '*Chest Pain*', '*Coughing of Blood*', '*Fatigue*', '*Weight Loss*', '*Shortness of Breath*', '*Wheezing*', '*Swallowing Difficulty*', '*Clubbing of Finger Nails*', '*Frequent Cold*', '*Dry Cough*', and '*Snoring*'*". Among the twenty-four input attributes, the *Patient-Id* is of type string and the rest of the attributes is of type integer. The *Patient-Id* attribute was removed at the beginning of the work as it was considered to be irrelevant for the identification of prognostic markers. The target variable has multiclass labels which represents records of healthy, benign, and malignant cases. The percentage of healthy class is 30.3%, benign class is 33.2%, and malignant class is 36.5%. Figure 1 shows the pie chart representation of each class of data in the data set.

Fig. 1 Percentage of each classes

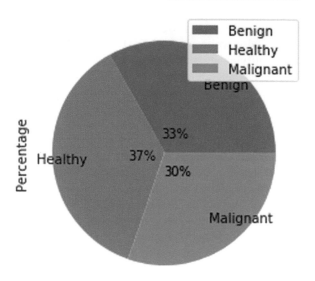

2.2 Data Cleaning and Feature Engineering

No missing values are found in the data set and there are no out-of-the bound values observed in the data set. Feature selection of attributes is done to improve the analysis outcomes. With smaller number of features, the classifier models can be more interpretable and achieve faster training. Since our data set has attribute values that represents different scale of categories we used Pearson Chi-square test for identifying feature relevance; further, selection of features was done through feature filtering method. All the attribute values in the data are of the same scale, and hence, there was no need for data transformation process.

2.3 Machine Learning Models

We started with the support vector machine (SVM) classifier for classification because of its capability to explore both linear and nonlinear classification based on its kernel selection. We observed that the radial basis function (rbf) kernel gives better performance over the linear kernel. This observation is an indication of a nonlinear relationship between the input features and the output variable. Hence, we used one of the best nonlinear classifiers, the random forest (RF), to classify the data. We explored the classification performance of RF with both binary classification and multiclass cases. In the case of binary classification, the target class was modified to risky and non-risky classes. The risky class represented the cancerous tumour groups, and non-risky class represented the benign and healthy group, respectively. Figure 2 shows the workflow of the research study.

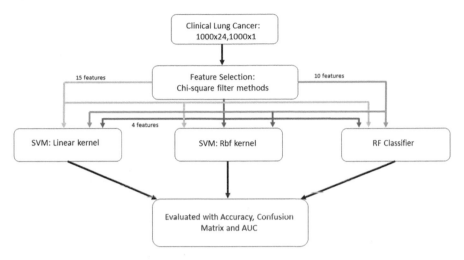

Fig. 2 Workflow of the research study

The performance under various number of features was explored using the filtering method on features, selected using the Pearson Chi-squared ranking. It was observed that from a total of 23 features, a minimum of four prominent features can effectively classify the groups with high accuracy.

2.4 Model Evaluation

The metrics we used to evaluate the performance of the classifiers include accuracy, precision, recall, f1-score, and receiver operating characteristics-area under curve (ROC-AUC) score.

- Accuracy = True Positive/(True Positive + True Negative) × 100.
- Precision = TruePositives/(TruePositives + FalsePositives)
- Recall = TruePositives/(TruePositives + FalseNegatives)
- F1-score = (2 × Precision x Recall)/(Precision + Recall)
- ROC-AUC curve plots the graph between sensitivity v/s 1-specificity.
- Sensitivity = True Positives/(True Positives + False Negatives)
- Specificity = True Negatives/(True Negatives + False Positives)

3   Results

This section describes the observations and results obtained in this research. For feature selection, the Chi-square scores obtained for the variables are 6.284,3.5,173.884, 518.093 354.673, 304.089, 323.916, 164.548, 215.19, 411.465, 94.193, 252.622, 251.709, 357.015, 319.856, 206.541, 311.004, 166.498, 113.034, 234.386, 168.197, 99.441, and 91.34, respectively. Based on these scores, the top fifteen attributes were filtered initially, which include "*Dust Allergy, Smoking, Fatigue, Occupational Hazards, Chronic Lung Diseases, Weight Loss, Wheezing, Genetic Risk, Chest Pain, Coughing of Blood, Frequent Cold, Obesity, Shortness of Breath, Alcohol Use*, and *Dry Cough*". These features were further reduced with filtering in successive steps until the classification accuracy started to deteriorate. Thus, the number of features was reduced to four influential features that decided the classification of a record. For multiclass classification, the SVM-linear kernel classifier achieved 87% accuracy, and the SVM-rbf kernel classifier achieved 96% accuracy, whereas the RF classifier achieved 100% accuracy, respectively. Tables 1, 2, and 3 give the classification performance metrics of SVM-linear, SVM-rbf, and RF classifiers, respectively.

Table 1 Multiclass classification accuracy of SVM (linear)

	Precision	Recall	F1-score	Support
0	0.90	0.73	0.80	96
1	0.77	0.84	0.81	106
2	0.93	1.00	0.97	128
Accuracy			**0.87**	

Table 2 Multiclass classification accuracy of SVM (rbf)

	Precision	Recall	F1-score	Support
0	0.93	0.94	0.93	96
1	0.94	0.93	0.94	106
2	1.00	1.00	1.00	128
Accuracy			**0.96**	

Table 3 Multiclass classification accuracy of RF classifier

	Precision	Recall	F1-score	Support
0	1.00	1.00	1.00	96
1	1.00	1.00	1.00	106
2	1.00	1.00	1.00	128
Accuracy			**1.00**	

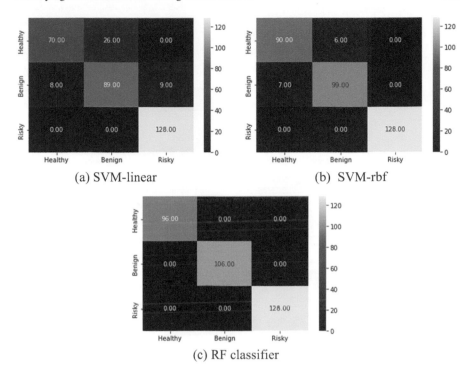

(a) SVM-linear

(b) SVM-rbf

(c) RF classifier

Fig. 3 Confusion matrices obtained for multiclass classification by different classifiers

Next, we obtained the confusion matrices with true positives, false positives, true negatives, and false negatives, as shown in Fig. 3. It can be observed from the confusion matrix of SVM-linear classifier that the model had more ambiguity is classifying between healthy and benign group.

In the next section, we evaluate the performance of the models using binary classification of the data as risky (cancerous) and non-risky (non-cancerous) groups. In order to execute this, we combined the healthy and benign groups of data into a single non-risky class. Figure 4 shows the confusion matrices obtained by the three classifier models for binary classification. It can be observed that the SVM-linear classifier had increased false positives and SVM-rbf classifier had increased false negative. When it comes to cancer detection, false negative has graver impacts, and hence, it should be minimized.

The ROC-AUC values obtained for both SVM-linear kernel and rbf kernel classifiers are 0.96, whereas for RF classifier is 1. Figure 5 shows the ROC-AUC curves for each classifier.

The maximum value for AUC is 1, and in this work, the RF model has achieved it. Therefore, there is no need of further investigation of performance by advanced models like neural networks. From the obtained results, it can be inferred that the multiclass classification tends to exhibit an inclination to nonlinear relationship

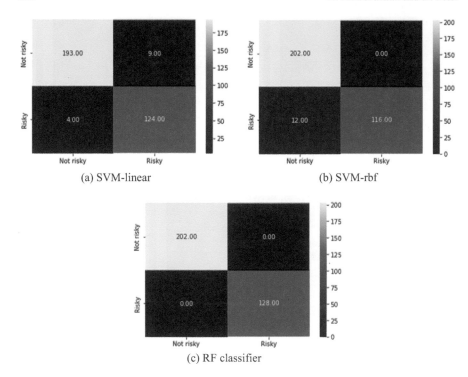

(a) SVM-linear   (b) SVM-rbf

(c) RF classifier

Fig. 4 Confusion matrices obtained for binary classification by SVM and RF classifiers

between input attributes and the output variable. This is evident when the classification accuracy improves while using the nonlinear classifier SVM-rbf over the SVM-linear. The RF model is giving the best performance of 100% accuracy. In the case of binary classification, both SVM classifiers are giving the same performance of 96% accuracy. The RF classifier here again giving an accuracy of 100%.

Figures 6 and 7 show the feature importance ranking based on information gain derived by the RF classifiers, with four features and 15 features, respectively. According to this outcome, the top ranked features are *Coughing of Blood, Obesity, Smoking, and Chest pain*. This result is different from the Chi-square relevant feature ranking. When we explored the classification accuracy with the feature ranking obtained from the RF model, the classification accuracy of the SVM-linear model decreased to 81%, and the SVM-rbf model accuracy decreased to 93%, respectively. Therefore, we recommend to choose the four critical features obtained through Pearson Chi-square test such as *Dust Allergy, Smoking, Fatigue*, and *Occupational Hazards*, as the primary predictors for the classification model.

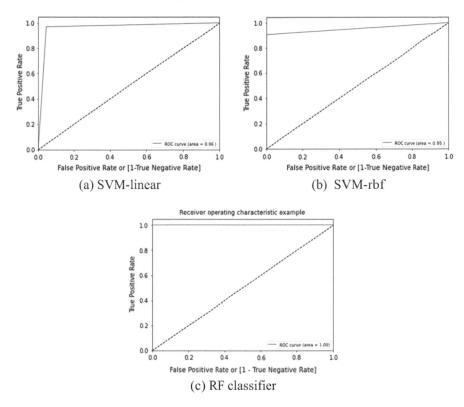

(a) SVM-linear

(b) SVM-rbf

(c) RF classifier

Fig. 5 ROC-AUC scores obtained for binary classification by SVM and RF classifiers

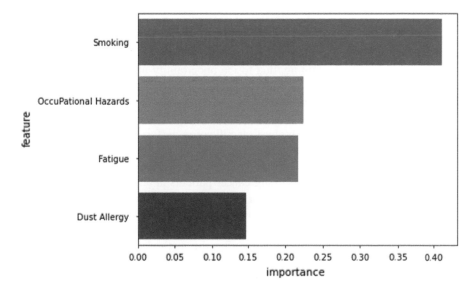

Fig. 6 Feature ranking based on information gain for four-feature data set

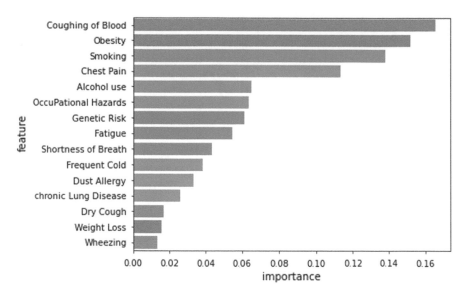

Fig. 7 Feature ranking based on information gain for fifteen-feature data set

4 Conclusion

This research was done with an aim to identify essential predictors from the clinical lung cancer data. For this, we experimented different machine learning models on the data to accurately classify the cancer patients. The clinical data used in this study had no missing values. So, the significance of the features was computed directly using Pearson Chi-square test on the categorical data. The selected attributes were further reduced to four attributes using data filtering method for feature selection. These attributes were selected as the essential predictors to rightly classify the records of the given data. In the beginning, due to the uncertainity about the separability of the data, we applied SVM with both linear and nonlinear kernels separately. Since the nonlinear kernel exhibited improved performance, we decided to experiment the popular nonlinear RF classifier to evaluate the performance. From the results, the RF classifier obtained the highest classification performance with an accuracy of 100%. In this work, the essential prognostic markers for lung cancer identified from the data are dust allergy, smoking, fatigue, and occupational hazards. These markers can be used to diagnose lung cancer patients at an early stage, which can help increase the survival rate through timely medical intervention. However, the real-world clinical data may not be quantified in the same scale as in this data, and also, there are possibilities of missing or erroneous data. This can be a limitation in deploying this inference into practice. But the methodology used in this paper to identify the markers can be reapplied to the relevant clinical lung cancer data to obtain desired results. In order to ensure the generalizability of our findings, it is necessary to experiment with more clinical data sets. This is proposed as a topic for future research.

References

1. Ferlay, J., Soerjomataram, I., Dikshit, R., Eser, S., Mathers, C., Rebelo, M., Bray, F., et al. (2015). Cancer incidence and mortality worldwide: Sources, methods and major patterns in GLOBOCAN 2012. *International Journal of Cancer, 136*(5).
2. Key Statistics of Lung Cancer, American Cancer Society (2022). https://www.cancer.org/can cer/lung-cancer/about/key-statistics.html
3. SEER Cancer Statistics Factsheets: Lung and Bronchus Cancer. National Cancer Institute. Available at http://seer.cancer.gov/statfacts/html/lungb.html. Accessed April 6, 2020.
4. Elizabeth, S., (2020). Deep learning delivers early detection citation. Nature Article.
5. Lim, W., Ridge, C., Nicholson, A., & Mirsadraee, S. (2018). The 8th lung cancer TNM classification and clinical staging system: review of the changes and clinical implications. *Quantitative Imaging In Medicine and Surgery, 8*(7), 709–718. https://doi.org/10.21037/qims.2018.08.02
6. Samuel, M. (2020). Early diagnosis of Lung Cancer with Probability of Malignancy Calculation and Automatic Segmentation of Lung CT scan Images. *Journal of Innovative Image Processing (JIIP), 2*(04), 175–186.
7. Sungheetha, A, & Rajesh, S. R. (2020). Comparative study: Statistical approach and deep learning method for automatic segmentation methods for lung CT image segmentation. *Journal of Innovation of Image Process, 2,* 187–193.
8. Vijayakumar, T. (2020). Posed inverse problem rectification using novel deep convolutional neural network. *Journal of Innovative Image Processing (JIIP), 2*(03), 121–127.
9. Smys, S., Chen, J. L. Z., & Shakya, S. (2020). Survey on neural network architectures with deep learning. *Journal of Soft Computing Paradigm (JSCP) 2*(03), 186–194.
10. Thawani, R., McLane, M., Beig, N., Ghose, S., Prasanna, P., Velcheti, V., & Madabhushi, A. (2017). Radiomics and radiogenomics in lung cancer: A review for the clinician. *Lung Cancer*.
11. Mascalchi, M., & Sali, L. (2017). Lung cancer screening with low dose CT and radiation harm from prediction models to cancer incidence data. *Annals of Translational Medicine, 5*(17).
12. Midthun, D. E. (2016). Early detection of lung cancer. F1000 Research, 5.
13. Kadir, T., & Gleeson, F. (2018). Lung cancer prediction using machine learning and advanced imaging techniques. *Translational Lung Cancer Research, 7*(3), 304–312. https://doi.org/10. 21037/tlcr.2018.05.15
14. Sebastian, A. M. & Peter, D. (2019). Evaluating the performance of machine learning techniques for cancer detection and diagnosis. *Lecture Notes on Data Engineering and Communications Technologies, 46*.
15. Syed, L., Jabeen, S., & Manimala, S. (2017). Telemammography: A novel approach for early detection of breast cancer through wavelets based image processing and machine learning techniques. In *Advances in Soft Computing and Machine Learning in Image Processing* (pp. 149–183).
16. Hornbrook, M., Goshen, R., Choman, E., O'Keeffe-Rosetti, M., Kinar, Y., Liles, E., & Rust, K. (2017). Early colorectal cancer detected by machine learning model using gender, age, and complete blood count data. *Digestive Diseases and Sciences, 62,* 2719–2727.
17. Zhang, C., Leng, W., Sun, C., Lu, T., Chen, Z., Men, X., Wang, Y., Wang, G., Zhen, B., & Qin, J. (2018). Urine proteome profiling predicts lung cancer from control cases and other tumors. *EBioMedicine, 30,* 120–128.
18. National Institutes of Health & National Cancer Institute. (2015). Risk Factors for Cancer. United States. https://www.cancer.gov/about-cancer/causes-prevention/risk
19. DataWorld Repository. https://data.world/cancerdatahp/lung-cancer-data

Dim Recognition in Images, Videos and Real-Time Video Streams

M. Kathiravan, R. G. Sakthivelan, S. Saravanan, S. G. Hari Hara Sudhan, M. Vasanth, and C. S. Abishek

Abstract This paper proposes an obscured image identification and examination system for naturally recognizing video transfers and images. This research study has developed few haze highlight models by using image tone, slope and range data, and a quick fourier transformation is performed to identify the obscure in recordings and images. Furthermore, the proposed research study has utilized a fast fourier transform calculation, which is worked into NumPy as the reason for the proposed procedure. Extensive examinations show that the proposed strategy works acceptably on testing the image information, which lays out a specialized starting point by taking care of a few PC vision issues, like movement investigation and picture rebuilding, utilizing the haze data.

Keywords Blur detection techniques · Fourier transform · Hair wavelet transform · Modified laplacian · Tenengrad

1 Introduction

This paper revolves around recognizing and dissecting to some extent with obscured images and propose a smart technique to naturally distinguish the obscured images, remove conceivable obscured locales, and further group them into two classes, i.e. close isotropic haze and directional movement obscured. The proposed strategy endeavours to handle two significant issues. One is obscure identification with concurrent extraction of obscured locales. The outcome of this progression gives

M. Kathiravan (✉) · S. G. Hari Hara Sudhan · M. Vasanth · C. S. Abishek
Computer Science and Engineering, Hindustan Institute of Technology and Science, Chennai, India
e-mail: mkathiravan@hindustanuniv.ac.in

R. G. Sakthivelan
Department of Information Technology, Rajalakshmi Engineering College, Chennai, India

S. Saravanan
Computer Science and Engineering, Faculty of Engineering and Technology, SRM Institute of Science and Technology, Vadapalani Campus, Chennai, India

© The Author(s), under exclusive license to Springer Nature Singapore Pte Ltd. 2023 703
S. Shakya et al. (eds.), *Sentiment Analysis and Deep Learning*,
Advances in Intelligent Systems and Computing 1432,
https://doi.org/10.1007/978-981-19-5443-6_54

helpful significant level of territorial data, working with an assortment of locale-based picture applications, for example happy-based picture recovery, object-based picture pressure, video object extraction, picture improvement and picture division. It can also be used as one of the rules for estimating the quality of captured images. The second level of proposed strategy is to consequently order the identified haze districts into two kinds: close to isotropic haze (counting out-of-centre haze) and directional movement obscure. The images are classified into these two classes since they are most contemplated in image rebuilding. The haze-characterized images also effectively track down applications through moving examination and image reconstruction. The subjects of image obscure investigation stand out enough to be noticed lately, most past work centres around taking care of the de-obscuring issue. General haze identification, going against the norm, is only sometimes investigated and is still a long way from common sense. Regna et al. acquainted a learning technique with which to order hazy or non-foggy locales in a single image. This strategy depends on a perception that foggy areas are more invariant to low pass separating. In our tests, we observe that utilizing this data isn't adequate for planning a dependable classifier. Different haze measures ought to be consolidated to accomplish great haze identification. Likewise, this strategy doesn't recognize obscure sorts. Using the measurements of data inclination alongside various headings, where the technique fabricates an energy work by considering the surmised obscure part to section pictures into obscure/clarify layers. This strategy just reveals movement-obscured areas by construing directional haze pieces. Other haze assessment techniques, for example the one proposed in, just give a proportion of obscure degree, which can't be straightforwardly used to segregate foggy from non-hazy locales.

2 Literature Survey

Camera shake during transparency prompts horrible dark image and remnants in numerous images. Standard outwardly weakened de-convolution techniques typically expect repeat space objectives on images or unreasonably deal with parametric designs for the development way during camera shake. Veritable camera developments can follow 12 tangled ways, and a spatial region prior can promptly stay aware of obviously exceptional picture characteristics [1]. This procedure familiarize with disposing the effects of camera shake from genuinely clouded images. The strategy expects a uniform camera to darken over the image and an immaterial in-plane camera to upset. To measure the cloudiness from the camera shake, the client ought to decide on an image region without drenching impacts. This research work shows the results for a collection of mechanized images taken from individual image groupings [2].

Dark is one of the most broadly perceived bending types in picture obtainment. Picture de-blurring has been thought of by and large as a fruitful strategy to chip away at the idea of clouded pictures. In any case, little work has been done on the perceptual evaluation of picture de-blurring estimations and de-blurred pictures. In this paper, we lead both theoretical and objective examinations of a picture to defocus and de-blurring [3]. A defocus de-blurred picture informational index (DDID) is first created using state-of-the-art picture defocus de-blurring computations, and a passionate test is finished to assemble the human assessments of the photographs. Then the displays of the de-blurring computations are evaluated considering the enthusiastic scores. With the insight that the ongoing image quality estimations are confined to predict the idea of defocus blurred pictures, a quality improvement module is proposed by considering the gray-level co-occurrence matrix (GLCM), which is for the most part used to check the lack of surface ease achieved by de-blurring. Preliminary outcomes considering the DDID informational collection show the sufficiency of the proposed technique.

An image-based methodology [4] for vehicle speed determining proof is presented. Standard speed assessment techniques use radar-or laser-based devices, which are by and large more expensive and appear differently on a disengaged camera structure. In this work, a single picture with vehicle development is used for speed assessment. Due to the overall development between the camera and a moving thing during the camera's transparency time, dark development occurs in the novel region of the image. It gives a conspicuous hint to the speed assessment of a moving thing. A vague objective region is first isolated and dark limits are evaluated from the developed darkened sub picture. The image is then de-blurred and used to decide various limits. Finally, the vehicle is not entirely set in stone by the imaging math, camera act and dark degree in the image. Tests have shown the evaluated speeds to be within 5% of genuine rates for both close-by and street traffic.

A key advance in a movement obscure based speed estimation work is a computation for the level haze degree. This paper proposes an original plan for utilizing picture matting and the straight forwardness map to satisfy this computation when both a defocus obscure and a movement obscure occur, and for a moving item with unpredictable shape edges heartily and precisely [5].

Imaging examples thicker than the profundity of the field of a magnifying instrument produces low-quality pictures as just a part of the example is in the centre. Hence, even in the best-engaged picture, there are consistently protests that are out of focus and, in this manner, obscured. It is challenging to precisely quantify the size, shape and limit of an obscured object. Accordingly, a few optical segments are frequently expected to gauge precisely the whole force conveyance of an example. To overcome this issue, we present an original technique for expanding the profundity of the field by combining a few optical areas in the wavelet space by utilizing multi-scale point-wise item (MPP) measures. Most existing combination techniques depend on measures that are just because of edges and don't recognize signals from the commotion. Our MPP models guarantee that the sign substance, as opposed to the commotion, is gathered. Rather than straightforwardly melding optical areas, we pre-process the pictures by performing nearby plane de-blurring that eliminates obscured content and combines the in-centre items. The general plan furnishes unparalleled quality pictures with an expanded depth of field, but the combined pictures are unfeeling toward the commotion. The exploratory outcomes show both subjectively and quantitatively that our methodology outflanks existing plans in the writing [6].

Here, an adaptable information-driven application-free technique to expand the profundity of the field is introduced. The key commitment in this work is the utilization of elements removed by empirical mode decomposition [7], specifically intrinsic mode images, for combination. The info pictures deteriorate into inborn mode pictures, and the combination is performed on the extricated oscillatory modes through weighing plans that permit the accentuation of central districts in each information picture. The combined picture binds together data from every central plane while keeping up with the verisimilitude of the scene. To approve the combination execution of our strategy, we have contrasted our outcomes with those of area-based and multi-scale disintegration-based combination procedures.

In customary combination strategies for multi-centre pictures [8], the centre guide created by a centre measure would ordinarily be delicate to mis-enlistment and commotion, or produce gravely adjusted limits. While many best-in-class calculations use more perplexing techniques or systems to resolve this issue, we propose in this paper to gauge a centre guide straightforwardly from two-scale blemished perceptions (centre guides) acquired by utilizing a small and large range of centre measures. This would add to a more vigorous combination by exploiting the corresponding properties of the two-scale noticed centre guides, i.e. strength to mis-enrolment (and commotion) and the better-adjusted limits. The evaluation is initially demonstrated in a probabilistic perspective utilizing irregular strolls-based calculation, in which we attempt to address the probabilities that each pixel of the centre guide is related to the noticed ones. Then we observed that this technique is identical to tackling another goal work, empowering an incredible lift both in computational productivity and assessment results. Exploratory outcomes exhibit that the proposed technique is a strong yet effective contrast and cutting edge combination strategy.

This paper [9] presents an original district-based system for multifocal picture combinations. The central thought is to portion the in-centre districts from the info pictures and join them up to deliver an all-in-centre picture. To this end, we propose three instinctive limitations on the combined cycle and model them into three energy terms, i.e. remaking blunder, out-of-centre energy and perfection regularization. The three terms are then planned into an enhancement system issue to tackle a division map. We additionally propose an eager calculation to limit the goal work, which, on the other hand, refreshes every pixel in the division map using a coarse-to-fine procedure. The intertwined picture is at long last created by joining the portioned in-centre districts in each source picture using the division map. Our methodology can yield a consistent outcome with many fewer ringing ancient rarities. Our methodology beats the cutting edge strategies in light of different blended and genuine pictures.

A hazy foundation because of the shallow profundity of field is frequently wanted for photographs like representations. In any case, little easy to use cameras don't allow enough defocus due to the small width of their focal points. We present a picture handling method that expands the defocus in a picture to re-enact the shallow profundity of the field of a focal point with a bigger opening. Our method assesses the spatially-shifting measure of obscure over the picture and afterwards utilizes a straightforward picture-based procedure to increment defocus. We first gauge the size of the hazing part at the edges and afterwards engender this defocus measure over the picture. Utilizing our defocus map, we amplify the current fogginess, and that implies that we obscure foggy areas and keep sharp districts sharp. Rather than more troublesome issues, for example profundity from defocus, we don't need an exact profundity assessment and don't have to disambiguate texture in fewer districts [10].

During the most recent couple of years, the idea of a happy recovery has been the focus of many investigations. A tonne of frameworks were acquainted with each other all together to accomplish picture indexation. One of the most well-known techniques is to register a division and extricate various boundaries from areas. Nonetheless, this division step depends on low-level information without considering basic perceptual parts of pictures, similar to the haze. At the point when a picture taker chooses to zero in just on certain articles in a scene, he thinks about how distinctively these items are from the remainder of the scene. It doesn't address a similar measure of data. Picture recovery devices commonly consider foggy areas to be a specific situation rather than a data compartment. Our thought is then to concentrate on the correlation between pictures by limiting our concentration just to the non-hazy areas, utilizing this meta information [11]. Our point is to present various elements and an AI approach to arrive at an obscured distinguishing proof in scene pictures.

We address the issue of visually impaired movement de-blurring from a solitary picture, brought about by a couple of moving items. In such circumstances, just a piece of the picture might be obscured and the scene comprises layers obscured to various degrees. The vast majority of current visually impaired de-convolution research focuses on recovering a single obscuring portion for the entire image. Be that as it may, on account of various movements, the haze can't be demonstrated with a solitary part and attempting to de-convolve the whole picture with a similar portion will cause genuine curiosity. Hence, the undertaking of de-blurring needs to include the division of the picture into locales with various blurs. Our approach [12] is based on the belief that the insights of subsidiary channels in images differ fundamentally by obscure. We can narrow the search to one layered box channel obscure. This enables us to demonstrate the normal subsidiary conveyances as an element of the width of the hazing part. Those conveyances are shockingly strong in separating areas with various hazy spots. The methodology produces persuading de-convolution results on true pictures with rich surfaces.

In this paper [13], we propose a partially obscured image classification and investigation system for distinguishing pictures with obscured districts and perceiving the haze types for those regions without expecting to perform obscure portion assessment and picture de-blurring. We foster a few haze highlights demonstrated by picture tone, inclination and spectrum data, and our use includes boundary preparation to robustly characterise obscured pictures. Our haze recognition depends on picture patches, making district-wise preparation and classification in one picture effective. Broad tests show that our technique works adequately on testing image data, laying out a specialized starting point for dealing with a few PC vision issues, for example, movement investigation and picture rebuilding, utilizing haze data.

Numerous computerized pictures contain obscured districts that are brought about by movement or defocus. The automated discovery and organization of obscured image locations is critical for various media analysis projects. This paper presents a straightforward and successful programmatic picture obscured locale identification and grouping method [14, 15]. In the proposed method, obscured picture districts are first identified by ex-a mining solitary worth of data for each picture pixel. The haze types (for example movement obscure or defocus obscure) are then resolved in light of a specific alpha channel imperative that requires neither picture de-obscuring nor obscure bit assessment. Broad tests have been conducted over a dataset that comprises 200 obscured picture areas and 200 picture locales with no haze that have been removed from 100 advanced pictures. Trial results show that the proposed procedure distinguishes and orders the two sorts of picture obscures accurately. The proposed strategy can be utilized in a wide range of multimedia examination applications like picture division, profundity assessment and data recovery [16].

The obscure draws out a significant inquiry—what are viable elements to separate among obscured and clarified picture areas? We address it by concentrating on a couple of obscure component portrayals in picture slope, fourier area and information-driven neighbourhood channels. Unlike previous techniques, which spent a lot of time rebuilding components, our highlights are designed to improve discriminative power and adapt to different haze scales in images. To aid evaluation, we create another haze insight dataset with a large number of images with named ground truth. Our outcomes are applied to a few applications, including obscure locale division, de-blurring and obscure amplification [17].

Obscure exists in numerous computerized pictures. It is essentially classified into two types: defocus obscure, which is caused by optical imaging frameworks, and movement obscure, which is caused by general movement among camera and scene objects. In this letter, we propose a basic yet powerful program-obscured picture district identification technique. Because of the perception that obscures constricts high-recurrence parts of a picture, we present a haze metric in light of the log found at the middle value of the range leftover to get a coarse haze map. Then, at that point, a clever iterative refreshing instrument is proposed to refine the haze map from course to fine by taking advantage of the natural significance of comparative neighbouring picture districts [18]. The proposed iterative refreshing system can to some extent settle the issue of separating an in-centre smooth locale from an obscured smooth area. Likewise, the proposed iterative refreshing system can be incorporated into other picture obscured area recognition calculations to refine the eventual outcomes. Both quantitative and subjective exploratory outcomes exhibit that our proposed strategy is more dependable and effective compared with different cutting-edge techniques.

3 Methodologies

Proposed an obscured picture identification and examination system for naturally recognising video transfers and pictures. We develop a few haze highlight models using image tone, slope, and range data, and we use quick Fourier transformation to

Fig. 1 System architecture

identify obscure in recordings and images. Furthermore, we utilised a Fast Fourier Transform calculation worked into NumPy as the reason for our procedure. Extensive examinations show that our strategy works acceptably on testing picture information, which lays out a specialised starting point for taking care of a few PC vision issues, like movement investigation and picture rebuilding. Utilizing haze data, the process is described in Fig. 1.

4 Implementation and Result Analysis

The proposed strategy has attempted to handle two significant problems. One is the obscure location with synchronous extraction of obscured regions. The result of this progression gives valuable undeniable level of territorial data, working with an assortment of area-based picture applications, for example happy-based picture recovery, object-based picture pressure, video object extraction, picture improvement, and picture segmentation. It can likewise act as one of the models for estimating the nature of captured images. The second target of our technique is to group the distinguished haze districts into two kinds: close to isotropic haze (counting out-of-centre haze) and directional movement blur. We classify pictures into these two classes since they are most generally concentrated in picture rebuilding naturally. The haze-grouped images also effectively track down applications via moving examination and image reclamation. We arrange pictures into these two classes since they are most normally concentrated in picture restoration. The obscurely grouped images also effectively observe applications such as moving examination and image restoration. This technique depends on a perception that foggy districts are more invariant to low-pass filtering. Different obscure measures ought to be joined to accomplish excellent haze recognition. Using slope data and various bearings, the technique informs an energy work taking into account the gathered haze portion to section images into obscured or clarified layers. This technique just finds movement in obscured locales by surmising directional haze bits. Other haze assessment techniques, for example the one proposed in, just give a proportion of obscure degree, which can't be straightforwardly used to segregate foggy areas from non-hazy areas.

In this part, we examined the testing findings in order to compare the performance of the various picture blur detection techniques. Figure 1 depicts a sample result of a blurred picture after using blur detection techniques, whereas Fig. 2 depicts a sample result of a blurred video after using blur detection techniques. Furthermore, the accuracy of the proposed fast fourier transfer) (FFT) results are compared with other blur detection approaches such as hair wavelet transform (HWT), modified laplacian (MLAP), laplacian (LAP) and Tenengrad (TEN).

The percentage is seen here using the fast fourier transform. If the image is excessively blurry or dim, the values will be shown in red, and occasionally in negative values as well. If the picture file is less blurry or dim, the result will be shown in green. The result or outcome will be presented at the top left of the file.

The blur percentage is seen here in Fig. 3, using the fast fourier transform. If the live video is excessively blurry or dim, the values will be shown in red, and occasionally in negative values as well. If the live video is less blurry or dim, the result will be shown in green. The result or outcome will be presented at the top left of the file.

The confusion matrix findings of the performance comparison of several blur detection algorithms are shown in Table 1. According to the assessment findings, FFT leads the best outcomes in terms of accuracy rate, followed by HWT, TEN, LAP and MLAP in that order. In terms of execution time, LAP comes out on top, followed by FFT, TEN, HWT and MLAP in that order. And, Fig. 4 shows the accuracy report in graphical representation.

In Fig. 4, there will be a welcome message at the top, followed by five selections. To detect dim/blur, we can select detect from image, video or real time.

By choosing the choose file button in Fig. 5, we can select and upload the image file. After uploading, we must select the detect option to detect the picture file, and the result will be displayed in a pop-up window.

The percentage is seen here using the fast fourier transform. If the image is excessively blurry or dim, the values will be shown in red, and occasionally in negative

Fig. 2 Image detection output

Fig. 3 Live video detection output

Table 1 Performance analysis of various methods

Dim recognition	TN	FP	FN	TP	Accuracy (%)	Total time (s)
FFT	99	1	5	95	98	5.5037
HWT	94	6	6	98	94	1.2148
MLAP	82	0	22	76	83.70	0.9815
LAP	77	25	13	64	89	5.7934
TEN	89	3	7	82	91.80	3.8479

Fig. 4 Website's main page

Fig. 5 Image detection page

values as well. If the picture file is less blurry or dim, the result will be shown in green. The result will be presented at the top left of the file, Fig. 6.

Once you've completed all of the steps, this page will display. And, once the outcome has been processed or completed, this window will open with the statement "Operation Completed Successfully". If this notification arrives, the job is done. This message applies to both video and real life. There's also a time video, represented in Fig. 7.

In Fig. 8, by choosing the choose file button, we can select and upload the video file. After uploading, we must select the detect option to detect the video file, and the result will be displayed in a pop-up window.

Fig. 6 Image detection output

Fig. 7 Successful completion window

Fig. 8 Video detection page

Figure 9 represents live video detection. A pop-up will show after selecting the open live and detect buttons from the drop-down menu. And, it detects the blur/dim vividly and precisely. And, as with picture and video detection, the result will be shown in real time in the same pop-up window.

Figure 10 represents the output of the live video detection. The percentage is seen here using the fast fourier transform. If the live video is excessively blurry or dim, the values will be shown in red, and occasionally in negative values as well. If the live video is less blurry or dim, the result will be shown in green. The result or outcome will be presented at the top left of the file.

Fig. 9 Live video detection

Fig. 10 Video detection output

5   Conclusions

A blur picture recognition and assessment system has been provided in this research for automatically determining whether an image has blurred areas and what sorts of blurs are present without the requirement to execute an image de-blurring. It is possible to identify blurred photographs using a variety of blur parameters, including colour, gradation and spectrum information. Many tests have shown that our approach is effective even with difficult picture data and may be used to identify and segment partial blur in images. For numerous blur-oriented as well as region-based computerized vision problems, like picture content retrieval, augmentation, high-level classification and feature segmentation, our solution lays a solid foundation.

Acknowledgements Thanks to the Department of Computer Science and Engineering, Hindustan Institute of Technology and Science, Chennai, India.

References

1. Darncy., Ebby, P., & Jeena Jacob, I. (2021). Rain streaks removal in digital images by dictionary based sparsity process with MCA estimation. *Journal of Innovative Image Processing, 3*(3), 174–189.
2. Manoharan, J. S., & Jayaseelan G. (2020). Single image dehazing using deep belief neural networks to reduce computational complexity.
3. Fergus, R., Singh, B., Hertzman, A., Rowels, S. T., & Freeman, W. T. (2006). Removing camera shake from a single photograph. *ACM Translation Graphics, 25*(3), 787–794.
4. Li, L., Yan, Y., Fang, Y., Wang, S., Tang, L., & Qian, J. (2016). Perceptual quality evaluation for image defocus deblurring. *Signal Process Image Common, 48*, 81–91.
5. Lin, H. Y., Li, K. J., & Chang, C. H. (2008). Vehicle speed detection from a single motion-blurred image. *Image Visual Computer, 26*(10), 1327–1337.
6. Ting-Fa, X., & Peng, Z. (2010). Object's translational speed measurement using motion blur information. *Measurement, 43*(9), 1173–1179.
7. Choi, H., Cheng, S., Wu, Q., Castleman, K. R, & Bovid, A. C. (2000). Extended depth-of-field using adjacent plane deblurring and MPP wavelet fusion for microscope images. In *Proceeding 3rd IEEE International Sump. Biomedicen Image* (pp. 774–777).
8. Hariharan, H., Kushan, A., & Abide, M. (2008) Extending depth of field by intrinsic mode image fusion. In *Proceeding 19th International Conference Pattern Recogning* (pp. 1–4).
9. Ma, J., Zhou, Z., Wang, B., Miao, L., & Song, H. (2019). Multi-focus image fusion using boosted random walks-based algorithm with two-scale focus maps. *Neurocomputing, 335,* 9–20.
10. Duan, J., Meng, G., Xiang, S., & Pan, C. (2014). Multifidus image fusion via focus segmentation and region reconstruction. *Neurocomputing, 140*(9), 193–209.
11. Bae, S., & Durand, F. (2007). Defocus magnification. *Computer Graphics Forum, 26*(3), 571 579.
12. Regna, J. D., & Konak, H. (2003). Automatic blur detection for meta-data extraction in content-based retrieval context. *Proceeding SPIE, 15*, 285–295.
13. Levin, A. (2006). Blind motion deblurring using image statistics. In *Proceeding Advance Neural Information Processing System* (pp. 841–848).
14. Liu, R., Li, Z., & Jia, J. (2008) Image partial blur detection and classification. In *Proceeding IEEE Conference Computer Visual Pattern Recogning* (pp. 1–8). Anchorage, AK, USA.
15. Su, B., Lu, S., & Tan, C. (2011). Blurred image region detection and classification. In *Proceding 19th ACM International Conference Multimedia* (pp. 1397–1400).
16. Shi, J., Xu, L., & Jia, J. (2014). Discriminative blur detection features. In *Proceeding IEEE Conference Computer Visual Pattern Recogning (CPR)* (pp. 2965–2972).
17. Tang, C., Wu, J., Hou, Y., Wang, P., & Li, W. (2016). A spectral and spatial approach of coarse-to-fine blurred image region detection. *IEEE Signal Processing Letters, 23*(11), 1652–1656.
18. Smys, S., Iliyasu A.M., Bestak R., & Shi F. (eds) (2018). New trends in computational vision and bio-inspired computing. *ICCVBIC.*

The Voice and Gesture to Speech Conversion in Indian Regional Sign Language

Aneesha Bhattatiri, Anushka Chaubal, Meghana Dalal, and Gautam Borkar

Abstract This project helps mute people to interact with the rest of the world using sign language. Communication is an important aspect of human beings. Interaction between normal people and an impaired person is very difficult because of communication barriers. This work includes a voice-based and text-based interaction approach. An interpreter is responsible for helping impaired individuals understand what is being said in a variety of situations. The main feature of this project is that it can be used to predict sign language and translate it to voice so that normal people can understand sign language without the need of learning them. In this project, gesture to speech translation is done for Hindi as well as Tamil language, Hindi being the nation's most spoken language, and Tamil being the language used in most parts of southern India has increased the requirement of this model. For future scope, the model created in this project can be used to develop an application that involves not one but two-way communication.

Keywords Sign language · pyttsx3 · Google Translate · Convolutional neural network · OpenCV · Feature extraction · TensorFlow · Keras

A. Bhattatiri · A. Chaubal (✉) · M. Dalal · G. Borkar
Department of Information Technology, Ramrao Adik Institute of Technology, D.Y. Patil University, Nerul, Navi Mumbai, India
e-mail: anu.cha.rt18@rait.ac.in

A. Bhattatiri
e-mail: ane.bha.rt18@rait.ac.in

M. Dalal
e-mail: meg.dal.rt18@rait.ac.in

G. Borkar
e-mail: gautam.borkar@rait.ac.in

© The Author(s), under exclusive license to Springer Nature Singapore Pte Ltd. 2023 717
S. Shakya et al. (eds.), *Sentiment Analysis and Deep Learning*,
Advances in Intelligent Systems and Computing 1432,
https://doi.org/10.1007/978-981-19-5443-6_55

1 Introduction

Every gesture has a different identification in sign language. It is a language that is based on gestures. A deaf and mute person uses nonverbal forms of communication such as sign language while interacting with others. There can be image-based and sensor-based recognition for sign language. Image-based gesture recognition is more studied compared to sensor-based gesture recognition because image-based recognition does not require a person to wear data gloves, etc., for gesture identification. Sign recognition and detection are the two phases. Sign detection is the extraction of features of a certain thing concerning several parameters. Recognizing a certain posture or shape to differentiate that posture from other signs is sign recognition. The main objective of the project is to make the communication experience of mute people as normal as possible. This project deals with Indian Regional sign language. A lot of research has been done in this area, but the majority of the research is done in the English language. The main objective is to design a sign to speech conversion system for Hindi and Tamil languages. Firstly, a gesture is captured using the OpenCV live camera, trained and the model is created further, the model predicts the gesture, and the gesture is predicted by considering different angles, backgrounds, and shapes of the images so that each image remains different from the other. The grayscale format is used to remove the maximum interference of the background and to reduce the noise. Lastly, the equivalent text indicating the gesture is converted to speech which is in Hindi and Tamil. Comparisons were made among the different algorithms, and the one giving better feature extraction, having a better accuracy rate, and giving minimal loss was chosen for training the model.

2 Literature Survey

According to researchers, sign language has been recognized using different steps including capturing gesture images, followed by image acquisition in order to send images for processing, feature extraction is done for differentiation between the images. Different researchers have used different methods while collecting the input data and outputting the data.

Truong used AdaBoost and Haar-like classifiers, 28,000 samples of hand sign images were used to train the translator, ASL was used, 1000 positive hand sign images of different scales, illumination with a set of 11,100 collections of Negative images was taken. Input fed was live video. Output was in the form of speech and text. Signs were recognized with 98.7% accuracy. API 5.3 was used and was integrated into the Windows SDK of Microsoft to convert text to speech [1].

Jain et al. used for classification and training were SVM and CNN, after feature extraction and preprocessing, SVM with 4 kernels and CNN with 2 layers were applied for training the model. The input given was an image dataset. Accuracy between the two algorithms was compared. SVM gave 81.4% while CNN with 2 layers gave 98.58% accuracy [2].

Mohandes et al. compared the recognition accuracy of Arabic sign language, image-based alphabet sign recognition, isolated word sign recognition, and continuous sign recognition using KNN, HMM, and PCNN algorithms used for image classification and segmentation [3].

Dabre and Dholay Image processing and neural networks were used to identify the features of the hand from the images that were taken from a video through a Web camera. The mode used was ISL. Positive as well as Negative images were taken. The conversion was done for video of daily used sentences gesture into the text and to audio using speech synthesizers. Haar Cascade Classifiers were used for sign interpretation. Microsoft.NET framework containing methods for both speech synthesis and recognition were used. The accuracy found was more than 92% [4].

Rajam and Balakrishnan 32 signs indicating binary up and down for the five fingers were taken. It was developed for single users. Static images of mode ISL had been fed through the preprocessing using the feature point extraction and ten images for each sign were trained that they were transformed to grayscale images and were resized. Preprocessing was done for the conversion of static images into text by considering fingertip position as a parameter [5].

Kim et al. The paper consists of the work on KSL where a data glove and an input sensing device was used to detect hand and finger motions, and the fuzzy min–max neural network provided better accuracy hence was used for recognition [6].

Wang et al. 4800 vocabulary Chinese sign Language images were taken and raw data taken from the 3D tracker and data glove. HMM algorithm was used for sign recognition and synthesizer for speech conversion, and the fast matching algorithm was used for improving performance. Performance accuracy found was around 90% [7].

Panwar Features like orientation, shape, and area were used for hand gesture recognition also, the information about the number of fingers and thumb position were used for hand sign recognition, the mode of sign used was ASL, preprocessing was done for noise removal, K-means clustering was used for segmentation of hand images, and gesture recognized by the help of 7-bit binary sequence during classification. The hand gesture recognition rate was calculated after bit generation [8].

Ghotkar et al. Alphabets for ISL were recognized, hand signs were taken from the inbuilt camera, and steps involved for recognition were real-time hand tracking, segmentation, feature extraction, and gesture interpretation.

Camshift method HSV color model and genetic algorithm were used for hand tracking, segmentation, and gesture interpretation. The system supports single and double-handed gestures [9].

Poddar et al. Webcam was used to recognize the hand signs made using contour recognition, image database was created using a webcam, signs used were of ASL form, followed by image preprocessing thereby, converting it to grayscale and removing background noise, after recognition signs are converted to speech and text further wav file containing speech was created [10].

Hossen et al. One-handed static gesture images of the Bengali alphabet were taken, BSL was used, having 1147 images in different backgrounds, followed by preprocessing, feature extraction, a deep convolutional neural network for classification and training. Output was the predicted class [11].

Bhatnagar et al. ASL was used, data glove-based system using a static subset of data with 2250 samples, sensor used for detection of movement of the fingers in order to calculate the statistical value and match it with the threshold, after which speech output was given based on the recognition [12].

Ameen and Vadera CNN was used for the classification of fingerspelling signs of ASL with depth and image intensity data [13].

Ellis and Barca Classification of signals from sensor gloves was done using conditional template matching, 10 flex and tactile sensors were implanted into gloves. Used Australian sign language [14].

Kyatanavar and Futane The basic automatic SLT uses two approaches. First, the data glove has inbuilt sensors and is worn by the signer to detect hand posture. Second, visual-based where the camera is used to capture images of signers [15].

Goyal et al. The method was developed for single users. Real-time ISL images were captured followed by feature extraction to identify signs articulated by the user through SIFT algorithm. Matched key points found from the input image were compared to the image stored for a specific letter in the database [16].

Purva and Vaishali ANN algorithm was used for the classification of ISL sign gestures. The accuracy achieved was 98% [17].

Pravin and Rajiv ISL gestures were taken from multiple sources, gestures were outputted as text and speech, algorithms compared in the study were blob detection, thresholding, and neural network fuzzy min–max. GFMMNN showed higher accuracy among all [18].

Nandy et al. Video database containing ISL signs were created. A direction histogram was used for classification. Euclidean distance and K-nearest neighbor metrics were used for recognition [19].

Khan et al. Steps involved for gesture recognition systems included extraction method from the camera, sensor, feature extraction, HMM, NN, fuzzy c means clustering algorithm was used for classification [20].

Lai and Lai Gestures of one to nine numbers were taken to recognize eleven kinds of gestures from dynamic video. YCbCr color space detected the skin color and hand contour pattern from the complex image, fingertip positions, and angles were calculated to recognize the hand signs [21].

Huang et al. Real dataset collected with Microsoft Kinect was used, and the efficiency of the model over the traditional ones based on hand-crafted features was demonstrated. 3D CNN is used to extract information from raw video without prior knowledge [22].

Akmeliawati et al. Webcam images and custom-based gloves used for recognition, preprocessing, hand motion tracking, ANN classification used for Malaysian sign language gesture classification of words, numbers, alphabets of static and dynamic type, further translation of it to the English language [23].

Liwicki and Everingham Individual letter recognition included hand segmentation, appearance description, and classification, continuous word recognition includes HMM, and 1000 low-quality videos of 100 words were taken for dataset creation. Word recognition accuracy is 98.9%. Used British sign language [24].

Holden et al. The first phase was a tracking module that identified face and hand signs, the second was the feature extraction phase, and the last phase was recognition that involved HMM and grammar to recognize phrases. 97% sentence-level and 99% word-level accuracy were achieved. Used Australian sign language [25].

After studying the survey papers of different researchers, it is found that image-based or visual-based input methods provide much better accuracy and are less prone to errors when compared to the sensor-based methods. Image-based methods can be performed at a much more affordable rate. Grayscale images are found to have a better feature extraction and processing power when compared to RGB images. Grayscale is less complex than RGB. Gesture recognition and speech conversion tasks require image acquisition, preprocessing, feature extraction, and speech synthesis steps in order to achieve accuracy. CNN with three convolution layers gives better performance for image recognition. Thus, this project uses one hand gestures, image-based input, grayscale images, three-layer CNN algorithm for acquisition, preprocessing, feature identification, and prediction which are mostly used in the above-cited survey papers as in listed Table 1.

3   Methodology

Methods: Following steps are followed as shown in the Fig. 1 for predicting the gesture and converting the predicted gesture to speech in Tamil and Hindi.

3.1   Image Acquisition

Hand signs, data gloves, are examples of image acquisition input devices. In this system, live signs are acquired by using an OpenCV camera. The user signs through the computer's Web camera, and the sign/gesture is captured using OpenCV software. The dataset of images created in this project is completely new and original.

3.2   Image Preprocessing

Is done for recognizing signs with good accuracy. Because OpenCV only collects BGR scale images, the live BGR signs are taken from a camera and transformed into gray image signs. RGB images are the most widely used image format for taking images and storing them. But, RGB consists of three layers and processing them one

Table 1 Literature paper review

Author and Ref. No.	Method and protocol	Advantages	Drawbacks
Truong et al. [1]	AdaBoost and Haar-like classifiers were used, 28,000 images of ASL hand signs, 1000 Positive hand signs, and 11,100 collections of Negative images were taken	Recognition of gestures in complex backgrounds	Require a big dataset for high accuracy
Jain et al. [2]	SVM and CNN were used, after feature extraction and preprocessing, SVM with 4 kernels and CNN with 2 layers applied for training the model. The input given was an image dataset having ASL images	Increase in accuracy value with a double layer of CNN	Fixed filter size for CNN layers
Mohandes et al. [3]	Recognition accuracy of ArSL, image-based alphabet sign recognition, isolated word sign recognition, and continuous sign recognition were compared using KNN, PCNN, and HMM algorithms	With image-based ArSL, there was no need to use complex sensor devices	To maintain accuracy with time
Dabre and Dholay [4]	Conversion of video of daily used sentences gesture of ISL sign to text then to speech using speech synthesizers. Haar Cascade Classifiers were used for sign interpretation	Efficiency increased fast recognition rate	Speech synthesis was slow, due to large sentences
Rajam and Balakrishnan [5]	32 ISL images indicating binary up and down positions for the five fingers were taken. Applied for single users. Preprocessing was done, fingertip positions, and static images were converted to text	Angular measurement improved the sign recognition accuracy and percentage	Comparatively lower accuracy

(continued)

Table 1 (continued)

Author and Ref. No.	Method and protocol	Advantages	Drawbacks
Kim et al. [6]	Data glove, an input sensing device was used to detect hand and finger motions, the fuzzy min–max neural network was used for recognition. Used Korean sign language	Gave good performance accuracy for gesture recognition	Sensor errors resulted in wrong results
Wang et al. [7]	4800 Chinese vocabulary signs were taken, raw data was taken from a 3D tracker and data glove, HMM algorithms were used for sign recognition and synthesizer for speech conversion, and a fast matching algorithm was used for improving performance	Recognition performance and speed were improved	System for signer-dependent isolated signs
Panwar [8]	Hand sign recognition was done on the basis of features like orientation, area, shape, and K-means clustering was used for segmentation of ASL hand gestures, gesture recognized by the help of 7-bit binary sequence during classification	Computationally efficient, does not require training in data	Maximum parameters are defined based or assumptions
Ghotkar et al. [9]	Alphabets for ISL were recognized, Camshift method, HSV color model and genetic algorithm for hand tracking, segmentation, and gesture interpretation. Used single and double-handed gestures	Cost-effective, used single and double-handed signs, could recognize signs with bare hands	Restricted to alphabet recognition only
Poddar et al. [10]	Webcam was used to recognize the ASL hand signs made using contour recognition, after recognition wav file containing speech created	Easy conversion of signs into texts and speech	Requirement for more memory

(continued)

Table 1 (continued)

Author and Ref. No.	Method and protocol	Advantages	Drawbacks
Hossen et al. [11]	One-handed Bengali sign gestures consisting of 1147 images of the Bengali alphabet were taken, and a deep convolutional neural network was used for classification and training	Deep convolutional improved the performance unlike others when trained on the same data	Small dataset and misclassification
Bhatnagar et al. [12]	Data glove-based system using a static subset of data with 2250 samples of ASL, and the sensor was used for detection of movement of the fingers to calculate the statistical value followed by the speech output	Sign to speech conversion system developed was relatively affordable	Cost limitation
Ameen and Vadera [13]	CNN was used for the classification of fingerspelling signs of ASL with depth and image intensity data	Depth and intensity inputs showed improvements	Less accuracy due to noise
Ellis and Barca [14]	Classification of signals from sensor gloves was done using conditional template matching, 10 flex and tactile sensors were implanted into gloves. Used Australian sign language	Sensor gloves were used as feedback to children	The discrepancy between signs
Kyatanavar and Futane [15]	The basic automatic SLT uses two approaches. First, the data glove has inbuilt sensors and is worn by the signer to detect hand posture. Second, visual-based where the camera is used to capture images of signers	The vision-based approach was found to be a cheaper one	The ambiguity between the finger signs

(continued)

Table 1 (continued)

Author and Ref. No.	Method and protocol	Advantages	Drawbacks
Goyal et al. [16]	Method developed for single users. Real-time ISL images were captured followed by feature extraction to identify signs articulated by the user through SIFT algorithm. Matched Key points found from the input image were compared to the image stored for a specific letter in the database	Uses real-time image capture	Single user system, alphabet restricted
Purva and Vaishali [17]	ANN was used for the classification of ISL signs. The accuracy achieved was 98%	RGB images are easily captured from the camera	Restricted to a small dataset
Pravin and Rajiv [18]	ISL gestures were taken from multiple sources, gestures outputted as text and speech, algorithms were compared in the study are blob detection, thresholding, and neural network fuzzy min–max. GFMMNN showed higher accuracy among all	Performance was good when tested with video gestures forming sentences	Accuracy was lower due to fewer feature attributes
Nandy et al. [19]	Video database containing ISL signs was created. A direction histogram was used for classification. Euclidean distance and K-nearest neighbor metrics were used for recognition	KNN algorithm gave 100% accuracy for classification	Lower accuracy for dynamic ISL
Khan et al. [20]	Steps involved for gesture recognition systems included extraction method from the camera, sensor, feature extraction, HMM, NN, and fuzzy c means clustering algorithm was used for classification	Algorithms used for dynamic gesture recognition, and shape capturing were effective	Computing time and errors due to lighting were more

(continued)

Table 1 (continued)

Author and Ref. No.	Method and protocol	Advantages	Drawbacks
Lai and Lai [21]	Gestures of one to nine numbers were taken to recognize eleven kinds of gestures from dynamic video. YCbCr color space detected the skin color and hand contour pattern from the complex image, fingertip positions and angles were found to recognize the hand signs	Picture processing time was less	Comparatively lesser accuracy
Huang et al. [22]	Real dataset collected with Microsoft Kinect was used and the efficiency of the model over the traditional ones based on hand-crafted features was demonstrated. 3D CNN is used to extract information from raw video without prior knowledge	Deep learning extracts different pieces of knowledge	Misclassification is found in some cases
Akmeliawati et al. [23]	Webcam images and custom-based gloves were used for recognition, preprocessing, and hand motion tracking. ANN classification was used for Malaysian sign language gesture classification of words, numbers, alphabets of static and dynamic type, and further translation of it to the English language	The system could recognize both static and motion fingerspelling and sign gestures, and it was low cost	Accuracy decreased when similar signs were found

(continued)

Table 1 (continued)

Author and Ref. No.	Method and protocol	Advantages	Drawbacks
Liwicki and Everingham [24]	Individual letter recognition included hand segmentation, appearance description, and classification, continuous word recognition included HMM, and 1000 low-quality videos of 100 words were taken for dataset creation. Word interpretation accuracy was 98.9%. Used British sign	A simple HMM-based lexicon model could give high accuracy on large lex cons	Small dataset with inexperienced signers
Holden et al. [25]	First phase was a tracking module that identified face and hand signs, the second was the feature extraction phase, and the last phase was recognition that involved HMM and grammar to recognize phrases. 97% sentence-level and 99% word-level accuracy were achieved. Used Australian sign language	Accuracy of sentence and word recognition was relatively high with the usage of HMM and grammar	Difficult to identify signs in complex backgrounds

Fig. 1 Working flow
diagram

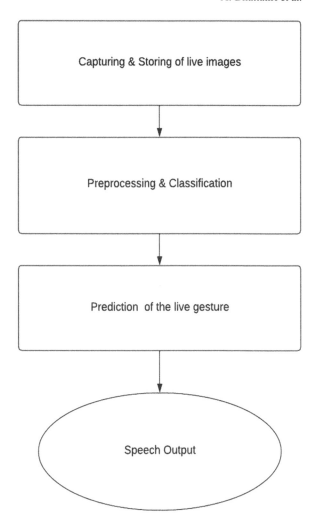

by one increases the computation cost and loading time. Therefore, OpenCV captures images in BGR format and converts them into RGB, this format being ineffective for classification and is hence converted to grayscale. Grayscale images are loaded faster with good accuracy because of their less complex behavior. They consist of black and white color images. After grayscaling, images are resized to decrease the training model's loading time. The images are processed into 64X64 dimensions making them consistent throughout the dataset.

3.3 Feature Extraction

Algorithms including CNN, DNN, ANN, and CNN with SVM were considered for extracting features after training the model. ANN and CNN are found to be used primarily in image classification in most of the image classification problems compared to DNN which is the least preferred. ANN when compared to CNN results in costly computation and storage, DNN on the other hand works inefficiently with images, and pattern recognition and extraction are difficult to achieve with DNN. ANN requires concrete data points, while CNN extracts features easily on its own without any human intervention from the input image. It can extract a hundred features effectively. CNN was chosen among all as the best fit for feature extraction and image recognition because it detects important features with good accuracy compared to the others.

3.4 Building the Training Model

CNN is majorly used as an application in computer vision. The working is considered to be similar to the perception that takes place in the human brain using neurons. The algorithm uses a filter for scanning through the entire pixel data of the image in addition it adds weight to detect features from it. CNN contains layers including a convolution layer, Max Pooling layer, flatten layer, dense layer, dropout layer, and a fully connected layer. These layers when combined make the identification of features effective. The layers at the beginning are used for identifying low-level features. Layer by layer gradually more complex features is recognized. The CNN model was built by adding three convolutional neural layers and the activation function used was "ReLu". To overcome the error of over-fitting, Max Pooling was done. Multi-class classification is being performed hence the activation function used in the output layer of the CNN model uses the "Softmax" function. The first layer is responsible for detecting low-level attributes like lines, with the input of 64 × 64 size of the grayscale image. It uses 32 filters in this layer. Max Pooling layer considers maximum values from the region hence detecting the desired feature increases through it. The second layer identifies features like angles, etc. This layer consists of 64 filters. Max Pooling is again used for this layer. The third layer identifies shapes, signs, etc. This layer consists of 128 filters. A dense layer is used in the model for expanding it with 256 elements. The other dense layer reduces the map to 15 elements in an array representing the considered classes. Every class has a specific probability equivalent to it, the class having the maximum probability value indicates itself to be the predicted gesture.

3.5   Gesture to Speech Conversion

After recognition, gesture signs are converted into speech for providing better communication for mute people. TensorFlow and Keras modules are used for training the collection of images using the chosen CNN algorithm by applying three convolution layers. Each class contained in a collection has 300–400 images for the training collection and 90–100 images for the testing collection, images are captured in a way that the hand portion gets visible thereby keeping the background interruption to a minimum. The images are taken in different backgrounds considering the background interference. After training the collection of different classes, the images are predicted using an OpenCV camera, and the live image is converted to the respective grayscale, which is compared to the collection of images. The predicted gesture is shown with its name and the word is converted into Hindi as well as the Tamil language. The translation of English words to Hindi and Tamil language is done using the Google Translate module containing the Translator class. The Hindi word is then transformed to its audio equivalent using the pyttsx3 module of Python. The same module is used for Tamil speech conversion. To choose a specific voice of Tamil and Hindi speech, go to settings, ease of access, then to the narrator of Windows 10, to add a new voice go to the start menu and choose windows registry editor, open a list of available voices, export the voices, modify the exported file, import the new file after editing, restart the PC to find the voice on the computer.

4   Result and Discussion

4.1   Comparison Among Different Algorithms

DNN. Figure 2 shows the accuracy after training the model is in the range of 0.4–0.5, showing that it is not the best choice to consider for image extraction and recognition, and a different algorithm could be taken into consideration, for dealing with this type of dataset.

ANN. Figure 3 shows the accuracy after training the model is in the range of 0.98–1.0, showing that it could be a better choice to consider for image extraction and recognition compared to DNN for dealing with this type of dataset.

CNN. Figure 4 shows the accuracy after training the model is in the range of 0.99-1.0, with 1.00 occurring the most, showing that it could be a better choice to consider for image extraction and recognition compared to DNN and ANN for dealing with this type of dataset.

CNN-SVM. Figure 5 shows, the accuracy after training the model is in the range of 0.99- 1.0, showing that it could be a better choice to consider for image extraction and recognition compared to DNN, and it gives similar accuracy when compared to

```
---,--- [                          ]  -- ---,---- ---- ------- -- ------,- ----
Epoch 31/50
220/220 [==============================] - 2s 10ms/step - loss: 1.1921e-07 - accuracy: 0.4695
Epoch 32/50
220/220 [==============================] - 2s 10ms/step - loss: 1.1921e-07 - accuracy: 0.4762
Epoch 33/50
220/220 [==============================] - 2s 10ms/step - loss: 1.1921e-07 - accuracy: 0.4918
Epoch 34/50
220/220 [==============================] - 2s 10ms/step - loss: 1.1921e-07 - accuracy: 0.5052
Epoch 35/50
220/220 [==============================] - 2s 9ms/step - loss: 1.1921e-07 - accuracy: 0.4591
Epoch 36/50
220/220 [==============================] - 2s 9ms/step - loss: 1.1921e-07 - accuracy: 0.4569
Epoch 37/50
220/220 [==============================] - 2s 10ms/step - loss: 1.1921e-07 - accuracy: 0.4732
Epoch 38/50
220/220 [==============================] - 2s 9ms/step - loss: 1.1921e-07 - accuracy: 0.4591
Epoch 39/50
220/220 [==============================] - 2s 9ms/step - loss: 1.1921e-07 - accuracy: 0.4695
Epoch 40/50
```

Fig. 2 DNN training accuracy

```
validation_steps=140)# No of images in test set

220/220 [==============================] - 2s 8ms/step - loss: 0.0234 - accuracy: 0.9927
Epoch 43/50
220/220 [==============================] - 1s 7ms/step - loss: 0.0066 - accuracy: 0.9982
Epoch 44/50
220/220 [==============================] - 2s 8ms/step - loss: 0.0063 - accuracy: 1.0000
Epoch 45/50
220/220 [==============================] - 2s 8ms/step - loss: 0.0217 - accuracy: 0.9936
Epoch 46/50
220/220 [==============================] - 2s 8ms/step - loss: 0.0049 - accuracy: 0.9991
Epoch 47/50
220/220 [==============================] - 2s 8ms/step - loss: 0.0058 - accuracy: 0.9973
Epoch 48/50
220/220 [==============================] - 2s 8ms/step - loss: 0.0165 - accuracy: 0.9936
Epoch 49/50
220/330 [==============================] - 2s 8ms/step - loss: 0.0159 - accuracy: 0.9964
Epoch 50/50
220/220 [==============================] - 2s 8ms/step - loss: 0.0065 - accuracy: 0.9982

Out[11]: <tensorflow.python.keras.callbacks.History at 0x1/8384415e0>

In [ ]:
```

Fig. 3 ANN training accuracy

```
Epoch 39/50
220/220 [==============================] - 4s 18ms/step - loss: 0.0091 - accuracy: 0.9984
Epoch 40/50
220/220 [==============================] - 4s 18ms/step - loss: 0.0111 - accuracy: 0.9982
Epoch 41/50
220/220 [==============================] - 4s 18ms/step - loss: 0.0017 - accuracy: 1.0000
Epoch 42/50
220/220 [==============================] - 4s 18ms/step - loss: 1.4247e-04 - accuracy: 1.0000
Epoch 43/50
220/220 [==============================] - 4s 18ms/step - loss: 0.0038 - accuracy: 0.9991
Epoch 44/50
220/220 [==============================] - 4s 18ms/step - loss: 0.0027 - accuracy: 0.9982
Epoch 45/50
220/220 [==============================] - 4s 18ms/step - loss: 0.0016 - accuracy: 0.9991
Epoch 46/50
220/220 [==============================] - 4s 18ms/step - loss: 0.0003 - accuracy: 0.9982
Epoch 47/50
220/220 [==============================] - 4s 20ms/step - loss: 1.3334e-04 - accuracy: 1.0000
Epoch 48/50
220/220 [==============================] - 5s 21ms/step - loss: 4.3132e-05 - accuracy: 1.0000
Epoch 49/50
220/220 [==============================] - 5s 23ms/step - loss: 0.0101 - accuracy: 0.9973
Epoch 50/50
220/220 [==============================] - 5s 21ms/step - loss: 0.1903e-04 - accuracy: 1.0000
```

Fig. 4 CNN training accuracy

ANN, but the CNN has given a better result out of all the considered algorithms. Hence, CNN is the best choice for dealing with this type of dataset.

```
validation_steps=140j# no of images in test set
220/220 [==============================] - 4s 18ms/step - loss: 0.9124 - accuracy: 0.9975
Epoch 40/50
220/220 [==============================] - 4s 19ms/step - loss: 0.9137 - accuracy: 0.9955
Epoch 41/50
220/220 [==============================] - 4s 19ms/step - loss: 0.9145 - accuracy: 0.9936
Epoch 42/50
220/220 [==============================] - 4s 19ms/step - loss: 0.9126 - accuracy: 0.9982
Epoch 43/50
220/220 [==============================] - 4s 19ms/step - loss: 0.9113 - accuracy: 1.0000
Epoch 44/50
220/220 [==============================] - 4s 20ms/step - loss: 0.9115 - accuracy: 0.9982
Epoch 45/50
220/220 [==============================] - 4s 20ms/step - loss: 0.9116 - accuracy: 0.9973
Epoch 46/50
220/220 [==============================] - 4s 20ms/step - loss: 0.9129 - accuracy: 0.9964
Epoch 47/50
220/220 [==============================] - 5s 21ms/step - loss: 0.9138 - accuracy: 0.9936
Epoch 48/50
220/220 [==============================] - 5s 22ms/step - loss: 0.9130 - accuracy: 0.9964

In [ ]:
```

Fig. 5 CNN with SVM training accuracy

Fig. 6 Gesture shown to predict the Hindi word for Hindi speech conversion

4.2 Hindi Language

4.3 Tamil Language

The trained model using the CNN algorithm has given the required output gesture, the word equivalent to the gesture is converted to Hindi and Tamil speech after prediction as shown in the validated results from Figs. 6, 7, 8, 9, 10 and 11. Considering the background interference, the gesture is predicted rightly in at most two–three attempts after which the voice is generated.

5 Conclusion

The aim of the system is to allow a deaf and a mute to communicate with others easily by allowing normal people to understand what the impaired person has said. The proposed approach converts hand sign gestures into speech. After comparing

```
if g>o and g>s and g>t and g>y:
    res=translator.translate("good",dest='hi',src='en')
    print('good-> "+res.text)
if s>b and s>g and s>t and s>y:
    res=translator.translate("sorry",dest='hi',src='en')
    print("sorry-> "+res.text)
if t>b and t>g and t>s and t>y:
    res=translator.translate("thankyou",dest='hi',src='en')
    print("thankyou-> "+res.text)
if y>b and y>g and y>s and y>t:
    res=translator.translate("you",dest='hi',src='en')
    print("you-> "+res.text)



cap.release()
cv2.destroyAllWindows()

you-> आप
```

Fig. 7 Code part for the Hindi speech translation

Fig. 8 Grayscale images inside the dataset

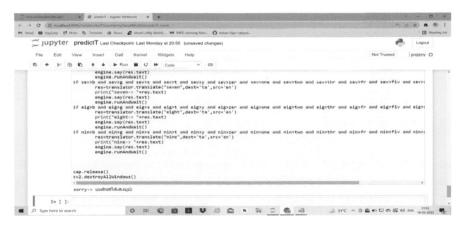

Fig. 9 Code for Tamil speech translation

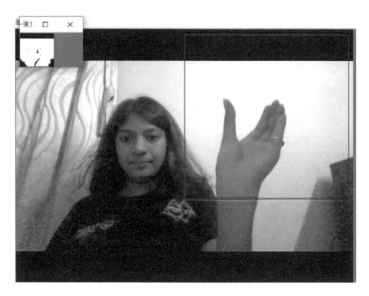

Fig. 10 Gesture shown to predict the Tamil word for Tamil speech conversion

Fig. 11 CNN model accuracy after training the set

several algorithms, the three-layer CNN was used to train the model, the classified grayscale image was successfully compared with the live grayscale gesture image, and the required speech output was produced effectively using the speech recognition system, installed on the PC. Thus, this system allows the mute to connect socially with others and overcome their inability to deliver signs to others. The model makes normal people understand the signs being communicated by the impaired effectively without the need of learning the signs. The project can be extended to other signs of different languages by the addition of their respective datasets and training their models with CNN. This model in the future can be used as a base for creating an application involving two-way interaction that is a gesture to speech as well as speech to gesture by which the impaired person will also be able to understand the speech produced by the normal person by seeing the gesture produced by the voice input.

References

1. Truong, V. N. T. (2016). Translator for American sign language to text and speech. In *Fifth IEEE Global Conference on Consumer Electronics*. Kyoto, Japan.
2. Jain, V., Jain, A., Chauhan, A., Kotla, S. S., & Gautam, A. (2021). An American sign language recognition using support vector machine and convolutional neural network. *International Journal of Information Technology*.
3. Mohandes, M., Liu, J., & Deriche, M. (2014). A survey of image-based Arabic sign language recognition. In *IEEE Eleventh International Multi-Conference on System, Signals and Devices* (pp. 1–4).
4. Dabre, K., & Dholay, S. (2014). A machine learning model for sign language interpretation using web camera images. In *IEEE International Conference on Circuits, System Communication and Information Technology Applications*. India.
5. Rajam, P., & Balakrishnan, G. (2010). Indian sign language recognition system to aid deaf and dumb people. In *IEEE International Conference on Computing Communication and Networking Technologies* (pp. 1–9).
6. Kim, J., Jang, W., & Bien, Z. (1996). A dynamic gesture recognition system for the Korean sign language. *IEEE Translation on System, Man, Cybernetics. B, 26*, 354–359.
7. Wang, C., Gao, W., & Xuan, Z. (2001) A real-time large vocabulary continuous recognition system for Chinese sign language. In *Proceeding IEEE Pacific Rim Conference on Multimedia* (pp. 150 157).
8. Panwar, M. (2012). Hand gesture-based interface for aiding the visually impaired. In *Proceeding IEEE International Conference on Recent Advance in Computer and Software System* (pp. 80–85).
9. Ghotkar, A. S., Khatal, R., Khupase, S., Asati, S., & Hadap, M. (2012). Hand gesture recognition for Indian sign language. In *IEEE International Conference on Computer Communication and Informatics* (pp. 1–4).
10. Poddar, N., Rao, S., Sawant, S., Somavanshi, V., & Chandak, S. (2015). Study of sign language translation using gesture recognition. *International Journal of Advanced Research in Computer and Communication Engineering*.
11. Hossen, M. A., Govindaiah, A., Sultana, S., & Bhuiyan, A. (2018). Bengali sign language recognition using deep convolutional neural network. In *Joint Seventh International Conference on Informatics, Electronics, and Vision*.
12. Bhatnagar, V., Magon, R., Srivastava, R., & Thakur, M. (2015). Cost-effective sign language to voice emulation system. In *Proceeding of the 2015 Eighth International Conference on Contemporary Computing* (pp. 521–525). Noida, India.
13. Ameen, S., & Vadera, S. (2017). Convolutional neural network to classify American sign language finger spelling from depth and color images. *Expert Systems*.
14. Ellis K., & Barca, J. (2012) Exploring of sensor gloves for teaching children sign language. *Advances in Human-Computer Interaction*.
15. Kyatanavar, R., & Futane, P. (2012) A comparative study of sign language recognition systems. *International Journal of Scientific and Research Publications, 2*(6).
16. Goyal, S., Sharma I., & Sharma S. (2013). Sign language recognition system for deaf and dumb people. *International Journal of Engineering Research & Technology*.
17. Purva, N., & Vaishali, K. (2014). Indian sign language recognition: a review. In *IEEE Proceedings on International Conference on Electronics and Communication Systems* (pp. 452–456).

18. Pravin, F., & Rajiv, D. (2011). "HASTA MUDRA" an interpretation of Indian sign hand gestures. In *Third International Conference on Electronics Computer Technology* (Vol. 2, pp. 377–380).
19. Nandy, A., Prasad, J., Mondal, S., Chakraborty, P., & Nandi, G. (2010). Recognition of isolated Indian sign language gesture in real-time. In *Communications Computer and Information Science book Series* (Vol. 70).
20. Khan, R., & Ibraheem, N. (2012). Hand gesture recognition: A literature review. *International Journal of Artificial Intelligence and Applications, 3*, 162–174.
21. Lai, H., & Lai, H. (2014). Real-time dynamic hand gesture recognition. In *IEEE International Symposium on Computer, Consumer, Control* (pp. 658–661).
22. Huang, J., Zhou, W., Li, H., & Li, W. (2015). Sign language recognition using 3D convolutional neural networks. In *IEEE International Conference on Multimedia and Expolore* (pp. 1–6).
23. Akmeliawati, R., Po-Leen Ooi, M., & Chow Kuang, Y. (2006). Real-Time Malaysian sign language translation using colour segmentation and neural network. In *IEEE on Instrumentation and Measurement Technology Conference Proceeding* (pp. 1–6). Warsaw, Poland.
24. Liwicki, S., & Everingham, M. (2009) Automatic recognition of fingerspelled words in British sign language. In *IEEE Computer Society Conference on Computer Vision and Pattern Recognition Workshops*. School of Computing University of Leeds.
25. Holden, E., Lee, G., & Owens, R. (2005). The Australian sign language recognition. *Machine Vision and Applications, 16*, 312–320.

Deep Learning-Based Quality Assurance of Silicon Detectors in Compact Muon Solenoid Experiment

Richard Joseph, Shashi Dugad, Ajay Singh Khalsa, Swarangi Dali, and Vaishnavi Ainapure

Abstract For the CMS experiment at CERN, Geneva, a substantial number of HGCAL sensor modules are manufactured at advanced laboratories across the globe. Each sensor module comprises around 675 checkpoints for visual inspection, making manual inspection practically unfeasible. In the industrial environment of manufacturing these sensor modules, this work is extremely challenging due to the various defect appearances, nonuniform, and ambiguous wire bond arrangements. Due to recent technological advances, there has been a rise in automated visual inspections and intelligent quality assurance systems in manufacturing. In order to simplify this, we propose a deep learning-based automated visual approach that processes and analyzes huge number of images for quality control and subsequent testing.

Keywords Compact muon solenoid · Object detection · Visual inspection · Scaled YOLOv4 · YOLOv4-tiny

1 Introduction

The Large Hadron Collider (LHC) is the world's most powerful and largest particle accelerator. The LHC is mainly composed of a 27-km ring of superconducting

R. Joseph · A. S. Khalsa (✉) · S. Dali · V. Ainapure
Vivekanand Institute of Technology, Mumbai, Maharashtra 400074, India
e-mail: 2018.ajay.khalsa@ves.ac.in

R. Joseph
e-mail: richard.joseph@ves.ac.in

S. Dali
e-mail: 2018.swarangi.dali@ves.ac.in

V. Ainapure
e-mail: 2018.vaishnavi.ainapure@ves.ac.in

S. Dugad
Tata Institute of Fundamental Research, Mumbai, Maharashtra 400005, India
e-mail: shashi@tifr.res.in

© The Author(s), under exclusive license to Springer Nature Singapore Pte Ltd. 2023   737
S. Shakya et al. (eds.), *Sentiment Analysis and Deep Learning*,
Advances in Intelligent Systems and Computing 1432,
https://doi.org/10.1007/978-981-19-5443-6_56

magnets with a variety of accelerating components along the route to amplify the energy of the particles. It accelerates protons to nearly the velocity of light and then collides them at four locations around its ring. To face the new challenges of the LHC, Physicists at the CMS collaboration are working on a completely new calorimeter the High-granularity Calorimeter which will delivers 10 times more collisions to its experiments to be installed in the endcaps of the detector. They are currently designing silicon sensors that are a key component of the HGCAL to withstand the high radiation dose. These fragile components must be visually inspected with a microscope and electrically tested to detect anomalies. Each sensor is divided into several small cells which will detect charged particles traversing the detector. To measure performance the silicon, sensor is 'contacted' with hundreds of pins or wire bonds.

The new High-granularity Calorimeter will be equipped with 28,000 silicon modules with 6 million readout cells [1].

The wire bonds must be parallel and in intact condition to function properly and currently, the quality inspection for these bonds are done manually.

Using an old-fashioned examination process, which has a lot of shortcomings. Manual inspection requires the presence of a person, who examines the object of concern and makes a decision based on certain pre-defined criteria or expertise. Visual inspection errors are estimated to be between 20 and 30% in most cases, according to studies. Human error causes some errors, while space constraints cause others. In manufacturing, visual inspection inaccuracies can take two forms: missing an existing problem or detecting a fault that doesn't exist. False alarms occur far less frequently than missed warnings, although misses might result in a loss of quality. Visual inspection has been proven to discover the majority of concealed faults throughout the production process time and time again [3].

With the emergence of machine learning, a visual inspection approach that automatically processes and analyze a vast number of gathered photos was developed. Traditional computer vision-based approaches, on the other hand, are readily disrupted by noise and clutter during real-time examination. Due to the increased manufacturing speed in today's environment, the goal is to reduce human intervention. Deep learning has recently been effectively applied to image recognition and image classification [4]. It can automatically detect defect patterns at a much faster rate and with better accuracy. Defects such as broken, crossing wire bonds of the sensor as in Fig. 1 can be recognized from images using a combination of deep learning and image processing technologies.

2 Literature Review

In regards to automated visual inspection, the authors Kai-kai Zhai et al. provide a deep learning-based concrete apparent defect classifier that can intelligently classify a single concrete defect picture, as seen in [4]. The deep residual network model was formed to obtain the classifiers for the four apparent concrete defects; second, image

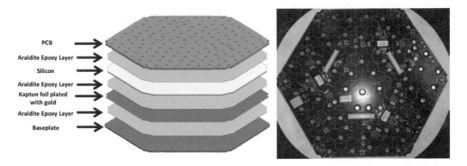

Fig. 1 Sensor board

processing technology was used to expand the image set of four types of apparent concrete defects, including general diseases, erosion and deposition, steel reinforcement exposed, and cracks; and finally, transfer learning was used to optimize the deep residual network model. The findings showed that the deep learning-based concrete apparent disease classifier created in this study can categorize images of a single concrete disease intelligently. The transfer learning accuracy rate was improved to 91.3%, and the identification accuracy of rebar exposed flaws was increased to 97.6%, meeting the demands of intelligent concrete defect detection in real-world applications.

In addition, for the Compressed Baryonic Matter (CBM) Experiment's Visual Inspection of Silicon Microstrip Sensors, custom-built optical inspection equipment was utilized to examine 1191 silicon microstrip sensors, with a machine learning-based technique for defect detection [5]. The surface of the sensor was also metro-logically regulated. The purpose of this research is to look at various sensor surface defects. Surface scrapes, dual metallization film defects, oxide layer defects, and dust particle detection are all instances of these. Convolutional deep neural networks were utilized to find flaws in the system (CDNNs). As a consequence, defective segments and defect agglomerations were discovered, and a two dimensional map of the defects based on their geometrical positions on the sensor was developed. Based on the total number of abnormalities discovered on the sensor's surface, a technique for determining the sensor's overall quality level was proposed.

Automated visual-based defect detection systems are applicable to a variety of materials, including metals, ceramics, and textiles, as shown in [6]. The work examines statistical, anatomical, and other techniques to textural defect diagnosis. They further discussed the state-of-the-art for detecting and classifying errors using supervised and non-supervised classifiers as well as deep learning. It is divided into three parts, the first of which covers the categorization of abnormalities that may arise on metal surfaces. The representation and batch processing of defects through pictures is covered in the following part, which is followed by the supervised and unsupervised classification techniques for image processing in the final step. The supervised classifiers used are K-nearest neighbor, NNs and Deep learning, SOM and

SVM while unsupervised classifiers used are Statistical/Novelty detection Gaussian mixture models.

3 Object Detection

Object detection is a technique of identifying and locating objects in photos and videos using computer vision [7]. It can be used to count objects in a scene, determine and track their precise locations, and precisely label them. It is required to accurately identify the exact categories of each object as well as the object's boundary box. Object detection algorithms as shown in Fig. 2 have become one of the research hotspots in the field of deep learning, thanks to the continued development of deep learning techniques in recent years. They are widely used in many real-world application scenarios, such as intelligent video monitoring, vehicle automatic driving, and robot environment perception.

Object detection is a technique of identifying and locating objects in photos and videos using computer vision [7]. It can count items in an environment, assess and analyze their exact positions, and label them appropriately. Each object's specific categories, as well as the object's bounding box, must be correctly identified. They're employed in a variety of real-world settings, facial recognition, industrial quality control, autonomous driving, and automobile detection.

There are two types of object detectors: single-step and dual-step detectors. Single-step approaches predict the entire image's classes and bounding boxes in a single instant, giving them a performance gain over dual-step algorithms. Single-stage models skips the region proposal stage of two-stage models and instead performs detection across a dense sample of locations. Inference is frequently faster with these models. It includes You Only Look Once (YOLO) and Single Shot MultiBox Detector (SSD) algorithms [8]. Dual-step techniques, on the other hand, involve a preliminary stage wherein important regions are identified, followed by classification to determine whether an object has been spotted in those areas. It includes or Mask R-CNN Faster or region-based convolutional neural networks (R-CNN) [9].

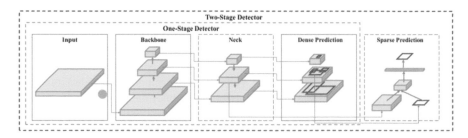

Fig. 2 Architecture of object detection algorithms

3.1 Yolo

YOLO is a neural network-based real-time object detection technique. The speed and precision of this algorithm are well-known. In a number of applications, it has been used to identify traffic signals, pedestrians, and animals. The approach just requires one forward propagation through a neural network to detect objects. This means that a single algorithm run may predict the whole image [10]. At the same time, the CNN is utilized to forecast various class probabilities and bounding boxes. There are several variants of the YOLO algorithm. Tiny YOLO, YOLOv4, and Scaled YOLOv4 [11, 12] are some of the popular examples.

The YOLO algorithm employs the following three techniques: residual blocks, bounding box regression, and intersection over union (IOU).

Residual blocks: To begin, the image is divided into grids. Each grid has a S x S dimension. The image below shows how an input image is divided into grids.

Bounding box regression: A bounding box is an outline that draws attention to an object in an image. Each bounding box present in the image consist of the following attributes:

- Probability of an object being present in the bounding box (pc),
- width (bw),
- the height (bh),
- Class (The letter c represents a class (for example, a person, a car, a traffic, light, and so on)).
- bounding box center (bx, by).

$$y = (\text{pc}, \text{bx}, \text{by}, \text{bh}, \text{bw}, c)$$

Intersection over Union(IOU): The object detection phenomenon of intersection over union (IOU) illustrates how boxes overlap. IOU is used by YOLO to produce a precise output box around the objects. The bounding boxes, as well as their confidence scores, are predicted by each grid cell. The outcome of IOU equals one if the expected and original bounding boxes are the same. Bounding boxes that aren't equivalent to the real box a removed using this approach.

3.2 YOLOv4

YOLOv4 algorithm developed by Bochkovskiy et al. is a significant overhaul over its predecessor with respect to speed and accuracy. In the MS COCO dataset, the YOLOv4 model (large) achieved revolutionary results. The YOLO-v4 network, as a one-step object detection algorithm, represents continued improvements to YOLO-v1 [10], YOLO-v2 [11], and YOLO-v3 [10, 11]. All YOLO series algorithms use the similar detecting mechanism. The region proposal network has been eliminated in comparison with the two-stage technique provided by Faster R-CNN [12], which considerably increases detection speed. The detection problem in YOLO-V4 is modified into a regression-based problem using a conventional one-stage approach.

YOLOv4 model runs in real time and is simple to train and execute on a single ordinary GPU. The YOLOv4 model, which is based on YOLOv3, has two types of optimization strategies: 'bag of freebies' and 'bag of specials'. The 'bag of freebies' approach, which may be construed as free presents, was presented by Mu Li et al. [14]. And its objective is to enhance object identification accuracy without affecting processing performance, using techniques including data augmentation, methods to solving the imbalanced data problem, and the bounding box (BBox) regression objective function. The term 'bag of specialties' refers to bargains, such as initiatives to improve the receptive field of CNNs, attention module, and improved activation function, all of which aim to raise the inference cost by a small amount while significantly improving object recognition accuracy. The author of the original model discovered the most appropriate ways from a significant number of enhancements approaches that can considerably improve the performance of the YOLOv4 model [15] through extensive comparative trials.

YOLOv4's structure as shown in Fig. 3 is mostly made up of three sections: the backbone, the head, and the neck. The object detector's backbone comprises a number of CNNs that have been pre-trained on ImageNet, such as VGG [16], ResNet [17], ResNeXt [18], or DenseNet [19]. The output of the object detector is the head, which is used to identify object classes and bounding boxes. To improve the object detector's ability to recognize objects at different sizes and extract integrated information, a module called the neck is placed between the backbone and the head. It employs a framework based on this technique, which includes the feature pyramid network (FPN) [20]. As a result, each feature map has both rich spatial and semantic information.

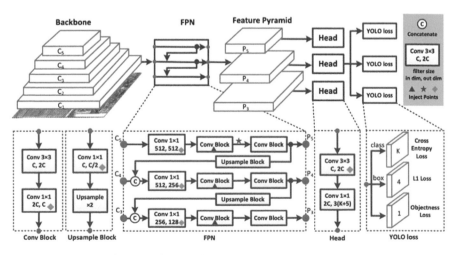

Fig. 3 Visualization of the YOLO network

3.3 Scaled Yolov4

Scaled YOLOv4 is an advancement of the YOLOv4 research developed by Chien-Yao Wang et al. [23]. Cross-stage Partial Networks are at the heart of it, allowing the network to increase its depth, breadth, resolution, and structure while retaining speed and accuracy.

Backbone: The architecture of CSPDarknet53 does not feature a residual block for down-sampling convolution for cross-stage processing [24]. As a result, it can be deduced that each CSPDarknet stage's computation time is.

$$whb^2(9/4 + 3/4 + 5k/2)$$

Only if $k > 1$ is met by the formula above will the CSPDarknet stage have a higher benefit in terms of computing than the Darknet stage. The number of residual layers held by each step in CSPDarknet53 is 1–2–8–8–4. We convert to achieve improved speed/accuracy equilibrium.

Neck: In YOLOv4, we CSP-ize the PAN [20] architecture [25] to effectively minimize the amount of computation. It essentially combines features from several feature pyramids before traversing two sets of inverted Darknet residual layers with no shortcut connections. This new update effectively reduces processing time by 40%. SPP: Originally, the SPP module was positioned in the first computation list group of the middle of the neck. As a consequence, the SPP module is placed in the initial computation to list group of the CSPPAN.

3.4 Yolov4-Tiny

The YOLOv4 model has laid the foundation for the YOLOv4-tiny object detection algorithm as in Fig. 4. For simplification of the network structure thereby improving network efficiency and reducing parameters algorithm, Yolov4-tiny was proposed. It works really well for real-time object detection as well.

For use on machines with limited computing power, the compressed version of YOLOv4, YOLOv4-tiny, is used. Deploying applications on mobile and embedded devices becomes easy due to the reduced parameters. The model weights are approximately 16 MB in size, allowing it to train 350 images in one hour on Tesla P100 GPU which gives the inference speed of 3 ms for YOLOv4-tiny which makes it one of the fastest object detection algorithm. YOLOv4-tiny has several modifications as compared to the original YOLOv4 network for attaining such a high efficiency. Network size reduction mainly leads to huge difference between YOLOv4-tiny and YOLOv4 in terms of speed. YOLOv4-tiny has only two YOLO heads, whereas there are three YOLO heads in YOLOv4 which further leads to formation of fewer YOLO layers and fewer anchor boxes for prediction. 29 pre-trained convolutional layers are used for building it, whereas 137 pre-trained convolutional layers are used in

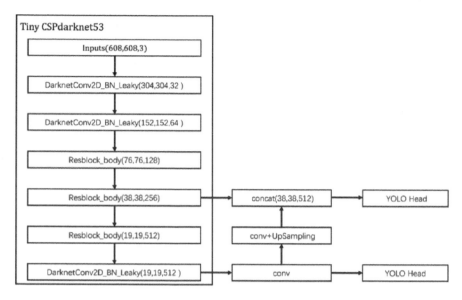

Fig. 4 YOLOv4-tiny architecture

YOLOv4 as a results the number of CSP backbone convolutional layers decreases [26].

YOLOv4-tiny is much faster than Scaled YOLOv4 and is preferred when the number of classes are fewer. It also has fewer layers and fewer anchor boxes for prediction as compared to Scaled YOLOv4.

4 Methodology

4.1 Data Collection

The dataset for training has been provided by the CMS experiment apparatus via Tata Institute of Fundamental Research (TIFR). Similar images were also provided by Texas Tech University (TTU) on which the pre-trained models were later tested. The difference being is the TIFR images consist of synthetic images while the TTU dataset comprises real-time images of silicon sensors used in the CMS experiment.

The detection of wire bonds is easier on the TIFR images (yellow base) since there appears to be clear contrast between the bonds and the base plate as compared to the TTU images (green base).

The images are of silicon detectors consisting of wire bonds in groups of three. A non-contact coordinate measuring machine (CMM) as shown in Fig. 5 is used to calculate the location, center position, and orientation of the silicon detectors with respect to reference pins which are marked in their mechanical mount. It is achieved

Fig. 5 Coordinate measuring machine

with an overall accuracy of 15 μm. The images created are in bmp format. The goal is to determine the quality of wire bonds.

4.2 Data Generation

There are 304 images in the dataset for the object detection algorithms. The resolution of these images are 1280 × 1024. In order to use the images for object detection, they have to be annotated and labeled with the appropriate class. The annotation is done by the integrated labeling tool similar to Labellmg software. It is based on Python and uses PyQt for graphical interface. It is used to create the rectangle bounding boxes to denote the region of interest to the object detection algorithms. To annotate the images, two different approaches were developed, the former focusing only on the quality of the bonds while the latter also focuses on the bond position and orientation.

In the first approach, a binary classification approach is used for annotation and object detection.

So, the classes are:

- three—three intact bonds
- not_three—broken or not-parallel wire bonds

In the second approach, 6 classes have been defined based on the position of the bonds in correspondence to the arms of a clock.

The classes are:

• 2_three	• 6_three	• 10_three
• 2_not_three	• 6_not_three	• 10_not_three

2_three denotes three intact 2 o'clock bonds, 2_not_three denotes broken 2 o'- clock bonds. Same applies for 6 o'clock and 10 o'clock bonds. After labeling the region of interest, annotations are exported as text file for YOLO algorithms. These files are the transformed version of images in object detection model format which includes the class, center x, center y, width, and height of the bounding rectangle. The final dataset consist of images and the corresponding annotation files. The dataset was then divided into 8:2 ratios as train and test.

4.3   Data Modeling

The models were trained on a combined dataset of TIFR and TTU comprising 230 images as large number of images were needed for training and showed promising results on the TTU images. The proposed generalized model is able to detect the quality of bonds in similar types of images. Because of limited images of TTU dataset, only 75 images were used for testing. Since each image have 3 labels the total classes tested are 225.

4.3.1   Scaled Yolov4

Scaled YOLOv4 (Scaling Cross-Stage Partial Network) is a CSP-based scaled-up YOLOv4 object detection neural network. It may be used on both small and large networks while retaining the optimal speed and accuracy. The base Scaled YOLOv4-csp model was trained for 300 epochs. Scaled YOLOv4 requires mish cuda activation. The batch size was 16 for training the model.

4.3.2   Yolov4-Tiny

YOLOv4-tiny is a compact version of YOLOv4 that may be used on systems with limited processing capacity. With a total of 29 pre-trained convolutional layers, tiny YOLOv4 compresses the number of convolutional layers in the CSP backbone. In addition, the number of YOLO layers has been lowered from three to two. For training the YOLOv4-tiny model, we have used yolov4-tiny-custom.cfg as we had a custom dataset. yolov4-tiny.conv.29 is the required weight file to train the YOLOv4-tiny model. The batch size is set to 64 and sub-divisions 16 for 6000 epochs.

5   Results

The final classification with labels of both approaches are illustrated in Fig. 6, and it also shows the decision boundary for the wire bonds.

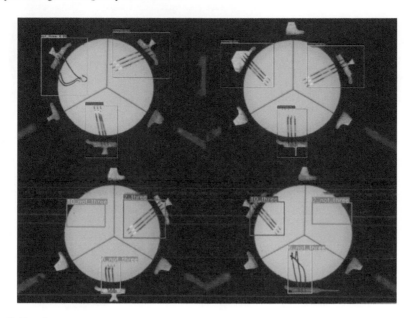

Fig. 6 Results

The results of training, i.e., Accuracy, Mean Average Precision (mAP), and False Positive rate of various algorithms are given in Table 1. mAP measures the performance of object detection models. It compares the ground-truth bounding box to the detected box and returns a score. False Positive rate denotes bad quality bonds detected as good quality is also calculated along with Accuracy and Mean Average Precision because it is an important evaluation metric in this case.

A confusion matrix summarizes the frequency of accurate and wrong predictions and breaks them down by class using count values. For both the algorithms, the number of testing images is 75 so in total, we have 222 bonds which are test for both the approaches. The vertical axis represents the predicted labels, and the horizontal axis represents the actual labels in the confusion matrix.

Figure 7a and b are the confusion matrices for approach I using Tiny YOLOv4 and Scaled YOLOv4 algorithm, respectively.

Figure 7c and d are the confusion matrices for approach 2 using YOLOv4 and Scaled YOLOv4 algorithm, respectively.

Table 1 Results for the TTU dataset

Algorithm	Accuracy (%)		mAP (%)		FP rate (%)	
	I	II	I	II	I	II
Scaled Yolov4	96.88	96.44	99.65	95.77	9.09	7.79
Yolov4-tiny	97.33	96.88	99.9	95.55	7.69	8.11

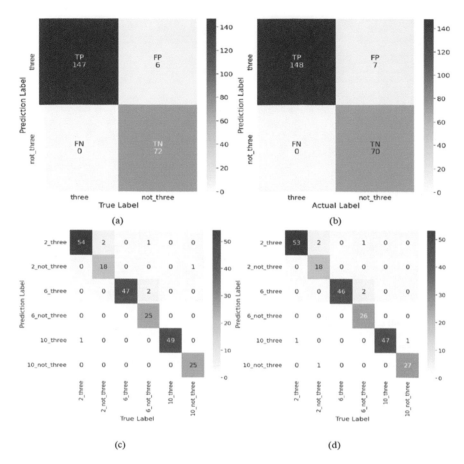

Fig. 7 Confusion matrix

For visual examination, many types of silicon detectors are available. Because we only utilized two types of detector in our dataset, the categorization of wire bonds in various detectors was erroneous. The dataset will be expanded in the future, and new photos with various types of detectors will be added, with the model being trained where it left off. However, for these sort of detectors, the model accurately detects the defects. The graphical of Accuracy, mAP of algorithms, and their approaches are shown in Fig. 8.

6 Conclusion

This project successfully uses deep learning-based techniques to develop a system for determining quality of silicon detectors by analyzing the images and detecting

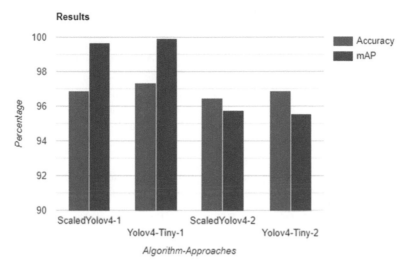

Fig. 8 Graph of Accuracy, mAP of algorithms, and their approaches

manufacturing defects. It can be a strenuous process to combine deep learning, image processing with the analysis of manufacturing techniques, but it's essential to ensure product quality. There can be no margin of error when it comes to quality control. Automated visual inspection (AVI) can help in recognizing patterns in manufacturing defects which can further help enhance manufacturing techniques to minimize damage. Defect detection has great practical significance, and the paper focuses on the deep learning-based approaches to automate manual inspection of silicon sensor bonds.

In this project, relevant literature was reviewed on traditional and intelligent AVI systems. Fully automated equipments in precision engineering and manufacturing can significantly decrease manufacturing costs while enhancing consistent production quality. However, extensive research into vision-based algorithms and practical AVI applications is vital.

The authors would like to express their gratitude and appreciation to the Tata Institute of Fundamental Research and Texas Tech University for providing us with the dataset and resources needed for this endeavor.

References

1. Ochando, C., & CMS collaboration. (2017). HGCAL: A high-granularity calorimeter for the endcaps of CMS at HL-LHC. In *Journal of Physics: Conference Series, 928*(1), 012025. IOP Publishing.
2. See, J. E., Drury, C. G. (2017). Ann speed, allison williams, and negar khalandi. "The role of visual inspection in the 21st century." In *Proceedings of the Human Factors and Ergonomics Society Annual Meeting* (Vol. 61, no. 1, pp. 262–266). Sage CA: Los Angeles, CA: SAGE Publications.
3. Newman, T. S., & Jain, A. K. (1995). A survey of automated visual inspection. *Computer Vision and Image Understanding, 61*(2), 231–262.
4. Huang, C., Zhai, K. K., Xie, X., & Tan, J. (2021). Deep residual network training for reinforced concrete defects intelligent classifier. *European Journal of Environmental and Civil Engineering,* 1–13.
5. Lavrik, E., Shiroya, M., Schmidt, H. R., Toia, A., & Heuser, J. M. (2022). Optical inspection of the silicon micro-strip sensors for the CBM experiment employing artificial intelligence. *Nuclear Instruments and Methods in Physics Research Section A: Accelerators, Spectrometers, Detectors and Associated Equipment, 1021,* 165932.
6. Czimmermann, T., Ciuti, G., Milazzo. M., Chiurazzi, M., Roccella, S., Oddo, C. M., Dario, P. (2020). Visual-based defect detection and classification approaches for industrial applications—A survey. *Sensors, 20*(5), 1459.
7. Hou, Q., Cheng, M. M., Hu, X., Borji, A., Tu, Z., & Torr, P. H. S. (2017). Deeply supervised salient object detection with short connections. In *Proceedings of the IEEE Conference on Computer Vision and Pattern Recognition* (pp. 3203–3212).
8. Liu, W., Anguelov, D., Erhan, D., Szegedy, C., Reed, S., Fu, C. Y., & Berg, A. C. (2016). Ssd: Single shot multibox detector. In *European Conference on Computer Vision* (pp. 21–37). Springer, Cham.
9. He, K., Gkioxari, G., Dollár, P., & Girshick, R. (2017). Mask r-cnn. In *Proceedings of the IEEE International Conference on Computer Vision* (pp. 2961–2969).
10. Redmon, J., Divvala, S., Girshick, R., & Farhadi, A. (2016). You only look once: Unified, real-time object detection. In *Proceedings of the IEEE Conference on Computer Vision and Pattern Recognition* (pp. 779–788).
11. Redmon, J., Farhadi, A. (2017) YOLO9000: Better, faster, stronger. In *Proceedings of the IEEE Conference on Computer Vision and Pattern Recognition* (pp. 7263–7271).
12. Ren, S., He, K., Girshick, R., & Sun, J. (2015). Faster r-cnn: Towards real-time object detection with region proposal networks. *Advances in neural information processing systems 28.*
13. Redmon, J. (2018) Darknet: Open source neural networks in c. (2013): 2018.
14. Zhang, Z., He, T., Zhang, H., Zhang, Z., Xie, J., & Li, M. (2019). Bag of freebies for training object detection neural networks. arXiv preprint arXiv:1902.04103
15. Zhang, Z., He, T., Zhang, H., Zhang, Z., Xie, J., & Li, M. (2019). Bag of freebies for training object detection neural networks. *arXiv preprint* arXiv:1902.04103
16. Sengupta, A., Ye, Y., Wang, R., Liu, C., & Roy, K. (2019). Going deeper in spiking neural networks: VGG and residual architectures. *Frontiers in Neuroscience, 13,* 95.
17. Bochkovskiy, A., Wang, C.-Y., & Mark Liao. H.-Y. (2020). Yolov4: Optimal speed and accuracy of object detection. *arXiv preprint* arXiv:2004.10934
18. Solawetz, J. (2022). Yolov4—an explanation of how it works. Roboflow Blog. Retrieved March 9, 2022, from https://blog.roboflow.com/a-thorough-breakdown-of-yolov4/
19. Iandola, F., Moskewicz, M., Karayev, S., Girshick, R., Darrell, T., & Keutzer, K. (2014). Densenet: Implementing efficient convnet descriptor pyramids. arXiv preprint arXiv:1404.1869
20. Iandola, F., Moskewicz, M., Karayev, S., Girshick, R., Darrell, T., & Keutzer, K. (2014). Densenet: Implementing efficient convnet descriptor pyramids. *arXiv preprint* arXiv:1404.1869

21. Wang, C.-Y., Yeh, I.H., & Mark Liao, H.-Y. (2021). You only learn one representation: Unified network for multiple tasks." *arXiv preprint* arXiv:2105.04206
22. Huang, X., Wang, X., Lv, W., Bai, X., Long, X., Deng, K., & Dang, Q. et al. (2021). PP-YOLOv2: A practical object detector. *arXiv preprint* arXiv:2104.10419
23. Wang, C.-Y, Bochkovskiy, A., & Mark Liao, H.-Y. (2021). Scaled-yolov4: Scaling cross stage partial network. In *Proceedings of the IEEE/cvf Conference on Computer Vision and Pattern Recognition* (pp. 13029–13038).
24. Thuan, D. (2021). Evolution of yolo algorithm and yolov5: The state-of-the-art object detection algorithm.
25. Guo, R., Li, S., & Wang, K. (2020) Research on YOLOv3 algorithm based on darknet framework. *Journal of Physics: Conference Series, 1629*(1), 012062. IOP Publishing, 2020.
26. Redmon, J., & Farhadi, A. (2018). Yolov3: An incremental improvement. *arXiv preprint* arXiv: 1804.02767

IoT-Enabled RFID-Based Library Management and Automatic Book Recommendation System Using Collaborative Learning

Maithili Andhare, Kishor Bhangale, Vijayalaxmi S. Kumbhar, Arti Tekade, Suyash Choudhari, Ajinkya Deshpande, and Sanket Chavan

Abstract Libraries are a valuable resource for learning. A person must acquire and retain knowledge. However, the current library system has several disadvantages, including manual management of books by librarians, tedious tasks of book searching, and difficulty to track the unreturned books. Thus, this article presents the automatic library management system (LMS) based on radio frequency identification (RFID) and the Internet of Things (IoT). The proposed system helps to identify and track a large number of tagged books using radio waves. The proposed LMS provides an interactive portal to display the availability of books in the library, book issue facility, tracking of issued books, book return facility, etc. The LMS portal also provides book recommendations based on the K-Means clustering algorithm and collaborative learning. The IoT helps to collect the data on the cloud platform so that remote access to the library system. This action simplifies the process of borrowing, updating, and returning books using RFID tags.

Keywords Internet of things · Library management system · K-means clustering algorithm · Cloud · NodeMCU · Node-Red · MySQL database · Grafana

1 Introduction

All educational institutions have a library, and the current library management system is a manual (registry) method or an ERP system that requires human labor. The current release of the library requires digitization. Although barcode technology is well known, it has several drawbacks. RFID technology can be used to eliminate human work and book mismanagement. It's just a scan-and-go system. It keeps track of the books and makes issuing and returning them more efficient [1, 2]. A database is being created to store information on the books available in the library so that users can search for books. Every book will be tracked by the system. A reminder email

M. Andhare · K. Bhangale · V. S. Kumbhar · A. Tekade · S. Choudhari (✉) · A. Deshpande · S. Chavan
Department of E&TC, PCCOER, Pune 412101, India
e-mail: suyash.choudhari27@gmail.com

that is sent automatically can be useful. As the world is shifting to Industry 4.0, this work uses modern no-code tools such as Node-RED operating in the cloud. MQTT protocol is being used which works on Publish and Subscribe model rather than a request–response model [3, 4].

To begin with, it is a type of non-contact data capture technology that allows for automatic identification and management. Second, any information may be written in, allowing it to keep up with the demand for an ever-increasing amount of data and faster data processing. Third, it is capable of rapidly recognizing targets. That is, once the targets enter magnetic fields, readers may quickly and simultaneously handle multiple label data, to achieve batch identification [5, 6]. Because high security cannot be overlooked, password protection can be configured to read and write tag data. Furthermore, it improves management and resolves issues with product information management in the motor process. In theory, the good information in the motor process. Due to its benefits of high security, high reliability, ease of operation, and the capacity to transport a huge amount of data, RFID technology has become a research hotspot in the field of indoor location technology [7].

IoT is widely used in many applications such as earthquake detection, landslide detection, industrial automation, energy management, remote application control, precision agriculture [8], security [9], building automation [10], border patrolling, robotics, transportation, and logistics. It provides a remote facility to collect the data using multiple devices, and process, and control the applications remotely. IoT has emerged as the prominent research area in library automation because of its flexibility, portability, low cost, efficiency, security, etc. [11].

The various automatic library system has been presented in the past. Radhika [12] provided a high-level overview of the components that make up a typical IoT-based library management system. The authors of [13] explained the whole architecture of a Digital Library System right from issuing and returning the books along with an anti-theft system. Feng et al. [14] compared and explained the advantages of RFID technology over barcode technology. The paper also explained the ease of use of RFID technology over barcode technology for a digital library-specific aspect. Cheng et al. [15] gave a whole overview of a digitized library system using RFID and Wi-Fi. Book issue and return system were also explained.

This article presents an IoT-enabled library management system based on RFID technology. The major contributions of the article are as follows:

- Library management system to automate the book issue, return, and maintenance process.
- Book recommendation system based on K-means clustering and collaborative learning.

The rest of the article is structured as follows: Sect. 2 provides the proposed methodology, Sect. 3 gives the information about experimental results and discussion, and Sect. 4 depicts the conclusion and provides the future scope of the work.

2 Proposed Methodology

The block diagram of the proposed LMS system is shown in Fig. 1. The NodeMCU Development Board is used in this system, which features a built-in ESP8266 Wi-Fi chip, a low cost, and a 2.5–3.6 V working voltage, making it perfect for IoT applications. In this study, IoT-based LMS is used for book authentication and issuance using RFID cards. To gain access to the library, the user must bring his ID card and complete an authentication process. Each book is tagged with an RFID tag placed on its cover. While scanning an RFID tag using the MFRC522 RFID reader, a specific tag ID with a user name has been saved in the database, which will be compared and authenticated. These are passive tags that are utilized in this system. The user's RFID identity card is scanned, and if it is legitimate, the user can access the books in that library's database. After the user has been authenticated, the rest of the process can begin. The pupils next use the library database to look up the availability of required books. If a book is available, then the user can issue it. The user scans the RFID tag of the book with the RFID reader, and then the reader reads the data from the tag and updates it into the database. Furthermore, the display will show the return date and the user leaves the library with the book. Thereafter user enters the library with a book to be returned. The user scans the RFID tag from the book to the RFID reader, and the database is updated. Users are free to leave the library after this process is completed.

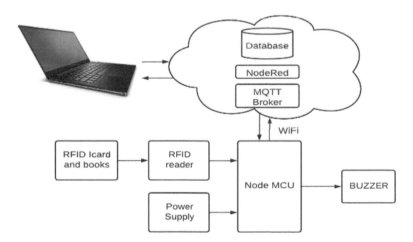

Fig. 1 Flow diagram of proposed IoT-based LMS

2.1 User Authentication Process

Figure 2 illustrates the user authentication process, and the steps are summarized as follows:

Step 1:   Issue RFID identity cards to the user.
Step 2:   Save the username and credentials in the database.
Step 3:   Scan the RFID identity card.
                    if valid user, then
                        Allow Book Issue
                        Allow Book Return
                    else
                        Invalid login
                        Contact Admin
                    end

The Node-RED authentication flow in Fig. 3 shows, if the user scans his card, then his password is retrieved from the database and stored in the global variable, and this password is checked with the password entered by the user. If both the passwords match, then the user will be able to issue, return, and search a book or else pop message is displayed indicating that the entered password is wrong.

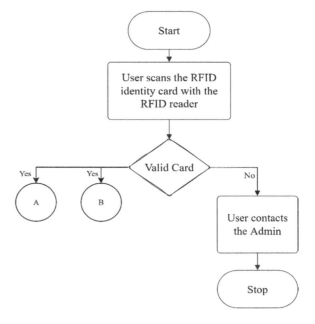

Fig. 2 User authentication process flowchart

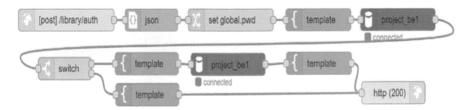

Fig. 3 Node-RED user authentication flow

2.2 Book Issue Process

The user can search for the availability of the book by typing keywords in the search box provided on the portal. These keywords are matched with the database and if any search results are found, then the user can take that particular book accordingly.

Figure 4 illustrates the book issue process, and the steps are summarized as follows:

Fig. 4 Flowchart for book issue process

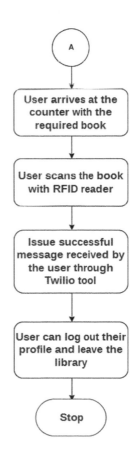

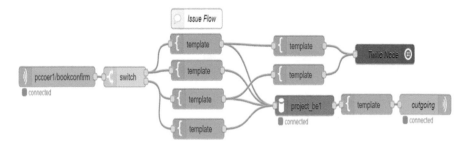

Fig. 5 Node-RED flow for book issue process

Step 1:   Search the library database for the availability of the required book.
Step 2:   If the required book is available, then bring that book to the system.
Step 3:   The user logs in their profile into the system.
            if valid user:
                Allow Book Issue
            else
                Invalid login
                Contact Admin
            end
Step 4:   Book issue successful message is sent to the user through the Twilio
            messaging tool which is integrated with Node-RED.
Step 5:   The user can leave the library.

After the process of authentication, if the user wants to take a book, he has to select the issue option after which he will be allowed to scan the particular book which he wants to take. Figure 5 depicts the flow. After this process is completed, the database will be updated accordingly, and the user will get the notification using the Twilio messaging tool node which is integrated into the Node-RED by providing a unique String Identifier (SID) and authentication token. The user will get the text message that the particular book has been issued to him.

2.3 Book Return Process

Figure 6 illustrates the book return process, and the steps are summarized as follows:

Step 1:   Bring the book which is to be returned to the system.
Step 2:   The user logs into their profile.
            if valid user:
                Allow Book Return
            else
                Invalid login
                Contact Admin

Fig. 6 Flowchart for the
book return process

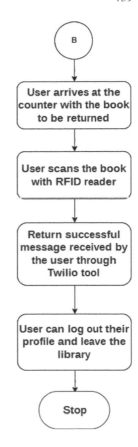

B

↓

User arrives at the
counter with the book
to be returned

↓

User scans the book
with RFID reader

↓

Return successful
message received by
the user through
Twilio tool

↓

User can log out their
profile and leave the
library

↓

Stop

end

Step 3:   The user scans the book's RFID tag/sticker with the RFID reader.

Step 4:   Book returned successful message will be sent to the user through the Twilio messaging tool which is integrated with Node-RED.

Step 5:   The user can leave the library.

After the process of authentication, if the user wants to return a book, he has to select the return option after which he will be allowed to scan the particular book which he wants to return. Figure 7 depicts the same Node-RED flow. After this process gets completed, the database will be updated accordingly.

2.4 Book Recommendation System

This portal provides the facility for user profile management and displays the library profile of the particular user using the K-means clustering algorithm. Initially, the K-means clustering algorithm is used to segregate the textbooks and reference books

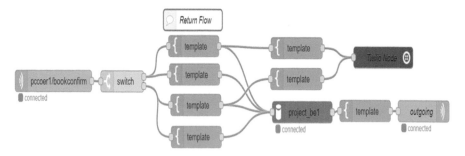

Fig. 7 Node-RED flow for the book return process

issued by the user. Then, the K-means clustering algorithm provides the domain-wise categorization of the books issued by the user such as analog domain, digital domain, machine learning domain, programming domain, and general books [16]. Based on the user history and K-means clustering algorithm, the system provides the recommendation of new unread books of the user's interest domain on the user portal. Based on the other user's accessibility of the book, its rating varies. Manual ratings are also available for the user, author, professors, and readers to improve the rating of books. Using the user's interest obtained from the K-means clustering algorithm, the proposed system provides book recommendations based on collaborative learning [17]. Collaborative learning combines the teaching–learning process with help of a group of students or feedback obtained from the students to complete the task. The flowchart of book recommendations is shown in Fig. 8.

3 Experimental Results and Discussion

3.1 Hardware Setup

The system components required for the implementation of the system, and its specifications are given in Table 1.

3.2 Software Setup

The software versions required for the implementation of the system, and its specifications are given in Table 2. The MySQL database is used to store the data of users and books. Node-RED orchestration tools are used to implement the flow of the proposed work. MQTT uses the publish–subscribe model. All the clients are connected to the network that can publish or subscribe to a topic. If a message is published on a particular topic, it will be distributed to all the clients who subscribe

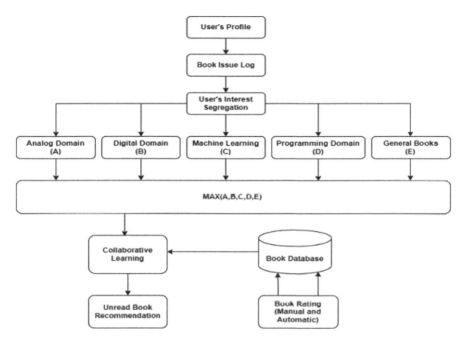

Fig. 8 Flowchart for book recommendation system

to that topic. In the same manner, the communication between the Node-RED and hardware components is done with the help of MQTT. Grafana, which is a dashboard tool, will represent the graph of library utilization. The proposed work is on the cloud, and so an Ubuntu machine is used on the cloud IAAS service.

3.3 RFID Card Detection by RFID Reader

The proposed RFID card scans in the range 0–3.5 cm, and evaluation of the same is represented in Table 3.

3.4 Database Generation

The master book database of all the books in the library is shown in Fig. 9. Every book has its unique book ID, which is a primary field in the database. Each book also has various fields such as Author, Publisher, Year of publication, Edition, Date of Registration, Issued Status, and Issued To in the database.

Table 1 Specifications of the components

Component	Parameters	Values
RFID card	Material	PVC
	Operating frequency	13.56 MHz
	Operating distance	Up to 100 mm
	Data transfer	106 kbit/s
	Data integrity	16 Bit
	Data storage time	minimum 10 years
	Operating temperature (°C)	−20 to 60
RFID 13.56 MHz scanner (MF522-ED)	Model	MF522-ED
	Working current at 3.3 V	13–26 mA
	Standby current at 3.3 V	10–13 mA
	Working frequency	13.56 MHz
	Card reading distance	0–60 mm
	Protocol	SPI
	Data communication speed	10Mbit/s Max
	Operating temperature (°C)	−20 to 80
Controller (NodeMCU ESP 8266)	Microcontroller	Ten silica 32-bit RISC CPU LX106 3.3 and 5 V MCU
	Operating voltage	
	Input voltage	7–12 V
	Digital i/o pins	16
	Analog i/o pins	1
	Flash memory	4 MB
	SRAM	64 KB
	UART	1
	I2C	1
	Clock speed	80 MHz
Buzzer (passive)	Operating voltage	3.3–5 V
	PCB dimension	3.1 × 1.3 cm

Table 2 Software specifications

Software	Version
Arduino IDE	1.8.19
MySQL	8.0.28
Node-RED	2.2.2
MQTT	5.0
Grafana	8.4.3
Cloud (Ubuntu)	18.04

Figure 10 provides the user master data plan of the proposed system. Every book has its unique book ID, which is a primary field in the database. Each book also has various fields such as RFID ID, First Name, Middle Name, Last Name, User Role, Mobile No., Academic Year, Year of admission, and Password in the database.

Table 3 Tag detection distance

S No.	Tag in detection range (cm)	RFID card detection status
1	0.4	Detected
2	0.8	Detected
3	1.2	Detected
4	1.6	Detected
5	2.0	Detected
6	2.4	Detected
7	2.8	Detected
8	3.2	Detected
9	3.6	Not detected
10	4.0	Not detected
11	4.4	Not detected
12	4.8	Not detected

Access	Book_RFID_ID	Title	Author	Publisher	Year_of_	Edition	RB_or_T	Date_of_Regi	Issued_stat.	Issued_To
1	66 C1 46 64	Internet of Things	Arshdeep Bahga, Vijay MAdisetti	Universities Pr...	2015	1	RB	2018-05-29	Available	NA
2	61 D2 20 90	8051 Microcontroll...	Muhammad Ali Mazidi	Pearson	2016	2	TB	2018-05-29	Issued	Akash S...
3	89 C2 17 P7	Wireless Communic...	Theodore S. Rappaport	Pearson	2013	2	TB	2018-05-29	Issued	Raj Jhoshi
4	78 CD 46 79	Electronics Devices...	Jacob Milman's, Christos C. Hal...	MCGraw Hill	2012	3	RB	2018-05-29	Available	NA
5	R7 4E 5R 4E	Communication Sy...	Simon Haykin	Wiley	2014	4	RB	2018-05-29	Available	NA
6	74 C7 95 64	Power Electronics	Mohammad Rashid	Pearson	2011	3	TB	2018-05-29	Issued	Amit Singh
7	7P 8F 9G 8R	Object Oriented Pr...	F Balgurusamy	MCGraw Hill	2016	6	RB	2018-05-29	Available	NA
8	4G 78 78 D8	Introduction to rob...	Saeed B. Niku	Wiley	2017	2	RB	2018-05-29	Available	NA

Fig. 9 Sample database of books

userid	userrfid	fname	mname	lname	userrole	mno	acadyear	yoj	pwd	recno
1001	22 00 81 5A	Suyash	S	Choudhari	user	9545530456	BE	2018	1	1
110	11 A1 38 36	Rahul	Ganpat	Mapari	user	8446485789	NA	2018	lib@4444	4
112	33 54 19 64	Vijayalaxmi	S	Kumbhar	user	8390455963	NA	2015	lib@6666	6
113	44 55 8F E3	Maithili	S	Andhare	user	9922435203	NA	2015	lib@2222	2
114	63 30 P8 65	Santosh	N	Randive	user	7276407032	NA	2015	lih@1012	12
115	45 30 4A 64	Rupali	R	Kawade	user	8956614785	NA	2016	lib@7777	7
116	36 C1 16 64	Arti	A	Tekade	user	9860687103	NA	2016	lib@5555	5
117	12 30 P8 64	Triveni	D	Dhamale	user	9422009032	NA	2016	lib@1011	11
118	12 A8 37 64	Snehal	B	Gholap	user	8600268023	NA	2016	lib@9999	9
119	38 5E 1B 64	Bhagyash...	L	Gawali	user	7219572032	NA	2017	lib@1010	10
120	11 9C 1B 64	Dipali	N	Dhake	admin	7588054035	NA	2017	lib@3333	3
121	77 72 3E 36	Kishor	B	Bhangale	user	7709710789	NA	2021	lib@8888	8

Fig. 10 Sample database of user

4 Conclusion

This article provides the effective implementation of the RFID and IoT-based library management system that automates the physical library system. The proposed system is very efficiently managed for books and user databases in terms of technology and is easy to use. Moreover, it consumes less time, automates the library, and reduces the workload of the library management authority. The main key benefit of this work is that all the activities including issuing, renewal, and return of books are digitized, and all these actions are modified in the database automatically. RFID in the library speeds up the process of book borrowing, tracking, and book search. In the future, automatic finding of book location using GPS-based system can be implemented. Furthermore, the proposed system can be implemented for online book libraries and book delivery systems.

References

1. Liang, X. (2018). Internet of Things and its applications in libraries: A literature review. *Library Hi Tech*.
2. Bhure, R. D. (2018). Application of internet of things in secured library management system. *International Research Journal of Engineering and Technology, 5*.
3. Singh, N. K., & Mahajan, P. (2014). RFID and it's use in libraries: A literature review. *International Journal of Information Dissemination and Technology, 4*(2), 117–123.
4. Gupta, P., & Margam, M. (2017). RFID technology in libraries: A review of literature of Indian perspective. *DESIDOC Journal of Library and Information Technology, 37*(1), 58.
5. Weinstein, R. (2005). RFID: A technical overview and its application to the enterprise. *IT professional, 7*(3), 27–33.
6. Finkenzeller, K. (2010). *RFID handbook: Fundamentals and applications in contactless smart cards, radio frequency identification and near-field communication*. John wiley & sons.
7. Su, J., Sheng, Z., Leung, V. C. M., & Chen, Y. (2019). Energy efficient tag identification algorithms for RFID: Survey, motivation and new design. *IEEE Wireless Communications, 26*(3), 118–124.
8. Mapari, R., Bhangale, K., Deshmukh, L., Gode, P., & Gaikwad, A. (2022). Agriculture protection from animals using smart scarecrow system. In *Soft Computing for Security Applications* (pp. 539–551). Springer.
9. Biradar, P., Kolsure, P., Khodaskar, S., & Bhangale, K. B. (2020). IoT based smart bracelet for women security. *International Journal of Resources Applied Science and Engineering Technologies (IJRASET), 8*(11), 688–691.
10. Sarraf, R., Ojha, S., Biraris, D., & Bhangale, K. B. (2020). IoT based smart quality water management system. *International Journal, 5*(3).
11. Shafique, K., Khawaja, B. A., Sabir, F., Qazi, S., & Mustaqim, M. (2020). Internet of things (IoT) for next-generation smart systems: A review of current challenges, future trends and prospects for emerging 5G-IoT scenarios. *IEEE Access, 8*, 23022–23040.
12. Radhika, V. (2021). Smart library access and management system. *Turkish Journal of Computer and Mathematics Education (TURCOMAT), 12*(7), 1495–1500.
13. Malipatil, N., Roopashree, V., Sanjana Gowda, R. H., Shobha, M. R., & Sateesh Kumar, H. C. (2020). RFID based library management system. *International Journal of Research in Engineering, Science and Management, 3*(7), 112–115.
14. Feng, C. (2010). Research for application of RFID in the library. In *2010 International Conference on Computer and Communication Technologies in Agriculture Engineering* (Vol. 1, pp. 262–264). IEEE.
15. Cheng, H., Huang, L., Xu, H., Hu, Y., & Wang, X. A. (2016). Design and implementation of library books search and management system using RFID technology. In *2016 International Conference on Intelligent Networking and Collaborative Systems (INCoS)* (pp. 392–397). IEEE.
16. Ahmad, A., & Dey, L. (2007). A k-mean clustering algorithm for mixed numeric and categorical data. *Data & Knowledge Engineering, 63*(2), 503–527.

17. Núñez-Valdez, E. R., Lovelle, J. M. C., Hernández, G. I., Fuente, A. J., & Labra-Gayo, J. E. (2015). Creating recommendations on electronic books: A collaborative learning implicit approach. *Computers in Human Behavior, 51,* 1320–1330.

An Effectual Analytics and Approach for Avoidance of Malware in Android Using Deep Neural Networks

Kapil Aggarwal and Santosh Kumar Yadav

Abstract Due to the rise in smartphone apps and increase in the usage of Android, there are a lot of security issues. Security issues need to be addressed in order to prevent vulnerabilities and identify them prior mishap. People who use smartphones are linked to a warning about the risks. Most people who use mobile phones do not have to think about a few negative possibilities when they install Android Package Kit (APK) files from different sources. It is important to have a system that can tell if the code in Android app is harmful. The first step in this work is to look at the Android APK datasets. Both good and bad APKs are analyzed, and the dataset is processed. The signatures that are hidden in the APKs are identified and extracted. This will make it easier to build a training dataset. Then, it is checked to see if each APK has the permissions it needs and how it affects the way it works. Once it's been cleaned, a dataset is made so that the model can be trained to predict malware. To finish the predictive analytics, any APK outside of the APKs is chosen and used. That's when it is possible to figure out how likely it is that the new APK will have corruptive code in it. Machine learning is used to track the results of different prediction measures, such as how long it takes, how accurate they are, and how much they cost. This article describes a machine learning technique to solve functional selection by safeguarding the selection and mutation operators of genetic algorithms. The proposed method is adaptable during population calculations in the training set. Furthermore, for various population sizes, it gives the best possible probability of resolving function selection difficulties during the training process. Furthermore, the work is combined with a better classifier in order to detect the different malware categories. The proposed approach is compared and validated with current techniques by using different datasets, wherein this approach shows elevated results in terms of accuracy.

K. Aggarwal (✉)
Department of Computer Science, Banasthali University, Aliyabad, Rajasthan, India
e-mail: kapil594@gmail.com

S. K. Yadav
Department of Computer Science, Shri Jagdishprasad Jhabarmal Tibrewala University, Churela, Rajasthan, India

© The Author(s), under exclusive license to Springer Nature Singapore Pte Ltd. 2023   767
S. Shakya et al. (eds.), *Sentiment Analysis and Deep Learning*,
Advances in Intelligent Systems and Computing 1432,
https://doi.org/10.1007/978-981-19-5443-6_58

Keywords Android malware · APK analytics · Android APK fingerprinting · Smartphone security

1 Introduction

The cybercriminals came up with a way to infect Android users with different banking Trojans, which are meant to get into people's account information. A report from ThreatFabric recently found that more than 300,000 Android users did not know they had downloaded malware with banking Trojan abilities, and that it was able to get around Google Play Store restrictions [1].

In the first step to submit apps to the Google Play Store, they had almost no malicious footprint and looked like useful apps, like QR code scanners, PDF scanners, cryptocurrency apps, or fitness apps. Afterward, these apps asked the user to update them, which was done outside the Google Play Store and installed malicious content on their Android phones. This was a way to install malicious content after the app has been installed. This way, it was completely unnoticeable to Google Play when the app was installed, so it was not dangerous [2].

The attackers were careful enough to submit an early version of their apps to the Google Play store that did not have any download or install features. Later, they updated the apps with more permissions, which made it easier for people to download and install the malware [3, 4]. They have also put limits on how the payload can be used. They used mechanisms to make sure the payload was only installed on real victims' devices and not in test environments, making it even more difficult to find.

Anatsa, Alien, Hydra, and Ermac are four different types of banking Trojans that have been found by ThreatFabric. Anatsa is the most common. Researchers at Zimperium have found a new virus that looks like a system update program, which makes it hard to find. Installed, it takes control of Android phones and steals data, texts, and pictures, for example. In the words of the researchers, hackers can record audio and phone calls, as well as take photos. They can also steal messages and files, as well as access instant messenger accounts. Hackers can also look into the user's browser, taking their search history and bookmarks. It is possible for them to see what the user is copying to the clipboard, as well as get information about the user's device. Android Package Kit (APK) is a type of file format used by the Android operating system. It is used for both installation and distribution of mobile apps. APK is an extension used by the Android operating system (OS) to identify application files [5–8]. The traditional structure of Android APK is shown in Table 1.

Android APK permissions as shown in Fig. 1, and vulnerability aspects are hereby underlined. Apps like QR code readers, scanners, fitness trackers, and cryptocurrency trading platforms are not always real [9, 10]. That's what researchers at ThreatFabric found when they looked at apps like these. Hackers have been able to make harmful versions of these apps that look just like the real ones. Figure 2 shows the market span of smartphone O.S. These apps would advertise what they do in the best way

Table 1 Traditional structure of Android APK

Description	APK file/folder
Optional folder for AssetManager	assets/
Compiled code of application	classes.dex
Resources without compilation	res/
Application resources	resources.arsc
Information of metadata	META-INF/
XML format manifest file	AndroidManifest.xml
Optional folder with compiled code	lib/

possible so that people do not suspect [11]. These ads convince people to download these apps, which makes them an easy prey for hackers.

The following are:

- REQUEST_DELETE_PACKAGES
- REQUEST_COMPANION_USE_DATA_IN_BACKGROUND
- GET_PACKAGE_SIZE
- TRANSMIT_IR
- SET_ALARM
- FOREGROUND_SERVICE

Fig. 1 Permissions in APK

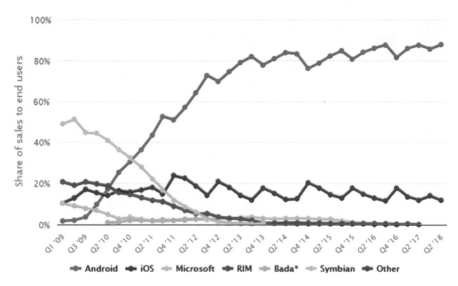

Fig. 2 Market span of smartphone O.S

- KILL_BACKGROUND_PROCESSES
- CHANGE_WIFI_MULTICAST_STATE
- MODIFY_AUDIO_SETTINGS
- CHANGE_NETWORK_STATE
- READ_SYNC_SETTINGS
- SET_WALLPAPER_HINTS
- ACCESS_NETWORK_STATE
- EXPAND_STATUS_BAR
- VIBRATE
- DISABLE_KEYGUARD
- REQUEST_COMPANION_RUN_IN_BACKGND
- INSTALL_SHORTCUT
- USE_FINGERPRINT
- CHANGE_WIFI_STATE
- ACCESS_LOCATION_EXTRA_COMMANDS
- MANAGE_OWN_CALLS
- BLUETOOTH
- SET_WALLPAPER
- READ_SYNC_STATS
- BROADCAST_STICKY
- WRITE_SYNC_SETTINGS
- REORDER_TASKS
- WAKE_LOCK
- ACCESS_NOTIFICATION_POLICY
- REQUEST_IGNORE_BATTERY_OPTIMIZATIONS

- ACCESS_WIFI_STATE
- NFC
- INTERNET
- BLUETOOTH_ADMIN
- RECEIVE_BOOT_COMPLETED

2 Malware Patterns and Machine Learning

Malware, which can come in many different forms, can be made to corrupt a computer system and achieve immoral goals, like destroying the computer, getting money, or getting into the computer in an illegal way, which could lead to less security or the leak of system information. Some harmful packets, including Trojans, rootkits, beasts, suspicious packers, scareware, evasion, backdoors, keyloggers, Trojan Spy, Trojan GameThief, etc., can be sent in different ways and can spread in two ways. Polymorphic malware changes its code each time it copies itself, but keeps its original coding and makes it look like it's always new. A lot of different IDS tools are out there, and some of them are free. They can both classify attacks (using PCAP files) and monitor network traffic [12–17].

Deep learning has a very low rate of mistakes, compared to machine learning which has a lot of mistakes. Deep learning, also known as deep structured learning and hierarchical learning as in Fig. 3, is when people learn by looking at a lot of different features or representations in a way that makes sense to them. The deep learning is whereby the advanced algorithms of neural networks and dynamic analytics are done [18–26].

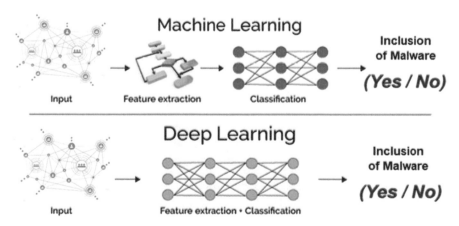

Fig. 3 Machine learning and deep learning

3 Research Statement and Goal

Since there are so many Android phones around the world, people get their apps from a lot of different places. While there are more and more problems in the Android ecosystem, if you have a way to show the model and technique, it's easier to find APK vulnerabilities [27–30].

Following are the major suspicious Apps released on different online repositories.

- com.family.cleaner—Cleaner: Safe and Fast
- com.op.blinkingcamera—Blinking Camera
- com.use.clever.camera—Clever Camera
- com.touch.smile.camera—Smile Camera
- com.just.parrot.album—com.qti.atfwd.core
- com.color.rainbow.camera—Rainbow Camera
- com.flappy.game.cat—FlappyCat
- com.bunny.h5game.parkour—Easter Rush
- com.op.blinking.camera—Blinking Camera
- cm.com.hipornv2—HiPorn

4 Experimental Results and Outcome

The first thing that is done is to get APK files from online repositories so that the permission-based dataset can be made. It is also taught how to use machine learning and deep learning.

The APKPure repository has a lot of Android apps that are analyzed with the APK Tool so that the analytics on their permissions and suspicious parameters can be looked at as in Figs. 4 and 5.

Fig. 4 APK repository of APKPure

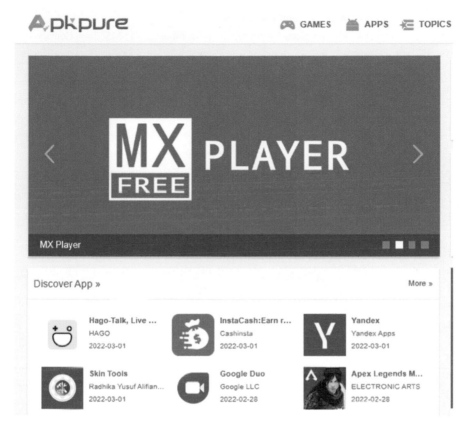

Fig. 5 APKPure apps repository

Following is the generation panel of dataset as in Table 2 from Android apps using Android AAPT. The toolkit of Android AAPT provides the instructions to fetch the permissions associated with Android app without installation in the smartphone. It is done so that detailed fingerprinting can be done.

The accuracy metrics on the traditional machine learning and the deep learning approaches are listed in Table 3.

Figure 6 shows how well deep learning and traditional machine learning perform based on time consumed. The execution time of deep neural network-based learning is found to be more accurate in performance than the classical approach of traditional machine learning in all assorted attempts. The implementation of deep learning-based algorithm raises the performance of the overall system in terms of accuracy as compared to that of the traditional approach.

Table 2 Formation of dataset for implementations

Call	Suspicious (1-malignant, 2-benign)	Bluetooth	SMS	Contact
0	1	0	1	0
1	2	1	1	1
0	1	0	0	1
1	2	1	1	1
1	1	1	1	1
0	1	0	1	1
1	1	0	0	0
1	1	1	0	0
1	1	1	1	0
1	1	1	1	1
0	2	1	0	1
1	1	1	1	1
1	1	1	0	1
1	1	0	1	1
1	1	1	0	1
1	1	0	0	1
1	1	1	0	0
0	1	1	0	1
0	1	1	1	1
0	1	1	1	1
1	1	0	0	1
1	2	0	1	1
1	1	0	1	0
1	1	1	1	0

Table 3 Evaluation of parameters

Traditional machine learning approach (accuracy)	Deep learning-based approach (accuracy)
84	93
82	93
75	95
81	94
86	97

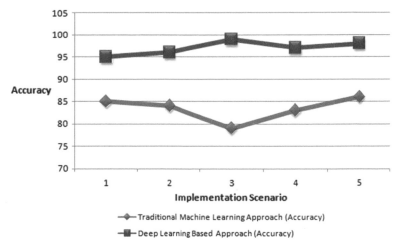

Fig. 6 Evaluation of execution time

5 Conclusion

This work discusses how to use deep learning to make better malware detection and predictions with a higher level of accuracy. Specialist methods are used to classify the Android APK, to predict with a high level of certainty by looking at the footprints and signs left by the APK. The methods help to get the results by taking into account a lot of different things while also making sure that each process knows how to work efficiently. It is possible to use blockchain technology in apps like Android APKs to make them more secure or private. Cryptocurrency has become one of the most important fields of study because of all the work being done on blockchains. Many digital coins are popular and have gained a lot of attention around the world, even though there have been some complaints and scandals. There are many different types of cryptocurrency, like Bitcoin, Ethereum, Litecoin, Gridcoin, Primecoin, Ripple, Nxt, Dogecoin, Namecoin, Auroracoin, etc. that work on blockchains, which means they do not record transaction details through a bank or payment service. And that is the main reason that many countries have banned cryptocurrency, despite the fact that these well-known and used digital currencies have very good security features. Dynamic cryptography is used to manage a blockchain's block of records. This means that all transactions can be encrypted, and the blockchain is resistant to hackers and sniffers. However, the proposed deep learning algorithm has a few drawbacks. The method described in this article is designed to solve the specific problem of optimizing functionality selection. The redundancy of the crossover operator is specific for this optimization problem category. Secondly, the selection problem for a dataset is optimized, and the number of samples and the dimension of data is the main factors influencing runtime. For future study, it is thus an essential job to use the method for large cases without adding computing complexity.

References

1. Vivekanandam, B. (2021). Design an adaptive hybrid approach for genetic algorithm to detect effective malware detection in android division. *Journal of Ubiquitous Computing and Communication Technologies, 3*(2), 135–149.
2. Jose, R. R., & Salim, A. (2019). Integrated static analysis for malware variants detection. *International Conference on Inventive Computation Technologies* (pp. 622–629). Cham: Springer.
3. Kumar, A. A., Anoosh, G. P., Abhishek, M. S., & Shraddha, C. (2020). An effective machine learning-based file malware detection—a survey. In *International Conference on Communication, Computing and Electronics Systems* (pp. 355–360). Springer, Singapore.
4. Deshotels, L., Notani, V., & Lakhotia, A. (2014). Droidlegacy: Automated familial classification of android malware. In *Proceedings of ACM SIGPLAN on Program Protection and Reverse Engineering Workshop 2014* (p. 3). ACM.
5. Yerima, S. Y., Sezer, S., & McWilliams, G. (2014). Analysis of Bayesian classification-based approaches for Android malware detection. *IET Information Security, 8*(1), 25–36.
6. Talha, K. A., Alper, D. I., & Aydin, C. (2015). APK Auditor: Permission-based Android malware detection system. *Digital Investigation, 13*, 1–14.
7. Faruki, P., Ganmoor, V., Laxmi, V., Gaur, M. S., & Bharmal, A. (2013). AndroSimilar: Robust statistical feature signature for Android malware detection. In *Proceedings of the 6th International Conference on Security of Information and Networks* (pp. 152–159). ACM.
8. Yerima, S. Y., Sezer, S., & Muttik, I. (2015). High accuracy android malware detection using ensemble learning. *IET Information Security, 9*(6), 313–320.
9. Cai, H., Meng, N., Ryder, B., & Yao, D. (2019). Droidcat: Effective android malware detection and categorization via app-level profiling. *IEEE Transactions on Information Forensics and Security, 14*(6), 1455–1470.
10. Adebayo, O. S., & Aziz, N. A. (2019). The trend of mobile malwares and effective detection techniques. In *Multigenerational Online Behavior and Media Use: Concepts, Methodologies, Tools, and Applications* (pp. 668–682). IGI Global.
11. Onwuzurike, L., Mariconti, E., Andriotis, P., Cristofaro, E. D., Ross, G., & Stringhini, G. (2019). MaMaDroid: Detecting android malware by building markov chains of behavioral models (extended version). *ACM Transactions on Privacy and Security (TOPS), 22*(2), 14.
12. Kim, T., Kang, B., Rho, M., Sezer, S., & Im, E. G. (2019). A multimodal deep learning method for Android malware detection using various features. *IEEE Transactions on Information Forensics and Security, 14*(3), 773–788.
13. Chen, S., Xue, M., Fan, L., Ma, L., Liu, Y., & Xu, L. (2019). How can we craft large-scale Android Malware? An automated poisoning attack. In *2019 IEEE 1st International Workshop on Artificial Intelligence for Mobile (AI4Mobile)* (pp. 21–24). IEEE.
14. Sharma, A., & Sahay, S. K. (2019). Group-wise classification approach to improve Android malicious apps detection accuracy. arXiv preprint arXiv:1904.02122
15. Su, D., Liu, J., Wang, X., & Wang, W. (2019). Detecting Android locker-ransomware on Chinese social networks. *IEEE Access, 7*, 20381–20393.
16. Saracino, A., Sgandurra, D., Dini, G., & Martinelli, F. (2018). Madam: Effective and efficient behavior-based android malware detection and prevention. *IEEE Transactions on Dependable and Secure Computing, 15*(1), 83–97.
17. Li, J., Sun, L., Yan, Q., Li, Z., Srisa-an, W., & Ye, H. (2018). Significant permission identification for machine-learning-based android malware detection. *IEEE Transactions on Industrial Informatics, 14*(7), 3216–3225.
18. Shen, F., Del Vecchio, J., Mohaisen, A., Ko, S. Y., & Ziarek, L. (2018). Android malware detection using complex-flows. *IEEE Transactions on Mobile Computing.*
19. Karbab, E. B., Debbabi, M., Derhab, A., & Mouheb, D. (2018). MalDozer: Automatic framework for android malware detection using deep learning. *Digital Investigation, 24*, S48–S59.

20. Zhu, H. J., You, Z. H., Zhu, Z. X., Shi, W. L., Chen, X., & Cheng, L. (2018). DroidDet: Effective and robust detection of android malware using static analysis along with rotation forest model. *Neurocomputing, 272*, 638–646.
21. Garcia, J., Hammad, M., & Malek, S. (2018). Lightweight, obfuscation-resilient detection and family identification of Android malware. *ACM Transactions on Software Engineering and Methodology (TOSEM), 26*(3), 11.
22. Vinayakumar, R., Soman, K. P., Poornachandran, P., & Sachin Kumar, S. (2018). Detecting Android malware using long short-term memory (LSTM). *Journal of Intelligent & Fuzzy Systems, 34*(3), 1277–1288.
23. Narayanan, A., Chandramohan, M., Chen, L., & Liu, Y. (2018). A multi-view context-aware approach to Android malware detection and malicious code localization. *Empirical Software Engineering*, 1–53.
24. Chen, S., Fan, L., Chen, C., Su, T., Li, W., Liu, Y., & Xu, L. (2019). StoryDroid: Automated generation of storyboard for Android apps. arXiv preprint arXiv:1902.00476
25. Wang, W., Li, Y., Wang, X., Liu, J., & Zhang, X. (2018). Detecting Android malicious apps and categorizing benign apps with ensemble of classifiers. *Future Generation Computer Systems, 78*, 987–994.
26. Hsien-De Huang, T., & Kao, H. Y. (2018). R2-d2: Color-inspired convolutional neural network (cnn)-based android malware detections. In *2018 IEEE International Conference on Big Data (Big Data)* (pp. 2633–2642). IEEE.
27. McLaughlin, N., Martinez del Rincon, J., Kang, B., Yerima, S., Miller, P., Sezer, S., Joon Ahn, G., et al. (2017). Deep android malware detection. In *Proceedings of the Seventh ACM on Conference on Data and Application Security and Privacy* (pp. 301–308). ACM.
28. Demontis, A., Melis, M., Biggio, B., Maiorca, D., Arp, D., Rieck, K., Roli, F., et al. (2017). Yes, machine learning can be more secure! a case study on android malware detection. *IEEE Transactions on Dependable and Secure Computing*.
29. Milosevic, N., Dehghantanha, A., & Choo, K. K. R. (2017). Machine learning aided Android malware classification. *Computers & Electrical Engineering, 61*, 266–274.
30. Feizollah, A., Anuar, N. B., Salleh, R., Suarez-Tangil, G., & Furnell, S. (2017). Androdialysis: Analysis of android intent effectiveness in malware detection. *Computers & Security, 65*, 121–134.

Comparative Analysis of RNN Variants Performance in Stock Price Prediction

Amit Raja Kalidindi, Naga Sudhakar Ramisetty,
Srikalpa Sankeerth Kruthiventi, Jayam Sri Harsha Srinivas,
and Lekshmi S. Nair

Abstract Stock price prediction is an important task for a trader to make good returns on investment. Several factors can control the price of the stock. To discover the underlying patterns and make the right predictions, researchers have been trying to make use of machine learning and deep learning for many years. The results of the previous work done in this area show that DL models such as recurrent neural network (RNN) and long short-term memory (LSTM) have better performance in forecasting stock prices when compared to standard ML models. In this work, we took 18 years of historical data of Tata Consultancy Services (TCS) stock price and analyzed the performance of different variants of RNN in forecasting the stock's closing price.

Keywords Deep learning · Recurrent neural networks · Time series prediction

1 Introduction

Stock exchange is a marketplace where investors can purchase and sell public company stock. The important aspect of stock trading is price prediction or trend

A. R. Kalidindi (✉) · N. S. Ramisetty · S. S. Kruthiventi · J. S. H. Srinivas · L. S. Nair
Department of Computer Science and Engineering, Amrita Vishwa Vidyapeetham, Amritapuri,
India
e-mail: karaja@am.students.amrita.edu

N. S. Ramisetty
e-mail: rnagasudhakar@am.students.amrita.edu

S. S. Kruthiventi
e-mail: sskruthiventi@am.students.amrita.edu

J. S. H. Srinivas
e-mail: jsharshasrinivas@am.students.amrita.edu

L. S. Nair
e-mail: lekshmisn@am.amrita.edu

S. Shakya et al. (eds.), *Sentiment Analysis and Deep Learning*,
Advances in Intelligent Systems and Computing 1432,
https://doi.org/10.1007/978-981-19-5443-6_59

prediction (bullish or bearish). Stock prices are predicted using two types of analysis, namely fundamental analysis and technical analysis. Different factors such as economy, the market for a product produced by the company, deals made by the company, and Year on Year (YoY) growth made by the company are used in fundamental analysis to predict the trends. In technical analysis, the stock prices are predicted by analyzing the trading activity such as volumes of stock getting traded, and past prices. Traders make use of the candlestick patterns as part of technical analysis to find the patterns and make stock price trend predictions. But, some of the underlying patterns can be missed naturally. The problem of stock price prediction comes under time series prediction [1]. Different techniques such as MA and ARIMA have been used in the past to predict stock prices [2]. ML models such as SVM and linear regression have also been used in stock price prediction [3, 4]. But in recent years, research has shown that DL models such as ANN, CNN, and RNN have better performance in stock price prediction when compared to models like ARIMA and standard ML models. Models like RNN can predict stock prices with good accuracy and they are suitable for time series prediction as output from previous step is fed as input to current step. But, it suffers from a vanishing gradient problem which makes it difficult to handle the long-term dependencies [5]. The vanishing gradient problem occurs when learning is hampered as a result of the gradients becoming very small. The introduction of LSTM has solved this problem. It is a kind of RNN that can have long-term memory. LSTMs have shown great performance in sequence prediction problems. The main objective of this work is to analyze the performance of different RNN variants in stock price forecasting. In this work, we analyzed the performance of simple RNN, LSTM, LSTM with peephole connections, bidirectional LSTM, gated recurrent unit (GRU), and bidirectional GRU in forecasting the closing price of a stock, given univariate and multivariate data as input.

2 Related Work

Researchers have proposed different techniques for time series forecasting. In [6], DL models were used to forecast the stock prices of companies listed on NSE. The performance of multilayer perceptron (MLP), RNN, LSTM, CNN and SVM was analyzed. CNN gave better performance when compared to other models. The results were also compared with the performance of ARIMA and it was found that all models performed better than ARIMA. Similar work was done in [7], the performance of the above models was tested on both NSE and NYSE data and similar results were obtained. In [8], a sliding window approach using deep learning models was proposed to forecast the stock prices on short term basis. The study [9] proposed CNNLSTM model to get good prediction accuracy in stocks. For the feature extraction from data, CNN was adopted. Using the retrieved features, LSTM was utilized to forecast the stock price. In [10], the Nifty 50 Index values data was used to evaluate different ML-based regression models and different LSTM-based models performances in forecasting the open values. The performance of ML-based regression models such as multivariate regression, decision tree regression, SVM regression and LSTM models such as classic LSTM and encoder-decoder LSTM were recorded and compared. The performance of LSTM models was tested on univariate input data and multivariate input data. LSTM models with univariate input performed better than models with multivariate input. The LSTM model with univariate input which uses the past 1 week's data was found to be more accurate when compared to other models. In [11], the performance of the LSTM model in forecasting the open prices of Google and Nike stock was observed. The performance of the model was analyzed by changing the number of epochs for training the model and by also changing the size of training data. It was observed that an increase in the number of epochs and training with fewer data improved the testing results. In [12], the performance of stateless and stateful LSTMs in forecasting the stock prices of companies listed in NSE was analyzed. The stateless model was found to be more stable than the stateful model. In [13], future prices of the stock market were predicted using GRU. Changes to the internal structure of GRU were made to improve efficiency. Mini-batch gradient descent was used and promising results were obtained. In [14], simple RNN, GRU, and LSTM were used to forecast the stock prices of companies in the SP500. LSTM was found to outperform other models. The study [15] proposed optimized GRU and LSTM models for stock price forecasting. PCA and LASSO were used for dimensionality reduction. Both GRU and LSTM were found to give good performance. The models that used LASSO performed better than those that used PCA. In [16], deep learning models using CNN, LSTM, GRU, and extreme learning machines (ELM) were built and evaluated in forecasting the stock prices of stocks in the BANKEX index. In [17], a deep learning approach using stacked LSTM was proposed to forecast wind speed. The stacked LSTM model with three layers was found to give best performance. In [18], an LSTM-based model was proposed to forecast the web traffic for Wikipedia. In [19], comparative analysis was done on LSTM and GRU in handling machine comprehension and it was observed that GRU has better performance.

3 Variants of RNN

3.1 Simple RNN

RNN as in Fig. 1 is a type of ANN in which the current step is fed with the output from the previous step. Usually, neural networks can have so many neurons and layers, where the input is fed through neurons and layers, and the output is calculated. Here, the input and output are unrelated. If the data has to be predicted based on a previous sequence or data then the data or sequence needs to be remembered. Thus, RNN came into existence with the idea of having hidden states where it is used to remember the sequence of data. The mathematical equation of RNN follows as

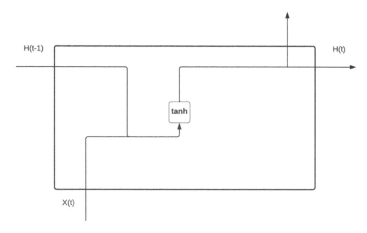

Fig. 1 Simple RNN

$$h_t = \tan h\left(w.[h_{t-1}, \ x_t] + b\right) \tag{1}$$

where h_t represents the current state, h_{t-1} represents the previous state and x_t represents the input, w represents the weights and b represents the bias.

RNN is trained by using a backpropagation algorithm. The weights are updated in such a way that the loss is minimal. Due to constant updation of weights using the gradient, the problem of vanishing gradient or exploding gradient occurs. The gradients become smaller and smaller which approach zeros that lead to the weights of the lower layers being unchanged and learning or training will not occur. This problem is known as the vanishing gradient. Sometimes, the gradients become larger and larger which leads to large weight updates and causes the gradient descent to be large. This situation is known as the exploding gradient. To avoid this scaling effect LSTM can be used. LSTM was built in such a way that the exploding or vanishing gradients problem is avoided [20].

3.2 Long Short-Term Memory (LSTM)

LSTM is a variant of RNN. Long-term dependencies can be learned by LSTM. The main difference between LSTM and RNN is that the hidden layer of LSTM is a gated unit or gated cell. The information is added or removed with the use of gates in LSTM. These gates consist of a sigmoid layer and a pointwise multiplication. As shown in Fig. 2, LSTM has three different kinds of gates that are used to maintain the state of the cell. The first gate is known as the forget gate. The sigmoid layer used for this output's values ranging from 0 to 1 decides the amount of data that should be eliminated from the past cell state. The second gate is known as the input gate. It controls the amount of data that must be added to the cell state. It has two layers. The sigmoid layer determines the values that must be modified and the tanh layer generates a set of new values. The third gate which is known as the output gate determines the output sequence. The tanh layer receives the updated cell state as input. The output from this layer is then multiplied by the output of the third sigmoid layer. The output from this multiplication becomes the new hidden state. The mathematical equation of LSTM follows as

$$i_t = \sigma \left(w_i . [h_{t-1}, \ x_t] + b_i \right)$$
$$f_t = \sigma \left(w_f . [h_{t-1}, \ x_t] + b_f \right)$$
$$o_t = \sigma \left(w_o . [h_{t-1}, \ x_t] + b_o \right)$$
$$\ddot{C}_t = \tan h \left(w_C . [h_{t-1}, \ x_t] + b_C \right)$$
$$C_t = f_t^* C_{t-1} + i_{t*} \hat{C}_t$$
$$h_t = o_t^* \tan h(C_t) \tag{2}$$

where i_t represents the input state, f_t represents the forget state, o_t represents the output state, w_x represents the weight for respective gates, b_x represents the bias, C_t represents the cell state, h_t represents the hidden state, and the $\tan h$ and σ represents the activation functions.

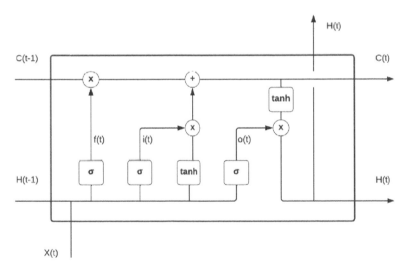

Fig. 2 LSTM

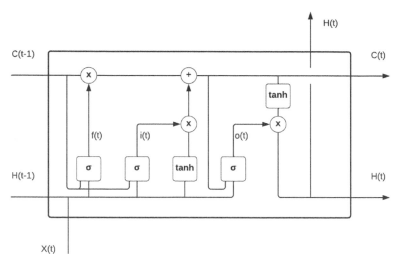

Fig. 3 Peephole LSTM

3.3 Peephole LSTM

LSTM with peephole connections is one of the variants of the basic LSTM architecture. As shown in Fig. 3, peephole connections are created in such a way that the gates are not only dependent on the hidden state but also on the cell state [21]. The LSTM peephole connection equation follows as.

$$i_t = \sigma\left(w_i.\left[C_{t-1}, h_{t-1}, x_t\right] + b_i\right)$$
$$f_t = \sigma\left(w_f.\left[C_{t-1}, h_{t-1}, x_t\right] + b_f\right)$$
$$o_t = \sigma\left(w_o.\left[C_t, h_{t-1}, x_t\right] + b_o\right) \tag{3}$$

Here, i_t represents the input state, f_t represents the forget state, o_t represents the output state, C_t represents the cell state, h_t represents the hidden state, w_x represents the weight for respective gates, σ represents the activation function, and b_x represents the biases.

3.4 Gated Recurrent Unit (GRU)

GRU is also one of RNN variations [22] which is being used in removing the problem faced by the exploding or vanishing gradient. Unlike LSTM, GRU has only two gates. They are reset gate and update gate. As shown in Fig. 4, the hidden state and cell state are not separate in GRU, but they are merged into a single state. The amount of knowledge passed into the future is controlled by the update gate. The amount

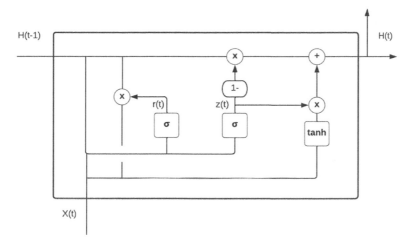

Fig. 4 GRU

of knowledge that needs to be forget is controlled by the reset gate. Due to the less number of gates and the less number of trainable parameters GRUs can train faster. The mathematical equation of GRU follows as

$$
\begin{aligned}
z_t &= \sigma\left(w_z.[h_{t-1}, x_t] + b_z\right) \\
r_t &= \sigma\left(w_r.[h_{t-1}, x_t] + b_r\right) \\
\hat{h}_t &= \tan h\left(w_{h..}[r_t * h_{t-1}, x_t] + b\right) \\
h_t &= (1-z_t) * h_{t-1} + z_t * \hat{h}_t
\end{aligned}
\tag{4}
$$

Here, x_t represents the input vector, h_t represents the output vector, $\hat{h}_t$ represents the candidate activation vector, z_t represents the update gate vector, r_t represents reset vector, w_x represents the weights, and b_x represents the bias.

3.5 *Bidirectional LSTM and Bidirectional GRU*

Bidirectional recurrent neural networks can be considered as a hybrid of two separate recurrent neural networks. In this model, the neural network will have access to the sequence information in both forward and backward direction [23]. As shown in Fig. 5, in Bidirectional LSTM and bidirectional GRU the inputs will flow in two directions: from past to the future or vice versa. The inputs that run from backward will allow bidirectional LSTM and bidirectional GRU to preserve information in the future and The inputs that run from forward will allow bidirectional LSTM and bidirectional GRU to preserve information in the past also. They are mainly used in NLP tasks such as translation entity recognition and classification. They are also

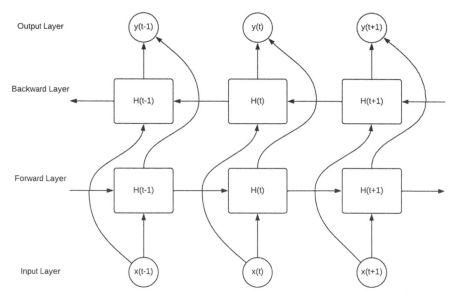

Fig. 5 Bidirectional LSTM or bidirectional GRU

applicable in speech recognition handwritten recognition and also protein structure prediction. They also have certain disadvantages as they takes more time to train compared to standard LSTM or GRU. The output can be calculated as

$$y = g\left(w_y.\left[a_f, a_b\right] + b_y\right) \tag{5}$$

Here, y represents the output, w_y represents the weights, b_y represents the bias, a_f is the activation in forward direction, and a_b is activation in backward direction.

4 Approach

In this work, we used historical data of Tata Consultancy Services (TCS) stock price and analyzed the performance of the above-mentioned variants of RNN in forecasting the stock's closing price. We used yahoo finance [24] to collect accurate data. We collected the stock data from Jan 1, 2003 to Dec 31, 2021. The available data was divided in 8:2 ratio for training and testing. The validation split was chosen as 0.2 of the training data. As we wanted to test the RNN variants performance on both univariate input and multivariate input we processed our data accordingly. For the models with univariate input, only the stock's closing price for the past 14 days is given as input. For the models with multivariate input; the low, high, open, close prices, and volume of stock traded for the past 14 days are given as input. The data was normalized using the Min–Max scaler. For the optimization of the models, we took mean squared error (MSE) as the criteria.

$$MSE = \frac{1}{n} \sum_{i=1}^{n} (y_i - \hat{y}_i)^2 \tag{6}$$

We took mean absolute percentage error (MAPE) as the evaluation criteria.

$$MAPE = \frac{1}{n} \sum_{i=1}^{n} \left| \frac{y_i - \hat{y}_i}{y_i} \right| * 100 \tag{7}$$

Initially, to find the best optimizer, we tested the performance of the above-mentioned RNN variants using different optimizers such as Adam, Rmsprop, stochastic gradient descent (SGD), and Adadelta. We used simple models for this purpose and they contained three layers, i.e., input layer, RNN variant layer, and dense layer. Hidden units were taken as 50 for each model. The number of epochs for training the models were taken as 50.

The Tables 1 and 2 show that for both univariate and multivariate input Adam optimizer performed better when compared to other optimizers, since it produced the lowest MAPE in the majority of cases. Adam optimizer has very fast computation time when compared to other optimizers. So, it can produce better results in less time. The memory used by it and the number of parameters taken are less when compared to many optimizers. So, we can say that Adam is the best optimizer based on the results obtained.

After identifying Adam as the effective optimizer, by using it as the optimizer we created and analyzed the performance of various models for each variant by varying the number of RNN variant layers and hidden units.

As mentioned above, the performance of the models has been tested on both univariate input and multivariate input. For the univariate input, the input shape is (14, 1) as only the closing price of the stock for past 14 days is taken as input. For the multivariate input, the input shape is (14, 5) as the low, high, open, close prices, and volume of stock traded for the past 14 days is given as input.

Table 1 Performance of RNN variants under different optimizers and with univariate input-MAPE values

RNN variant	Adam optimizer	Rmsprop optimizer	SGD optimizer	Adadelta optimizer
Simple RNN	2.05	2.37	9.11	11.16
LSTM	1.45	2.32	2.42	12.08
Peephole LSTM	1.76	2.34	2.34	8.97
BiLSTM	2.00	2.33	2.86	5.83
GRU	1.25	1.74	1.75	12.86
BiGRU	1.29	1.35	2.44	6.69

Table 2 Performance of RNN variants under different optimizers and with multivariate input-MAPE values

RNN variant	Adam optimizer	Rmsprop optimizer	SGD optimizer	Adadelta optimizer
Simple RNN	5.80	5.99	8.03	10.68
LSTM	1.42	2.24	2.94	12.09
Peephole LSTM	2.58	4.72	3.22	10.39
BiLSTM	2.95	1.88	3.46	12.68
GRU	2.16	2.57	1.60	3.65
BiGRU	1.77	2.39	2.72	4.00

5 Results

For each RNN variant we created three models which differ in number of RNN variant layers (1–3) used. Then, the performance of each model was tested by changing the number of hidden units (50, 75, 100, 125, and 150). As mentioned above MAPE was used to measure the performance.

$$\text{Average MAPE} = \frac{\text{MAPE}(50) + \text{MAPE}(75) + \text{MAPE}(100) + \text{MAPE}(125) + \text{MAPE}(150)}{5} \quad (8)$$

To get the overall performance of a model, the average of MAPEs given by each model was calculated when number of hidden units are varied.

The results demonstrate that LSTM and GRU versions outperform the standard RNN. As previously stated, RNN has the vanishing gradient problem, which limits its ability to retain long-term dependencies. The performance of RNN has decreased drastically for both univariate input and multivariate input as in Fig. 6 when number of layers increased. The existence of cell state and gates in LSTM and GRU variants help them to retain necessary information. So, LSTM and GRU variants performed better than the simple RNN.

The figures in Table 3 show that in the models with univariate input, GRU and Bidirectional GRU performed better when compared to LSTM variants when the number of layers are 1 and 2. But, when the number of layers was 3, classic LSTM and Peephole LSTM performed better than GRU and bidirectional GRU. In the overall performance, GRU performs best, followed by bidirectional GRU. The performance of bidirectional LSTM was comparatively less than other LSTM variants and GRU variants.

The figures in Table 4 show that in the models with multivariate input, GRU, and bidirectional GRU performed better when compared to LSTM variants in all cases. Significant decrease in the performance of LSTM and peephole LSTM was observed when number of layers increased. The performance of bidirectional LSTM is low when compared to other models, but no significant difference in performance was

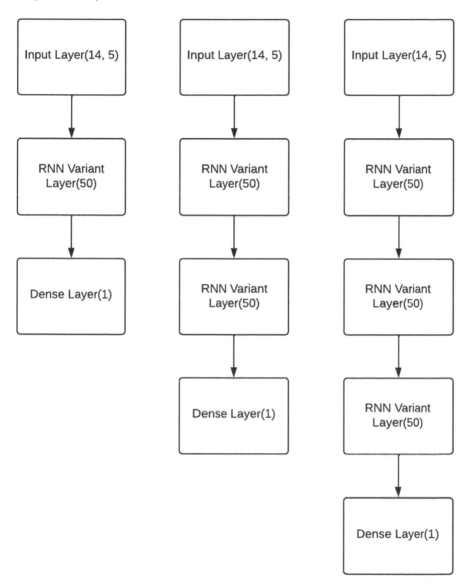

Fig. 6 Sample models with multivariate input and number of hidden units as 50

observed when number of layers changed. In the overall performance, bidirectional GRU performs best, followed by GRU.

The figures show how the variants perform in predicting the TCS stock's price when the number of layers is varied and the number of hidden units is set to 150. The models were trained for 50 epochs.

Table 3 Performance of RNN variants with univariate input

RNN variant	Average MAPE layers—1	Average MAPE layers—2	Average MAPE layers—3
Simple RNN	1.68	2.80	7.9
LSTM	1.46	1.82	1.65
Peephole LSTM	1.61	2.04	1.67
BiLSTM	1.65	1.73	2.72
GRU	1.39	1.53	1.75
BiGRU	1.43	1.58	1.84

Table 4 Performance of RNN variants with multivariate input

RNN variant	Average MAPE layers—1	Average MAPE layers—2	Average MAPE layers—3
Simple RNN	2.96	6.63	11.12
LSTM	1.70	2.19	3.37
Peephole LSTM	1.84	1.92	3.15
BiLSTM	2.61	2.54	2.80
GRU	1.68	1.76	2.78
BiGRU	1.66	1.61	1.81

Figures 7, 8, and 9 show the performance of RNN variants in forecasting the TCS stock's price using univariate input. Figures 10, 11, and 12 show the performance of RNN variants in forecasting the TCS stock's price using multivariate input. For both univariate and multivariate input, the figures show how the performance of simple RNN dropped with the increase in number of RNN layers. They also show that the predictions made by the GRU and bidirectional GRU are more accurate than predictions made by LSTM variants.

We have also tested the models on few other stocks listed in NSE to confirm our results. We collected the stock data of Infosys, Reliance, SBI, and Bajaj Finance from the date Jan 1, to Jan 4, 2022. We used the models that were trained using the TCS stock data to predict stock price of the above mentioned stocks. The models that were used contain an input layer, three RNN variant layers and a dense layer. The hidden units were set to 150. The models were trained for 50 epochs.

The figures from Tables 5 and 6 show that GRU and bidirectional GRU perform better when compared to other RNN variants. The LSTM variants also gave a descent performance in forecasting. The main advantage of GRU architecture over LSTM architecture is it's faster training time due to less number of weights or parameters. As the results are similar to the above results, when compared to the other LSTM variants, we can state that GRU and bidirectional GRU have an advantage in forecasting the stock price.

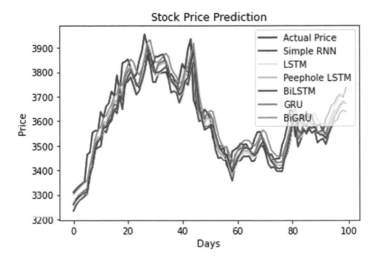

Fig. 7 Closing price prediction with univariate input and number of layers as 1

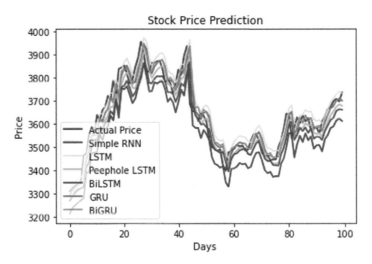

Fig. 8 Closing price prediction with univariate input and number of layers as two

6 Conclusions

In this work, we analyzed the performance of different RNN variants in forecasting the closing price of the stock. We used the TCS stock price data from yahoo finance. We evaluated different optimizers for finding the best optimizer and found Adam as the best optimizer for optimizing the models. For each RNN variant different models were built by varying the number of layers and hidden units. The performance was tested on both univariate and multivariate input. The LSTM and GRU variants outperformed the standard RNN. GRU and bidirectional GRU outperformed LSTM, peephole LSTM, and bidirectional LSTM. To confirm our results, the trained models were also tested on the stock data of Infosys, Reliance, SBI, and Bajaj Finance. Similar results were obtained, indicating that our results are promising and our approach is novel.

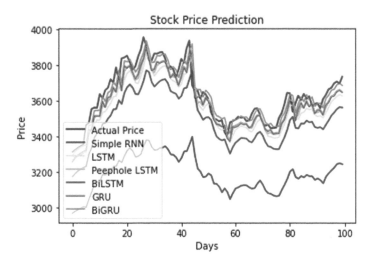

Fig. 9 Closing price prediction with univariate input and number of layers as three

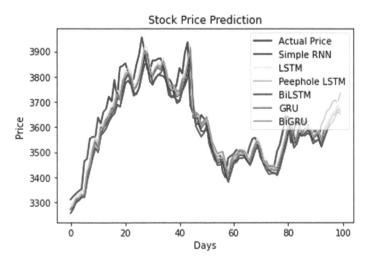

Fig. 10 Closing price prediction with multivariate input and number of layers as 1

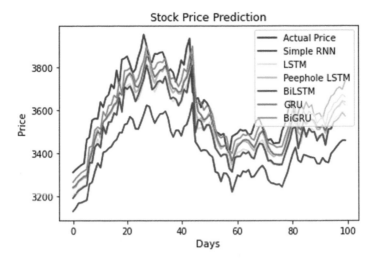

Fig. 11 Closing price prediction with multivariate input and number of layers as two

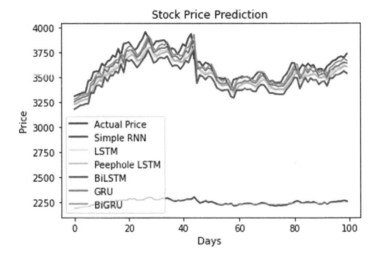

Fig. 12 Closing price prediction with multivariate input and number of layers as three

Table 5 Closing price prediction—univariate input

RNN variant	MAPE Infosys	MAPE Reliance	MAPE SBI	MAPE Bajaj Finance
Simple RNN	1.70	1.83	2.03	2.02
LSTM	1.42	1.73	1.72	1.96
Peephole LSTM	1.37	1.62	1.71	1.87
BiLSTM	1.49	1.72	1.75	2.00
GRU	1.39	1.46	1.70	1.82
BiGRU	1.37	1.49	1.71	1.82

Table 6 Closing price prediction—multivariate input

RNN variant	MAPE Infosys	MAPE Reliance	MAPE SBI	MAPE Bajaj Finance
Simple RNN	2.20	2.38	3.15	2.44
LSTM	1.4	1.63	1.77	1.83
Peephole LSTM	1.49	1.73	1.82	1.91
BiLSTM	1.66	1.88	1.91	2.03
GRU	1.43	1.56	1.75	1.78
BiGRU	1.43	1.62	1.73	1.77

References

1. Mehtab, S. S, & Jaydip. (2020). A time series analysis-based stock price prediction using machine learning and deep learning models.
2. Adebiyi, A. A., & Aderemi Ayo, C. (2014). Stock price prediction using the ARIMA model. In *Proceedings—UKSim-AMSS 16th International Conference on Computer Modelling and Simulation, UKSim 2014*. https://doi.org/10.1109/UK-Sim.2014.67
3. Karim, R. A., & Md Hossain, Md. (2021). Stock market analysis using linear regression and decision tree regression. 1–6. https://doi.org/10.1109/eSmarTA52612.2021.9515762
4. Vanukuru, K. (2018). Stock market prediction using machine learning. https://doi.org/10.13140/RG.2.2.12300.77448
5. Hochreiter, S. (1998). The vanishing gradient problem during learning recurrent neural nets and problem solutions. *International Journal of Uncertainty, Fuzziness and Knowledge-Based Systems., 6*, 107–116. https://doi.org/10.1142/S0218488598000094
6. Anand, C. (2021). Comparison of stock price prediction models using pre-trained neural networks. *Journal of Ubiquitous Computing and Communication Technologies (UCCT), 3*(02), 122–134.
7. Hiransha, M., Gopalakrishnan, E. A., Menon, V. K., & Soman, K. P. (2018). NSE stock market prediction using deep-learning models. *Procedia Computer Science, 132*, 1351–1362. https://doi.org/10.1016/j.procs.2018.05.050. ISSN 1877-0509.
8. Selvin, S., Vinayakumar, R., Gopalakrishnan, E. A., Menon, V. K., & Soman, K. P. (2017). Stock price prediction using LSTM, RNN and CNN-sliding window model. In *2017 International Conference on Advances in Computing, Communications and Informatics (ICACCI)* (pp. 1643–1647). https://doi.org/10.1109/ICACCI.2017.8126078
9. Lu, W. L., Li, J., Sun, Y., Wang, A., & Jingyang. (2020). A CNN-LSTM-based model to forecast stock prices. *Complexity, 2020*, 1–10. https://doi.org/10.1155/2020/6622927

10. Mehtab, S., Sen, J., & Dutta, A. (2021). Stock price prediction using machine learning and LSTM-based deep learning models. In: Thampi, S. M., Piramuthu, S., Li, K. C., Berretti, S., Wozniak, M., Singh, D. (Eds.) *Machine Learning and Metaheuristics Algorithms, and Applications. SoMMA 2020. Communications in Computer and Information Science* (Vol. 1366). Springer, Singapore. https://doi.org/10.1007/978-981-16-0419-58

11. Moghar, A., & Hamiche, M. (2020). Stock market prediction using LSTM recurrent neural network. *Procedia Computer Science, 170*, 1168–1173. https://doi.org/10.1016/j.procs.2020.03.049. ISSN 1877-0509.

12. Yadav, A., Jha, C. K., & Sharan, A. (2020). Optimizing LSTM for time series prediction in Indian stock market. *Procedia Computer Science, 167*, 2091–2100. https://doi.org/10.1016/j.procs.2020.03.257. ISSN 1877-0509.

13. Rahman, M. H., Junaid, M., Forhad, T.-S., Hossen, M., & Muhammad. (2019). Predicting prices of stock market using gated recurrent units (GRUs). *Neural Networks, 19*, 213–222.

14. Lee, S. I., & Yoo, S. J. (2020). Threshold-based portfolio: The role of the threshold and its applications. *The Journal of Supercomputing, 76*, 8040–8057. https://doi.org/10.1007/s11227-018-2577-1

15. Gao, Y., Wang, R., & Zhou, E. (2021). Stock prediction based on optimized LSTM and GRU models. *Scientific Programming, 2021*, Article ID 4055281, 8 p. https://doi.org/10.1155/2021/4055281

16. Balaji, J. A., Ram, Harish, D. S., & Binoy, B. N. (2018). Applicability of deep learning models for stock price forecasting an empirical study on bankex data. *Procedia Computer Science, 143*, 947–953.

17. Sowmya, C., Kumar, A. G., & Kumar, S. S. (2021). Stacked LSTM recurrent neural network: A deep learning approach for short term wind speed forecasting. In *2021 International Conference on Intelligent Technologies (CONIT)*.

18. Mettu, N., & Sasikala, T. (2019). Prediction analysis on web traffic data using time series modeling. *RNN and Ensembling Techniques*. https://doi.org/10.1007/978-3-030-03146-6_67

19. Viswanathan, S., Kumar, A. M., & Soman, K. P. (2019). A sequence-based machine comprehension modeling using LSTM and GRU. *Lecture Notes in Electrical Engineering, 545*, 47–55.

20. Hochreiter, S. S., & J. (1997). Long short-term memory. *Neural Computation, 9*, 1735–1780. https://doi.org/10.1162/neco.1997.9.8.1735

21. Gers, F. A., Schraudolph, N. N., & Schmidhuber, J. (2002). Learning precise timing with LSTM recurrent networks. *Journal of Machine Learning Research, 3*, 115–143.

22. Chung, J., Gulcehre, C., Cho, K., & Bengio, Y. Empirical evaluation of gated recurrent neural networks on sequence modelling. https://doi.org/10.48550/arXiv.1412.3555

23. Schuster, M. P., & Kuldip. (1997). Bidirectional recurrent neural networks. *Signal Processing, IEEE Transactions on., 45*, 2673–2681. https://doi.org/10.1109/78.650093

24. Yahoo Finance Website. https://finance.yahoo.com/

Deep Learning-Based Haze Removal System

Santosh Randive, Joel Joseph, Neel Deshmukh, and Prashant Goje

Abstract Machine learning and computer vision are some of the booming areas of applied artificial intelligence (AI). Through machine learning, we give machines the ability to make decisions on their own. Computer vision is a subset of AI where we give these machines the ability to see and we give these machines the ability to make sense of images and understand crucial parts of the images. Often these images get corrupted because of noise and atmospheric disturbances, and these noises and disturbances are identified as haze or atmospheric haze. The presence of haze in the image hides the key elements/subjects in the image, so the AI algorithm finds it difficult to identify the subjects in the image. In this paper, we propose to use a deep learning-based haze removal system that will prove to be one of the most important preprocessing steps before giving the data to your AI system.

Keywords Machine learning · Deep learning · Computer vision · Artificial intelligence · Haze removal · Image processing · Generative adversarial networks

1 Introduction

There are different methods in computer vision to identify what a neural network pays attention to when giving predictions. Paper [1] discusses the possible things that each layer of a convolutional neural network (CNN) might learn to identify. Paper [2] discusses a technique called class activation map that helps in highlighting the

S. Randive · J. Joseph (✉) · N. Deshmukh · P. Goje
PCET's, PCCOER, Ravet Pune, India
e-mail: joel.joseph211@gmail.com

S. Randive
e-mail: santosh.randive@pccoer.in

N. Deshmukh
e-mail: neeldeshmukh619@gmail.com

P. Goje
e-mail: prashantgoje8s@gmail.com

© The Author(s), under exclusive license to Springer Nature Singapore Pte Ltd. 2023   797
S. Shakya et al. (eds.), *Sentiment Analysis and Deep Learning*,
Advances in Intelligent Systems and Computing 1432,
https://doi.org/10.1007/978-981-19-5443-6_60

parts of images that the network is paying attention to. These attention maps give an understanding of our dataset and also our neural networks. Now, these maps do not provide much information if the image contains noise and atmospheric haze. The network does not understand the part of the image it should pay attention to because of these disturbances in pixels, resulting in poor performance of the neural network. Consider an object detection algorithm, the algorithm is trained to classify objects like bus, person, car, etc., in the image, now if these images contain haze, the algorithm will find it difficult to identify the objects in the image. This algorithm is heavily used in self-driving cars, and if the algorithm gets a hazy image, the algorithm might find it difficult to classify the objects in front and thus would result in unwanted behavior and would result in some dangerous accidents. To avoid this, we need to add a preprocessing block that can remove the haze and unwanted noise from the image and then pass a clear image to the object detection algorithm. Initially, some image processing techniques were used to improve the image quality, but these systems are slow and are prone to errors. With the rise of deep learning and computer vision, we can now achieve better results compared to these traditional methods. Here, we propose a similar system that can be used as a preprocessing step to remove haze and then pass the haze-free images to the further machine learning blocks.

2 Proposed Method

2.1 Working

To tackle this problem, we applied various standard deep learning techniques like down-sampling the input image followed by softening the activation of the deep layers and then up-sampling the image. The network works well but often tends to lose necessary pixels from the image. In this paper, we propose to use a generative adversarial network (GAN) [3]-based network that can take a hazy image as input and obtain the de-hazed image of the same. A GAN network consists of two networks: a generator and a discriminator. The generator tries to generate new data, and the discriminator tries to identify whether the image fed is fake or real. The goal of the generator is to fool the discriminator such that its output is a probability score of 0.5, meaning that the data provided can be either real or fake. GAN's end goal is to predict features given a label, instead of predicting a label of given features.

A simple training process of GAN is shown in Fig. 1.

2.2 Generator Architecture

The generator of a Pix2Pix cGAN is a modified U-Net [5]. Just like the U-Net, the generator model consists of an encoder block that contains a series of down-sampling

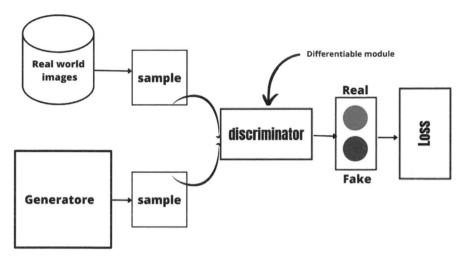

Fig. 1 Block diagram of generative adversarial networks

layers followed by a decoder block that contains a series of up-sampling layers. Each encoder block consists of convolution, batch normalization [6], and activation (Leaky ReLU) layers, whereas the decoder block comprises transposed convolution, batch normalization, dropout (applied to the first three blocks), and ReLU layers.

2.3 Training the Generator

As discussed earlier, the goal of the generator is to make artificial haze-free samples of image preserving the important information from the input image. The objective of the discriminator is to identify these fake images generated by the generator from the real images. Figure 2 shows the block diagram of training the generator.

In simple GAN, network (the GAN loss) learns to adapt to the data over the training steps, whereas the conditional loss of the GAN network tries to keep the overall structure of the image. In Pix2Pix GAN, the GAN is optimized over the following loss function:

(i)   The generator loss is then calculated as the sigmoid cross-entropy of the generated image and an array of ones.

(ii) An L1 loss, i.e., mean absolute error (MAE) is also calculated between the generated image and the target image. This is done so that the generated image preserves the objects/subjects in the image, and the generated image is structurally similar to the target image.

(iii) The overall generator loss is then calculated as follows: *gan_loss* + *lambda* * *l1_loss*, here *lambda* = *100*. It is a weight factor to the *l1_loss*.

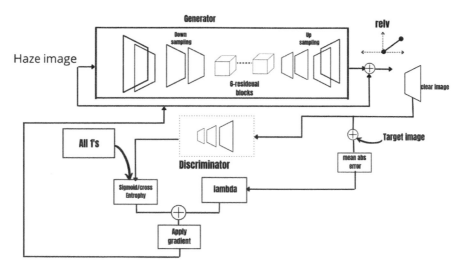

Fig. 2 Block diagram of generator training

(iv) After calculating the loss, the gradients are calculated, and using standard optimizers like Adam [7], the model weights are then updated.

2.4 Discriminator Architecture

The discriminator block is just a simple convolutional neural network. It consists of a series of down-sampling blocks, and each block consists of convolution and batch normalization layers activated by a Leaky ReLU function. The discriminator tries to classify every image into real or fake class.

2.5 Training the Discriminator

As discussed earlier, the goal of the discriminator is to classify between real and fake samples of the images. The discriminator network gets the ground truth image and the generated image from the generator network. The main objective of the discriminator is to see how similar these images are and try to classify the generated image as a fake image. The discriminator block diagram is shown in Fig. 3.

The total discriminator loss is calculated as the sum of real loss and generated loss, where

(i) The real loss is the sigmoid cross-entropy of the real images and an array of ones, as these images are real images and should be labeled as ones.

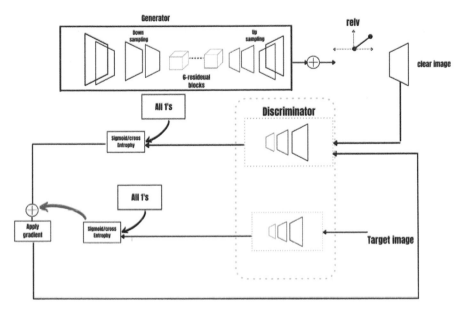

Fig. 3 Block diagram of discriminator training

(ii) Generated loss is the sigmoid cross-entropy of the generated image and an array of zeros, as these are fake images that should be labeled as zeros.
(iii) The total loss is then calculated as the sum of real loss and generated loss.

After calculating the loss, the gradients are calculated, and using standard optimizers like Adam, the model weights are then updated.

2.6 The Overall Training Overview

The training set consists of hazy images and the de-hazed version of the same image which is considered as the ground truth image. The hazy image is given to the generator, and at first, the generator outputs a random noise, overtraining the generated image looks more like the ground truth image. The discriminator gets the generated image and the ground truth image. The loss for both the networks is calculated, and the gradients are updated by backpropagation to both the networks.

3 Results

Over time, the network learns to generate new de-hazed images of the given hazy image. Figures 4, 5, 6, and 7 show some of these images generated by the network.

Fig. 4 Given the hazy image (left) to the generator, the de-hazed image (right) is obtained, and ground truth image (middle) is used for comparison

Fig. 5 Given the hazy image (left) to the generator, the de-hazed image (right) is obtained, and ground truth image (middle) is used for comparison

Fig. 6 Given the hazy image (left) to the generator, the de-hazed image (right) is obtained, and ground truth image (middle) is used for comparison

On calculating the MSE between the ground truth image and the generated image, we attain an average MSE of 0.0221. The comparison with other systems is given in Table 1.

4 Future Scope

The following system can be integrated into today's automated machine learning systems like self-driving cars, image classification systems, etc. The generic block diagram of a system like this is shown in Fig. 8.

Fig. 7 Given the hazy image (left) to the generator, the de-hazed image (right) is obtained, and ground truth image (middle) is used for comparison

Table 1 Comparison table based on different approaches

Parameters	ATM [9]	BCCR [10]	DCP [11]	Our approach
MSE	0.0689	0.0243	0.0172	0.0221
PSNR	26.751	37.173	40.628	38.122

Fig. 8 Block diagram of the haze removal system

The image feed block is just a simple image capturing device (camera), the captured image is passed to our trained ML model to remove haze, the output of this block is a haze-free image, and this haze-free image can then be passed to further machine learning algorithms for further classifications.

5 Conclusion

The presence of haze in our data can result in unwanted behaviors in our machine learning models. Getting rid of these atmospheric anomalies while keeping the structural information of our input image is of cardinal essence. To solve this issue, a conditional GAN as discussed above will be really helpful. The GAN learns to generate haze-free images while keeping the structural information intact.

References

1. Zeiler, M. D., & Fergus, R. (2013). Visualising and understanding convolutional networks. arXiv: 1311.2901, 12 November 2013
2. Zhou, B., Khosla, A., Lapedriza, A., Oliva, A., & Torralba, A. (2015). Learning deep features for discriminative localization. arXiv: 1512.04150, 14 December 2015
3. Goodfellow, I. J., Pouget-Abadie, J., Mirza, M., Xu, B., Warde-Farley, D., Ozair, S., Courville, A., & Bengio, Y. (2014). Generative adversarial networks. arXiv: 1406.2661, 10 June 2014
4. Isola, P., Zhu, J. -Y., Zhou, T., & Efros, A. A. (2016). Image-to-image translation with conditional adversarial networks. arXiv: 1611.07004, 21 November 2016
5. Ronneberger, O., Fischer, P., & Brox, T. (2015). U-Net: convolutional networks for biomedical image segmentation. arXiv: 1505.04597, 18 May 2015
6. Ioffe, S., & Szegedy, C. (2015). Batch normalisation: accelerating deep network training by reducing internal covariate shift. arXiv 1502.03167, 11 February 2015
7. Kingma, D. P., & Ba, J. (2017). Adam: a method for stochastic optimization. arXiv: 1412.6980, 30 January 2017
8. Cai, B., Xu, X., Jia, K., Qing, C., & Tao, D. (2016). DehazeNet: an end-to-end system for single image haze removal. arXiv: 1601.07661, 17 May 2016
9. Sulami, M., Glatzer, I., Fattal, R., & Werman, M. (2014). Automatic recovery of the atmospheric light in hazy images. In *IEEE International Conference on Computational Photography (ICCP)* (pp. 1–11).
10. Meng, G., Wang, Y., Duan, J., Xiang, S., & Pan, C. (2013). Efficient image dehazing with boundary constraint and contextual regularization. In *IEEE International Conference on Computer Vision (ICCV)* (pp. 617–624).
11. He, K., Sun, J., & Tang, X. (2011). Single image haze removal using dark channel prior. *IEEE Transactions on Pattern Analysis and Machine Intelligence, 33*(12), 2341–2353.

A One-Stop Service Provider for Farmers Using Machine Learning

K. Vidya Sree, G. Sandeep Kumar Reddy, R. Dileep Varma, P. Mihira, and S. Remya

Abstract Agriculture is one of the most important sectors that affect a country's economic development. Agriculture employs about 65% of the Indian population and accounts for about 22% of gross domestic product (GDP), yet they earn only a few percent of the country's GDP. This is due to the lack of knowledge and awareness about which seeds to sow and how to deal with pests, both of which can have a significant impact on productivity. Hence, this paper proposes a method for combining AI with agriculture to develop a smart agricultural system that makes it easier for farmers to produce and maximize their output. Therefore, FarmEasy, a website that makes it simple for farmers to make the greatest decisions at the right time has been developed. The proposed method implements four major features, namely, crop recommendation based on nutrients present in soil and weather in that area, fertilizer recommendation based on nutrients in soil and crop, disease prediction based on a picture of the affected crop, and a Chatbot for farmers' assistance to clear their queries and inquiries. According to the observations, FarmEasy has the ability to effectively satisfy the information demands of farmers on a large scale.

Keywords Agriculture · Crop recommendation · Fertilizer · Plant disease · AI
Chat-bot · Smart agriculture · Machine learning · Artificial neural networks ·
Agricultural technology

K. Vidya Sree (✉) · G. Sandeep Kumar Reddy · R. Dileep Varma · P. Mihira · S. Remya
Department of Computer Science and Engineering, Amrita Vishwa Vidyapeetham, Amritapuri, India
e-mail: kvidyasree@am.students.amrita.edu

G. Sandeep Kumar Reddy
e-mail: gskreddy@am.students.amrita.edu

R. Dileep Varma
e-mail: dvrraju@am.students.amrita.edu

P. Mihira
e-mail: pmihira@am.students.amrita.edu

S. Remya
e-mail: remyas@am.amrita.edu

S. Shakya et al. (eds.), *Sentiment Analysis and Deep Learning*,
Advances in Intelligent Systems and Computing 1432,
https://doi.org/10.1007/978-981-19-5443-6_61

1 Introduction

Indian agricultural progress may be examined from two angles: institutional and technical. The institutional method, defined by land reforms and the transformation of agricultural relations, was implemented in the mid-1950s. As an agricultural development engine, it was a flop. A technical approach to agricultural expansion arose around the mid-1960s, and it had a substantial impact on agricultural production and productivity [28], apparently in locations with the basic prerequisites for switching to new technology [3]. There has been a global crisis for the past few decades in which no farmer has been able to earn a fair price for his produce. Traditional agricultural approaches can often lead to overproduction of the same crop, resulting in a crop surplus. Because this influence occurs in a certain crop each year, most farmers do not choose to cultivate that crop the next year. As a result, there is a significant shortage of the product, which has resulted in a significant price increase. The underlying cause of these accidents is a long-standing and ongoing agricultural communication imbalance [4]. The ability to anticipate a crop's yield based on prior trends will be tremendously useful in forecasting the farmer's earnings in this situation.

So, in order to address the aforementioned challenges that farmers face, this paper describes a method in the hopes of maintaining agriculture's long-term viability. The goal of this research is to provide an interface that sends all crop production data and fertilizer recommendations [23] on a regular basis, as well as notifying farmers about crop output in order to satisfy future requests and assisting farmers with their concerns via Chatbot help. Artificial intelligence (AI) and machine learning (ML) are two rapidly growing technologies. The perspective is to provide agriculture with the huge benefits of this new power with this goal in mind. All potential benefits of machine learning and artificial intelligence (AI) in agriculture have been investigated and the ML pipeline to manage them has been designed.

Crop recommendation is implemented using six different classifiers, ChatBot is performed using artificial neural networks (ANN) models, and disease prediction is carried out using seven different models and the best model for each has been picked after comprehensive comparison research to carry out real-time testing and utilization of these characteristics. Fertilizer recommendation is estimated by comparing ideal plant values to the inputted plant values and providing appropriate recommendation using the rule-based technique.

This paper further intends to address the previous works carried out in Sect. 2, related to all the four features: crop recommendation, fertilizer recommendation, disease prediction, and Chatbot (Fig. 1). The detailed proposed model is described in Sect. 3. A comprehensive analysis of models implemented, and their performance results are discussed under Sect. 4. Furthermore, Sect. 5 articulates the paper and provides a thorough understanding of the founding principle.

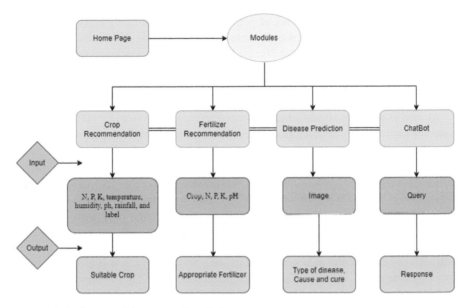

Fig. 1 Block diagram of the proposed method

2 Related Work

Agriculture plays a critical role in a country's growth and development. For this progress, farmers must be certain of the types of crops to be harvested for certain soil and weather conditions, the varieties of fertilizers to be used without affecting soil and crop quality, and the diseases to be diagnosed and preventative measures to be taken. Chatbots serve a supportive role in delivering assistance to farmers by ensuring communication. There has been a lot of background study done in these areas.

In design and implementation of fertilizer recommendation system for farmers, Subramanian et al. [6] have addressed a way to pick reasonable fertilizer for the land and determining the supplement amounts in the soil using lab hardware can be incredibly expensive, especially in developing countries. They conducted soil analysis, which included monitoring the flow of ions in the soil using a microcontroller named Latter, which was used to track the nutrients in the soil. This approach assisted farmers in maximizing their output for a specific crop cycle while minimizing the impact on the land and soil qualities. It also ensured that healthy crops were grown by lowering the risk of overfertilization.

Niranjan et al. [7] have proposed employing content-based image retrieval for leaf recognition and disease detection. Their suggested solution was a straightforward piece of software that focused on two parts of the process flow. The first was leaf recognition, which required determining which plant group or type it belongs to. It includes image processing techniques such as pre-processing, feature extraction, building a training and testing set of leaf images, and eventually classification; the second was disease detection, which included identifying leaf illnesses in the query leaf image. Pre-processing of the diseased leaf, image segmentation, feature extraction of the ROI chosen by the user, i.e., diseased region, creation of a training and testing set of damaged leaf images, and classification of the affected portions into their respective diseases were all used to do this.

Ahmed et al. [8] analyzed a nutrient recommendation system for soil fertilization based on evolutionary computation. They presented nutrient recommendations based on time-series sensor data and an enhanced genetic algorithm. The method assisted in determining the region to assess crop adaptability at various nutrient levels and provided insight into agricultural nutrient evaluation in a changing environment. In data mining, scheduling problems, and sequence planning, evolutionary computation has been used to solve a variety of tasks. An enhanced GA approach for recommending optimum nutrients to address optimization, resulting in increased yield production was proposed. The method provided a novel way to determine the best nutrients for increasing yield while maintaining soil fertility. It was able to improve local optimization to prevent the local premature individual in the population.

Mahendra et al. [25] proposed crop prediction using machine learning approaches where the system suggested the best suitable crop for a particular land based on content and weather parameters. It also included information on the number and type of fertilizers needed, as well as the seeds needed for growing. After pre-processing the data, the SVM algorithm was used to predict rainfall, and the model was trained using an SVM classifier. After fitting and testing the model, it was eventually able to calculate the required annual rainfall. One of the input parameters for the crop prediction system was forecasted rainfall. The decision tree technique was used to implement crop prediction. Various elements such as humidity, temperature, projected rainfall, and soil PH are taken into account when predicting the crop. As a result, a farmer-friendly algorithm that forecasts the optimal crop for a specific piece of land was developed.

Vijayalakshmi and PandiMeena [10] proposed Agriculture TalkBot using AI. This Chatbot incorporated NLP methods to interpret user queries. The user requests were tokenized into words, then stop words such as a, is, and the like are eliminated, and the remaining words are transformed to their root words. These words are first translated to a bag of words, then to a vector format so that the classification system can process them effectively. After then, the training dataset was used to train the bot. Then, using the gradient descent approach, a neural network was built, and the error was minimized. The test data set goes through the identical pre-processing, classification, and neural network development steps as the training data set. To obtain accurate findings, the class with the highest probability was iterated.

Manjula and Narsimha [11] applied data mining techniques for crop recommendation and yield prediction in agriculture This study suggests crops and fertilizers to farmers in order to boost agricultural productivity. It also provided crop farming tactics such as diversified farming, spacing, irrigation, sow processing, and so on, as well as fertilizer and pesticide recommendations. Nitrogen (N), phosphorus

(P), and potassium (K) readings were used to suggest. It suggested using content-based methods and collaborative filtering algorithms. The content-based system recommends items based on their attributes, whereas collaborative filtering methods propose elements based on similarity metrics between users and/or elements.

Niranjan et al. [7] have performed an intelligent question answering conversational agent using naive Bayesian classifier which automated a question answering system. This system responds to the student's questions in a more interactive manner. The teacher can access and modify the QA knowledge base, and he or she could also be aware of the areas where students are more likely to have doubts. Because it is a probabilistic categorization method, the naive Bayesian technique is applied. The system was evaluated using a single category and a self-made knowledge base, and it was discovered that the system was capable of detecting the proper match the majority of the time. The system was set up in such a manner that it automatically applied the Bayesian algorithm to the data provided by the teacher and categorizes it.

Yashaswini et al. [12] gathered data from a variety of sources, including government websites and libraries. Noise and unrelated data were filtered out. The information was then safely saved in a database. The KNN technique was used to train the database using machine learning in tensor flow architecture. They used NLP, and the training data were evaluated for correctness on a regular basis with a small set of validation data. After going through all the steps of data gathering, preprocessing, cleaning, training, and testing, the system was finally ready to use and is sent to the server. The system was then NLP trained, which indicates that it does not need to be taught on a regular basis. It learns on its own and improves its system with more sophisticated responses.

3 Proposed Method

Each model is developed independently and a website that included them collectively has been built as shown in Fig. 1. The first step of this research is the in integration of artificial intelligence with agriculture [27]. It includes, a Chatbot for inquiries, an integrated crop and fertilizer recommendation system, and the capacity to detect plant disease spread among other things. The software pipeline for the project may be thought of as a mapping of three sections: front-end development, feature design, and integration are all performed with HTML and CSS. Python is also used for machine learning, deep learning, model training, saving, and deployment of inferences.

3.1 Crop Recommendation

It is the most crucial module of all. In agriculture, it is critical that suggestions are provided accurately and precisely since mistakes can result in severe capital loss and material loss. This module's goal is to assist farmers in increasing agricultural output by proposing the best crop for their soil composition and climatic circumstances. The module is created with the help of a model pipeline that accepts the N: Ratio of Nitrogen content in the soil, P: Ratio of Phosphorus content in the soil, K: Ratio of Potassium content in the soil, temperature, humidity, pH, and rainfall from the user and recommends the crop according to its learning. Different ML models used for training are as follows.

A. *K*-Nearest Neighbors Classifier

This is one of the most fundamental and straightforward classification approaches, and it should be one of the initial options for classification research when there is little or no prior information about the distribution of data. The necessity for regression analysis when trustworthy parametric estimates of probability densities are unavailable or impossible to calculate led to the development of *K*-nearest neighbor classification. To make the algorithm operate optimally on a given dataset, the most appropriate distance measure is selected. There are many different distance measures to choose from, but the most popular ones are considered. The Euclidean distance function, which is the default in the SKlearn KNN classifier package in Python is used in this work. *K*-nearest neighbor is based on Euclidean distance as given in Eq. 1.

$$d(a_x, a_y) = \frac{q}{\left(a_{x_1} - a_{y_1}\right)^2 + \left(a_{x_2} - a_{y_2}\right)^2 + \left(a_{x_p} - a_{y_p}\right)^2} \tag{1}$$

B. SVM

Support vector machines (SVMs) are a set of supervised learning algorithms for classification, regression, and outlier detection. It's a categorization system. Each data item is depicted as an extension in an *n*-dimensional architecture (where *n* is the number of alternatives), with the value of each feature equal to the value of a single coordinate, using this algorithmic rule. It is a discriminative classifier with an appropriately bounded separation hyperplane. To put it in another way, the algorithmic approach generates the optimum hyperplane for categorizing new samples given in the labeled coaching data (supervised learning) [17].

C. Gaussian Naive Bayes

It is a set of algorithmic rules rather than a single algorithm. The naive Bayes classifier is a simple probabilistic classifier that combines Bayesian statistics theorems with persistent naïve maximum likelihood estimation. It is a classification method based on the Bayes theorem and the hypothesis of predictor independence. A naive Thomas Bayes classifier, in simple terms, believes that the existence of one character in an extreme class does not imply the presence of another feature. A fruit is termed an apple if it is red, round, and has a diameter of roughly two inches [17]. A naïve Bayes classifier would consider each quality to contribute to the chance that this fruit is an apple on its own, even if they are dependent on one another or the existence of additional traits. These learners predict the class label for each of the training data sets. The training data set's class label is chosen using the majority voting approach, which is used to vote on the class label that is predicted by most of the models. The rules are generated using ensembled models. Similarly, decision tree classifier (Sect. 3.1C) and logistic regression (Sect. 3.2C) are adopted to analyze which of the models is most appropriate to incorporate crop recommendations. In addition, a comparison (Sect. 4B) is undertaken between the models in order to select one for final implementation (Fig. 2).

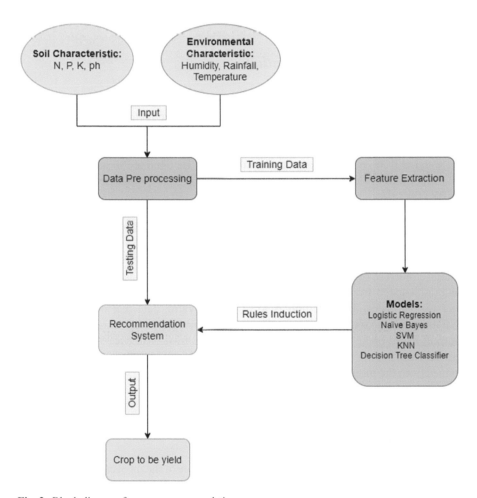

Fig. 2 Block diagram for crop recommendation

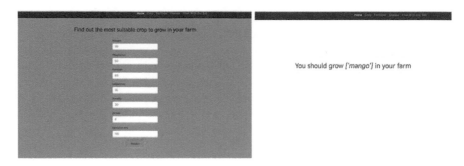

Fig. 3 a, b Sample implementation of crop recommendation

3.2 Fertilizer Recommendation

Fertilizers are used in agricultural cultivation to increase crop output and quality. These have had a significant impact on agricultural production and farming profitability all over the world. If fertilizers are not provided, the soil would deteriorate, making plant development exceedingly difficult since they cannot survive just on water. All the nutrients provided in them may be obtained in nature. The most common nutrient sources in mineral fertilizers are nitrogen, potassium, and phosphate (NPK) [9, 18].

The fertilizer recommendation system (Fig. 4) is developed using a model pipeline that takes pH values, soil type, such as the composition of NPK present in the soil, and crop type as inputs, and outputs the optimal fertilizer to apply. The dataset was received online, and it includes the usual NPK and pH values required for the individual crops. The rule-based technique is used to create this module. These algorithms take knowledge from the categorization model in the form of easy-to-understand and explain rules. This method is ideal for analyzing data that has both numerical and qualitative characteristics. There is a distinct dictionary of recommendation outputs based on key values. As a consequence, correct fertilizer recommendations are received taking into account all characteristics [15] needed by matching key values from the dictionary (Fig. 3).

3.3 Disease Prediction

When it comes to using technology in agriculture, one of the most confusing difficulties is detecting plant diseases. Despite the fact that deep learning and neural networks have been used to detect whether a plant is healthy or diseased, new technologies are always being developed.

In this work, to determine whether a plant leaf is healthy or unhealthy, a traditional machine learning algorithm and image processing to pre-process the data are used. Following the loading of the user's image from the user interface (UI), it is

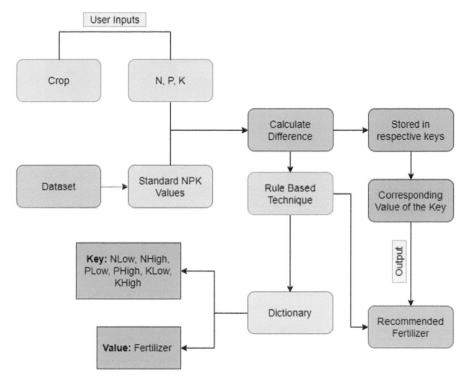

Fig. 4 Block diagram for fertilizer recommendation

transformed from RGB to BGR and subsequently from BGR to HSV. Image segmentation is used to extract colors from images. This is done to separate the leaf from the backdrop so that the color of the leaf can be retrieved (Fig. 6). The color, shape, and texture can be extracted from an image by using the global feature descriptor (Fig. 5).

A. Decision Trees Classifier

The image's features are retrieved and layered together using the NumPy function. This model has now been trained using seven different machine learning models, namely, logistic regression, linear discriminant analysis, decision trees, naive Bayes (Sect. 3.1C), K-nearest neighbors (Sect. 3.1A), random forest, and support vector machine (Sect. 3.1B).

A prominent data mining strategy for constructing prediction algorithms is the decision tree methodology. A population is split into splinter groups, which form an inverted tree with root, internal, and leaf nodes. When the sample size is sufficiently high, research data may be separated into training and validation datasets [14].

B. Logistic Regression

Fig. 5 a, b Sample implementation of fertilizer recommendation

Logistic regression is a classification algorithm. It predicts a binary result using a collection of independent factors. A binary outcome is one in which there are only two possible outcomes, i.e., the event occurs (1) or it does not occur (0). Independent variables are those variables or factors that have the potential to impact the outcome (or dependent variable). So, when working with binary data, logistic regression is the ideal form of analysis to apply. When the outcome or explanatory variable is binary or categorical in nature, i.e., if it fits into one of two categories (such as "yes" or "no," "pass" or "fail", etc.), it is binary (Fig. 7).

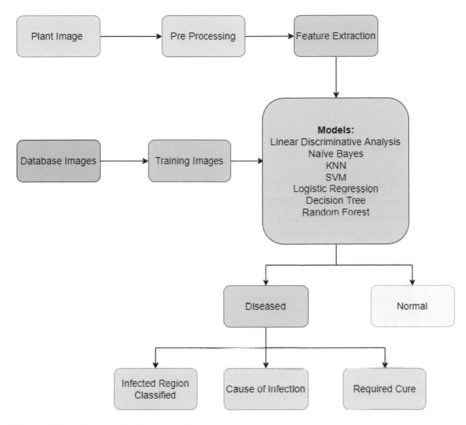

Fig. 6 Block diagram for disease prediction

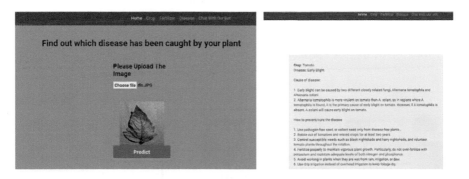

Fig. 7 a, b Sample implementation of disease prediction

Similarly, linear discriminant analysis, naive Bayes (Sect. 3.1C), K-nearest neighbors (Sect. 3.1A), random forest, and support vector machine (Sect. 3.1B) are used to determine which model is most suited for constructing disease prediction-based image processing. Moreover, a comparison (Sect. 4B) of the models is performed in order to pick one for the final development.

3.4 Chatbot

The most important task of a Chatbot [26] is to extract relevant components by analyzing and determining the objective of the user's request. Following the completion of the analysis, the user receives an appropriate answer. Chatbots employ three techniques of categorization, matching patterns, NLU, and NLP (Fig. 8). Chatbots driven by artificial intelligence boost organizational operational efficiency and save expenses while also delivering convenience to customers. FAQs can be automated, decreasing the need for human intervention. Bots are a cleverer way of ensuring that clients receive the quick answer they require without having to wait in line [21]. Gaussian NB (Sect. 3.1C) and sequential models are used to accomplish these features and functions.

Sequential Classifier
Keras is the most often used deep learning framework. A model is created by combining multiple layers. The characteristics are sent into the dense layer, which is a completely linked layer. There are two dense layers followed by the sigmoid layer [22]. Taking into account two datasets (Sect. 4A) unique rows are obtained for query, intent datasets, and preprocess the dataset to produce the intent corpus. The intents are then vectorized and encoded, and the dictionary mapping labels are returned as integer values. Tags, patterns, and responses are used to generate appropriate JSON.

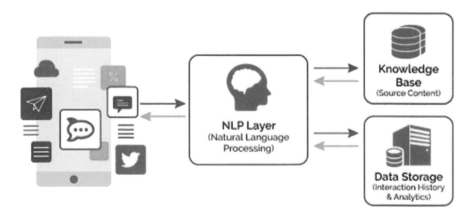

Fig. 8 Workflow of Chatbot

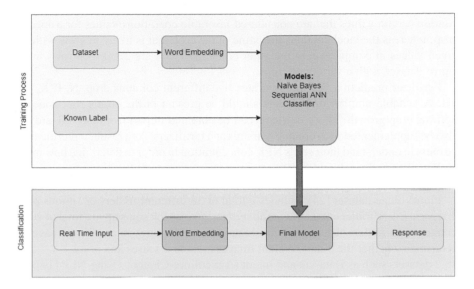

Fig. 9 Block diagram for Chatbot

Now, ANN is set up with one input layer, two hidden layers, and one output layer. The ANN model is compiled and fit to the training set. Similar techniques are performed on the data-tag dataset, and trained entity models are prepared to extract entities. The intent model is saved and used to forecast user inquiries by extracting things and assigning them to intents based on query phrases. Finally, the response from JSON is retrieved and shown on the UI. In addition, a voice chat function is incorporated alongside a conventional Chatbot and operates on the same paradigm as text chat. Therefore, the text is collected from speech and delivered to our stored intent model (Fig. 9).

4 Experimental Analysis and Results

This study presents a brief description of the observations carried out to evaluate the developed methodology.

4.1 Dataset Collection

For the proposed modules, a total of four datasets are used. Crop recommendation dataset [13] comprises of eight different columns of data, i.e., N, P, K, temperature, humidity, pH, rainfall, and label. N, P, K are considered to be the most important nutrients for any crop. Other columns, such as temperature, humidity, pH, and rainfall, contain certain values that are considered favorable conditions/values for a specific crop, whereas the label contains the name of the crop that is likely to grow with the given values in comparison to the other columns. There are 2200 such rows in the entire dataset, with a variety of crops.

Fertilizer prediction [14] incorporates five different columns crop, N, P, K, and pH. A suitable number of nutrients should be present in the crop's root zone for optimal crop growth. Those nutrients can be obtained in part from the soil and can also be supplemented with organic manures and fertilizers. As a result, it is critical for farmers to understand their soil's NPK concentration in order to determine how much N, P, and/or K they should supply using organic or mineral fertilizers to maximize crop growth, production, and yield.

PlantVillage dataset [24] includes a total of 38 different folders of various plant diseases. Every folder contains a number of images, which sums up to a total of 20.6k images in all folders. This assists machine vision technology in acquiring images to determine whether the obtained plant images include diseases and pests.

2 datasets such as Data-tag and Intent (Agriculture-Chatbot-Using-NLP [19]) are chosen for implementing the Chatbot model. The intent dataset has two columns namely, Query and Intent, which include a number of preconfigured sample queries and their related intents. For instance, Hello!, Greeting. In this example, 'Hello!' refers to a query while 'Greeting' represents the intention (Intent). The data-tag dataset consists of words and tags as columns. Similar to the intent dataset, it has numerous words with their respective tags. These two datasets are then combined using feature extraction into a simple JSON file with three fields: tag/intent, pattern, and response.

4.2 Performance Analysis

This sub-section evaluates the performance of the suggested research methodology. Here, the proposed research method for the crop recommendation system (Table 1) is compared to existing research methodologies, namely, logistic regression, K-nearest neighbors classifier, Gaussian NB, support vector machine (SVM), and decision tree classifier.

Table 1 Performance comparison results for crop recommendation model

Model	Train accuracy (%)	Val accuracy (%)
Logistic regression	97.5	96.4
KNN classifier	98.4	97.7
Gaussian NB	99.5	99.3
SVM	98.8	97.7
Decision tree classifier	95.4	98.9

Table 2 Performance comparison results for disease prediction model

Model	Accuracy (%)
Linear discriminant analysis	90.3
Naive Bayes	86.1
K nearest neighbors classifier	92.2
SVM	77.4
Random forest classifier	95.9
Decision tree classifier/CART	92.3
Logistic regression	92.03

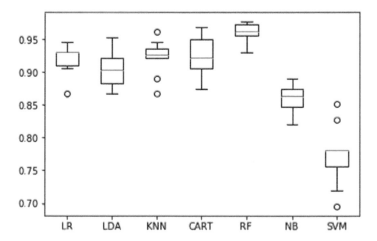

Fig. 10 Boxplot algorithm comparison

For disease prediction, a total of seven models have been implemented, and their performances are compared against their respective accuracies (Table 2). In addition, a boxplot algorithm comparison (Fig. 10) is performed for a more in-depth examination.

For the Chatbot model, two algorithms have been implemented, GuassianNB (Sect. 3.1), and sequential ANN algorithm. Their performances have been measured by their accuracies, which are 93.5% and 95.8%, respectively.

4.3 Results

All four features are implemented using machine learning techniques and then incorporated into a website with a customizable user experience. When the project is launched, it first displays a landing page (Fig. 11) from which it can be navigated to any of the four features based on the requirements.

Fig. 11 Home page of FarmEasy

The models created (Table 1) are performed to obtain a crop recommendation system. This is accomplished by developing a user interface (UI). The UI (Fig. 3a) is implemented as a web portal. The model trained using the training dataset, is predicted against user inputs. The source code will communicate to any test scenario that suggests a crop (Fig. 3b). If the test case fails to match any of the predictions, a no-match result is generated.

The designed module proposes appropriate fertilizer to be used for user supplied NPK levels (Fig. 5a), which are checked against their base values using the dictionary technique. The source code will communicate to any test scenario that suggests a fertilizer (Fig. 5b). A no-match result is issued if the test case fails to match any of the predictions.

Third feature uses image processing and classification with a random forest classifier to determine whether a plant has been infected or not, depending on the image uploaded by the user (Fig. 7a). After processing the image, it predicts the kind of disease, the source of the infection, and its prevention or cure (Fig. 7b).

Furthermore, fourth feature accepts user queries and responds appropriately using a sequential classifier and the intent corpus. It also enables voice chat typing in addition to this (Fig. 12).

5 Conclusion

The developed project, FarmEasy is a one-stop platform for farmers in rural India, to provide an interface that delivers all crop production data and fertilizer suggestions on a regular basis, as well as to alert farmers about crop output in order to meet future demands and to aid farmers with their issues via Chatbot service. Chatbot with voice chat feature which facilitates the farmers with seamless live communication comes very handy. The main novelty of this work is integrating farmers with technology from the phase of sowing seeds to getting the harvest by maximizing their yield and improving the quality of the crop. FarmEasy has been developed keeping in mind all the complex constraints.

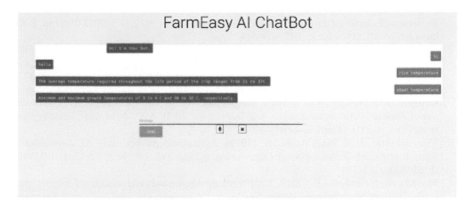

Fig. 12 Sample display of ChatBot

Future study will focus on periodically updating datasets in order to generate reliable predictions, and the processes can be automated. Another feature that will be implemented is the ability to deliver the appropriate fertilizer for the crop and region. To put this into action, a thorough investigation of available fertilizers and their effects with soil and climate is required. It is necessary to conduct a statistical data analysis.

Acknowledgements This project would not have been feasible without the help and involvement of many individuals. We'd like to appreciate everyone who contributed to the project's implementation, both directly and indirectly. First and foremost, we would like to express our gratitude to Dr. Remya S., Assistant Professor, Department of CSE, Amrita School of Engineering, Amritapuri, for her great guidance, encouragement, and timely assistance during the entire Project Phase. We appreciate our project coordinators/panel members, Assistant Professor Deepthi L. R., Assistant Professor, and Assistant Professor Sandhya Harikumar, for their active engagement and supervision.

References

1. Chand, R., Saxena, R., & Rana, S. (2015). Estimates and analysis of farm income in India, 1983–84 to 2011–12. *Economic and Political Weekly, 50*(22):139–145. http://www.jstor.org/stable/24482496
2. Deshpande, T. (2017). State of agriculture in India. *PRS Legislative Research, 53*(8), 6–7.
3. Chadha, G. K. (2003). Indian agriculture in the new millennium: Human response to technology challenges. *Indian Journal of Agricultural Economics, 58*(1), 1–31.
4. Mishra, S., & Narasimha Reddy, D. (2011). Persistence of crisis in Indian agriculture: Need for technological and institutional alternatives. In *India development report* (pp. 48–58).
5. Ben Ayed, R., & Hanana, M. (2021). Artificial intelligence to improve the food and agriculture sector. *Journal of Food Quality, 2021.*
6. Subramanian, K. S. (2020). Design and implementation of fertilizer recommendation system for farmers.
7. Niranjan, M., Saipreethy, M. S., & Kumar, T. G. (2012). An intelligent question answering conversational agent using Naïve Bayesian classifier. In *IEEE 2012 International Conference on*

Technology Enhanced Education (ICTEE), Amritapuri, India, 2012.01.3–2012.01.5 (pp. 1–5). https://doi.org/10.1109/ictee.2012.6208614

8. Ahmed, U., Lin, J. C.-W., Srivastava, G., & Djenouri, Y. (2021). A nutrient recommendation system for soil fertilization based on evolutionary computation. *Computers and Electronics in Agriculture, 189*, 106407. ISSN 0168-1699. https://doi.org/10.1016/j.compag.2021.106407

9. Pai, M. L., Suchithra, M. S., & Dhanya, M. (2020). Analysis of soil parameters for proper fertilizer recommendation to increase the productivity of paddy field cultivation. *International Journal of Advanced Science and Technology, 29*(3), 4681–4696. http://sersc.org/journals/index.php/IJAST/article/view/5683

10. Vijayalakshmi, J., & PandiMeena, K. (2019). Agriculture talk bot using AI. *International Journal of Recent Technology and Engineering, 8*(2S5), 186–190. https://doi.org/10.35940/ijrte.b1037.0782s519

11. Manjula, A., & Narsimha, G. (2019). Crop recommendation and yield prediction for agriculture using data mining techniques. *Journal of Emerging Technologies and Innovative Research, 6*(3), 4. www.jetir.org, http://www.jetir.org/

12. Yashaswini, D. K., Hemalatha, R., & Niveditha, G. (2019). Smart chatbot for agriculture. *International Journal of Engineering Science and Computing, 9*(5).

13. Ingle, A., Patel, Y., & Patel, R. (2021, May). Crop Recommendation Dataset [Version 1]. https://www.kaggle.com/datasets/atharvaingle/crop-recommendation-dataset

14. Kerneler, K., Bagga, N., & Arudhra, V. (2022). Fertilizer Prediction [Version 1]. https://www.kaggle.com/datasets/gdabhishek/fertilizer-prediction

15. Suchithra, M. S., & Pai, M. L. (2019). Improving the prediction accuracy of soil nutrient classification by optimizing extreme learning machine parameters. *Information Processing in Agriculture.*

16. Peterson, L. (2009). K-nearest neighbor. *Scholarpedia, 4*(2), 1883. https://doi.org/10.4249/scholarpedia.1883

17. Reddy, D. A., Dadore, B., & Watekar, A. (2019). Crop recommendation system to maximize crop yield in Ramtek region using machine learning. *International Journal of Scientific Research in Science and Technology*, 485–489. https://doi.org/10.32628/IJSRST196172

18. Hochmuth, G., & Hanlon, E. UF/IFAS Southwest Florida Research and Education Center, UF/IFAS Extension, Gainesville, FL 32611.

19. umangpatel00. (2020, November), Agriculture-Chatbot-Using-NLP, MIT Licensed. https://github.com/umangpatel00/Agriculture-Chatbot-Using-NLP/tree/master/datasets

20. Song, Y. Y., & Lu, Y. (2015). Decision tree methods: Applications for classification and prediction. *Shanghai Archives of Psychiatry, 27*(2), 130–135. https://doi.org/10.11919/j.issn.1002-0829.215044

21. Tamrakar, R., & Wani, N. (2021). Design and development of CHATBOT: A review

22. Dharani, M., Jyostna, J. V. S. L., Sucharitha, E., Likitha, R., & Manne, S. (2020). Interactive transport enquiry with AI chatbot. In *2020 4th International Conference on Intelligent Computing and Control Systems (ICICCS)* (pp. 1271–1276).

23. Kumar, M., Tomar, M., Singh, S., Chandran, D., Punia, S., Dhumal, S., et al. (2022). *Plant-based proteins and their multifaceted industrial applications.* Elsevier.

24. Siyah, B., Emmanuel, T. O., & Medicherla, S. (2019). Plant Village Dataset [Version 1]. https://www.kaggle.com/datasets/emmarex/plantdisease

25. Mahendra, N., Vishwakarma, D., Nischitha, K., Ashwini, & Manjuraju, M. R. (2020). Crop prediction using machine learning approaches. *International Journal of Engineering Research and technology, 9*(08), 1–4. https://doi.org/10.17577/ijertv9is080029

26. Samyuktha, M., & Supriya, M. (2020). Automation of admission enquiry process through chatbot—A feedback-enabled learning system. In V. Bindhu, J. Chen, & J. Tavares (Eds.), *International Conference on Communication, Computing and Electronics Systems. Lecture Notes in Electrical Engineering* (Vol. 637). Springer. https://doi.org/10.1007/978-981-15-2612-118

27. Vasudevan, A., Nair, N. S., Benny, A. S., Shabana, K. M., Shenoy, A., & Dutta, M. (2011). KARSHIK: Agricultural information monitoring and reference based on wireless networks.

In *ACWR 2011 Proceedings of the International Conference on Wireless Technologies for Humanitarian Relief* (pp. 537–539), Amritapuri.

28. Rekha, P., Venkat Rangan, P., Ramesh, M. V., & Nibi, K. V. (2017). High yield groundnut agronomy: An IoT based precision farming framework. In *2017 IEEE Global Humanitarian Technology Conference (GHTC)*.

29. Sivaganesan, D. (2021). Performance estimation of sustainable smart farming with blockchain technology. *IRO Journal on Sustainable Wireless Systems, 3*(2), 97–106.

30. Chen, J. I.-Z., & Yeh, L.-T. (2020). Greenhouse protection against frost conditions in smart farming using IoT enabled artificial neural networks. *Journal of Electronics, 2*(04), 228–232.

Developing a Mood Meter to Predict Well-Being Through Machine Learning

Mehar Gupta and Urvi Latnekar

Abstract This study reveals how the everyday routine affects human emotions. The proposed study has developed an application to predict happiness by using a linear regression and neural network model. The primary goal of this research work is to discover the variables that causally impact human mood/emotion to take appropriate actions. In this perspective, the proposed application has developed a meta app with an ability to extract data from gadgets to predict the emotion. The authors experiment various visualization approaches to handle the data more precisely and quickly. The proposed app strive for transparency of information via p-values and prediction intervals and established the app as an open-source project. The data utilized in this research are derived from low-cost consumer devices and passive services, which requires only minimal cost and effort.

Keywords Machine learning · Visualization · Mood · Mobile · Quantified-self

1 Introduction

It is known that our environment and actions substantially affect our mood, health, intellectual, and athletic performance. However, there is less certainty about how much our environment (e.g., weather, air quality, noise) or behavior (e.g., nutrition, exercise, meditation, sleep) influences our happiness, productivity, sports performance, or allergies. Sometimes, it is surprising that people are less motivated, their athletic performance is poor, or disease symptoms are more severe.

This paper focuses on daily mood. Although negative moods have essential regulating functions like signaling the need for help or avoiding harmful behavior like going on buying sprees, taking risks, or making foolish investments [1], other studies

M. Gupta (✉)
Thapar Institute of Engineering and Technology Patiala, Punjab, India
e-mail: mehargupta98@gmail.com

U. Latnekar
Bennett University Greater Noida, Uttar Pradesh, India

© The Author(s), under exclusive license to Springer Nature Singapore Pte Ltd. 2023   825
S. Shakya et al. (eds.), *Sentiment Analysis and Deep Learning*,
Advances in Intelligent Systems and Computing 1432,
https://doi.org/10.1007/978-981-19-5443-6_62

show that bad moods can also have unfavorable consequences like less resistance to temptations, especially to unhealthy food [2], impaired learning capabilities [3], and inhibited creative thinking [4].

Our ultimate goal is to know which variables causally affect our mood to take beneficial actions. However, causal inference is generally a complex topic and not within the scope of this paper. Hence, the authors started with a system that computes how past behavioral and environmental data (e.g., weather, exercise, sleep, and screentime) correlate with mood and then use these features to predict the daily mood via multiple linear regression and a neural network. The system explains its predictions by visualizing its reasoning in two different ways. Version A is based on a regression triangle drawn onto a scatter plot, and version B is an abstraction of the former, where the slope, height, and width of the regression triangle are represented in a bar chart. The authors created a small A/B study to test which visualization method enables participants to interpret data faster and more accurately.

The data used in this paper come from inexpensive consumer devices and services which are passive and thus require minimal cost and effort to use. The only manually tracked variable is the average mood at the end of each day, which was tracked via the app.

2   Related Work

This section provides an overview of relevant work, focusing on mood prediction Sect. 2.1 and related mobile applications with tracking, correlation, or prediction capabilities. Section 2.3.

2.1   Prediction of Mood

In the last decade, affective computing explored predicting mood, well-being, happiness, and emotion from sensor data gathered through various sources.

V. Suma's work [5] on computer vision presents the high potential that this field holds in applications relevant to AI, deep learning, and neural networks.

Harper and Southern [6] investigate how a unimodal heartbeat time series, measured with a professional EGC device, can predict emotional valence when the participant is seated.

Choudhury et al. detect major depressive disorder of Twitter users who posted more than 4500 tweets on average with an average accuracy of ' 70% [7].

Several studies estimated mood, stress, and health with data from multimodal wearable sensors, a smartphone app, and daily manually reported behaviors such as academic activities and exercise, claiming maximum accuracy of 68.48% [8], 74.3% [9], 82.52% [10], all with a baseline of 53.94. Another study scored 78.7%, with a baseline of 50.4% [11].

2.2 Limitations

All the studies mentioned above are less practical for non-professional users committed to long-term everyday usage because expensive professional equipment, time-consuming manual reporting of activity duration, or frequent social media behavior is needed. Therefore, the authors focus on cheap and passive data sources, requiring minimal attention in everyday life.

One study meeting these criteria shows that mood can be predicted from passive data, specifically, keyboard and application data of mobile phones with a maximum accuracy of 66.59% (62.65% if without text) [12]. However, this project simplifies mood prediction to a classification problem with only three classes. Furthermore, compared to a high baseline of more than 43% (due to class imbalance), the prediction accuracy of about 66% is relatively low.

2.3 Related Apps

Several apps allow users to track their moods but lack correlation and prediction features [13–20]. Some health apps allow correlating symptoms with food and behavior but still do not allow for prediction [21, 22].

Apps capable of prediction are, and [23] which estimates the activity of the sympathetic nervous system from heart rate and heart rate variability and [24] which calculates stress, energy, and productivity levels from heart data as well [24]. Further, FitBit allows for logging how the user feels and computes a 'stress management' score taking the manually logged feeling, data about sleep, electro-dermal activity, and exercise into account [25].

2.4 Limitations

While these apps are capable of prediction, they are specialized in a few data types, which exclude mood, happiness, or well-being.

The product description of the smartwatch app 'Happimeter' states to 'get your body signals to predict your mood with machine learning' [26]. The authors could not test the app as it requires the operating system wearOS, and the app has a user rating of only 1.5 of 5 stars on Google Play and was not updated for more than a year [26].

3 Data Sources

This project aims to use non-intrusive, inexpensive sensors, and services that are robust and easy to use for a few years. Meeting these criteria,the authors tracked one person with a FitBit sense smartwatch, indoor and outdoor weather stations, screentime logger, external variables like moon illumination, season, day of the week, manual tracking of mood, and more.

4 Data Exploration and Processing

This section describes how the data processing pipeline aggregates raw data, imputes missing data points, and exploits the past of the time series. Finally, the authors explore conspicuous patterns of some features.

4.1 Pre-processing

The sampling rates of the raw data typically vary between five minutes (e.g., heart rate) to about weekly (e.g., Bodyweight and VO_2Max).

Data Aggregation The goal is to have a sampling rate of one sample per day. In most cases, the sampling rate is greater than $1/24h$, and the authors aggregate the data to daily intervals by taking the sum, fifth percentile, 95th percentile, and median. We use these percentiles instead of the minimum and maximum because they are less noisy and found them more predictive.

Data Imputation The sampling frequency of bodyweight and VO_2Max is usually $< 1/24h$. Because bodyweight and VO_2Max represent physical entities that change relatively slowly, the authors assume a linear change, allowing linear interpolation of consecutive measurements to obtain the 24 h frequency. If there are days or features where many values are missing, the authors drop these days or features, respectively. Otherwise, data imputation fills missing values with the feature's average.

Time series As the dataset is a time series, and yesterday's features could also affect today's mood, the authors added all of yesterday's features to the set of today's predictors. We also include the mood of the last days until there is no new significant information about autocorrelation, given the mood of the previous days. As shown in Fig. 1, computing the partial autocorrelation [27] determines these days, when including all days from the left until the first insignificant day. In our case, this means the values of one to four days ago.

Standardization Standardization rescales the features to have a mean of 0 and unit variance.

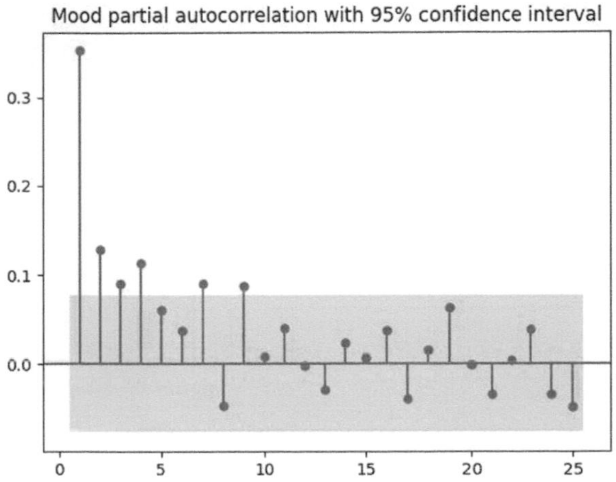

Fig. 1 Partial autocorrelation of mood. Values outside of the blue are within the 95% confidence interval, thus statistically significant

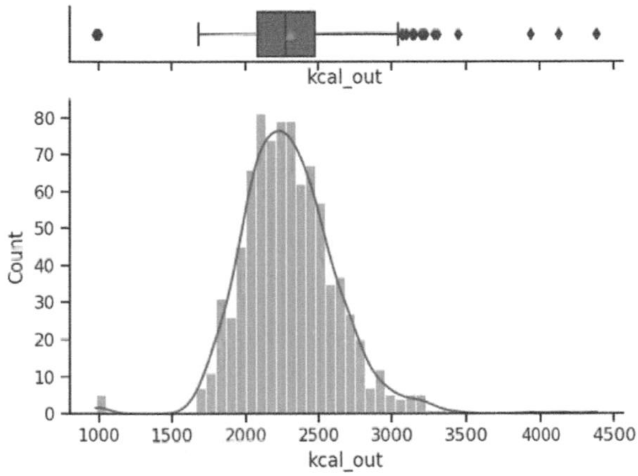

Mood partial autocorrelation with 95% confidence interval

Fig. 2 Distribution of the estimated metabolic energy output. Values below 1600 kcal kcal and above 3500 kcal kcal are outliers

4.2 Data Exploration

The dataset has many outliers because the sensors and services are cheap consumer devices. For example, the estimated metabolic energy output, shown in Fig. 2, has values at about 1000 kcal kcal and above 4000 kcal kcal.

Moreover, Fig. 3 shows a suspicious CO_2 spike at 5000 ppm. A closer look into the raw sensor data depicted in Fig. 4 indicates an improbable plateau at 5000 ppm. The causal explanation is an ending sensor range at 5000 ppm, which falsely counts all values greater than 5000 to 5000 ppm.

The distribution of the wakeup time looks Gaussian except for one suspicious spike at 320 min after midnight. However, an alarm clock at 5:20 am indicates the plausibility of this spike (Fig. 5).

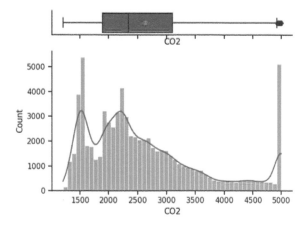

Fig. 3 Distribution of CO_2 data. Note the frequent occurrence of 5000ppm as it is the sensor's maximum range

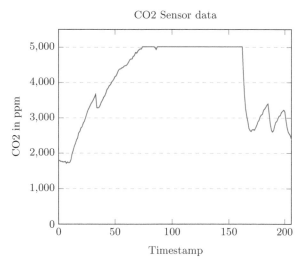

Fig. 4 Actual CO_2 level is higher than 5000 ppm, but the sensor's maximum is 5000 ppm

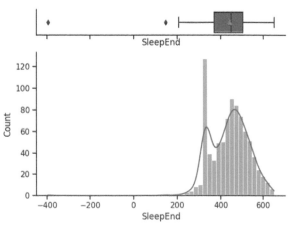

Fig. 5 Wakeup time in minutes after midnight. Note the spike at 320, which is caused by the alarm clock

Improbable values in the dataset are not corrected manually because the authors do not have access to data in the actual mobile application due to our strict privacy policy. Instead, the authors exploit robust statistics by aggregating the data via the fifth and 95th percentile instead of the maxima. Our experiments have shown that these percentiles are more predictive than the maxima.

4.3 Proposed App

The authors presented a meta app that can import data from consumer devices and services and allows for manual tracking. It allows the user to explore data via plotted time-lines and the relationship between variables through scatter plots and correlation coefficients. The app computes the Pearson correlation coefficient and p-values between all attributes. Because comparing each attribute with every other, the authors correct the p-values according to the Benjamini-Hochberg procedure [28] to control the false discovery rate due to multiple testing. We declare a result as significant for $p < 0.05$. Users can visually explore the data via a plotted time series with a seven-day moving average and manually inspect the relationship between two variables through scatter plots. The following sections eludidate the various approaches used to develop the app.

4.4 Train-Test-Split

Because there is a temporal dependency between observations, standard cross-validation, which assigns samples randomly to the train or test set, would lead to using some data from the future to forecast the past, which is not possible in the real-life application. We use time-series-splits to avoid this fallacy. However, many splits are inefficient w.r.t. data use because only the last split uses all training data. Ultimately, the authors create only one split, which is a simple train test split, where the training set contains old data points, and the test splits the most recent ones.

4.5 Experiments

Multiple Linear Regression We have used the training dataset while applying elastic net regularization and sample weighting to estimate the parameters for multiple linear regression.

Regularization We use a combined L_1 and L_2 weight penalty called elastic net regularization [29] with

Fig. 6 Exponential sample
weight decay discounting
older data as they might be
less valuable to predict
today's mood

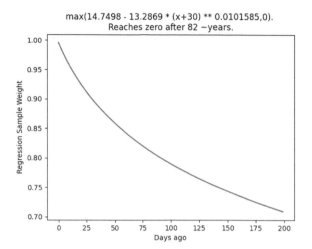

max(14.7498 - 13.2869 * (x+30) ** 0.0101585,0).
Reaches zero after 82 ~years.

$$\min_{w} \frac{1}{2n_{\text{samples}}} \|Xw - y\|_2^2 + \alpha\rho\|w\|_1 + \frac{\alpha(1 - \rho)}{2}\|w\|_2^2 \tag{1}$$

being the objective function to minimize, where ρ is the L_1 ratio, α the strength
of the penalty terms, w the weights, X the features, and y the predictions. Specific
cases are for $\rho = 0$, which simplifies to ridge regularization and $\rho = 1$ to Lasso
regularization. The authors search for the optimal ρ and α via cross-validation.

Sample weighting We assume that the factors influencing a person's mood change
over time. To account for it, the authors use exponential sample weight decay, as
shown in Fig. 6. The formula is

$$sample\_weight_i = max(14.7498 - 13.2869(i + 30)^{0.0101585}, 0), \tag{2}$$

which is exponential decay fitted to a value of 1 for today's datapoint and reaches
zero after about 82 years. $max(x, 0)$ ensures that the sample weight never becomes
a negative value.

Neural Network The neural network has two fully connected hidden layers, each
with a leaky rectified linear unit [30] as an activation function. The first and second
hidden layers have 16 and 8 neurons, respectively. Because the authors want to
regress on a scalar, the output layer has one unit. AdamW [31] is the optimizer,
minimizing the mean squared error for 4125 epochs with a learning rate of 10^{-4}
and a weight decay of one. The number of epochs is determined by early stopping
using cross-validation [32]. We searched for the best neural network architecture and
hyperparameters manually through cross-validation.

5 Results

It is observed that of 198 variables, 77 correlate significantly with mood.

5.1 Multiple Linear Regression

Elastic net regularization with a penalty strength $\alpha = 0.12$ and L_1-ratio $\rho = 1$ leads to the best prediction performance on our dataset.

Table 1 shows all regression weights w for $w \neq 0$. Note that Lasso regularization selected only nine features to predict the mood. The 95% prediction interval is ± 2.3 on the original scale from 1 to 9 and ± 1.2 on the standardized scale with unit variance. The average mean squared error on the standardized test set is 0.45, meaning 55% of the original variance can be explained. The effect of sample weighting is negligible.

5.2 Neural Network

The average mean squared error is 0.50 on the standardized test set, meaning half of the original variance can be explained.

5.3 Explainability, Visualization, and A/B Test

Screenshot (a) of Fig. 7 shows how the app visualizes each feature via time series to allow the user to spot changes over time and trends through a seven-day moving average. Screenshot (b) of Fig. 7 shows an example of a scatter plot enabling explo-

Table 1 Regression weights. All other features have a weight of zero

Feature	Regression weight
HumidInMax()	− 0.116
HeartPoints	0.101
CO_2 Median()	− 0.092
MoodYesterday	0.065
NoiseMax()Yesterday	− 0.053
PressOutMin()Yesterday	0.050
BodyWeight	0.046
VitaminDSup	0.032
DistractingScreentime	− 0.009

 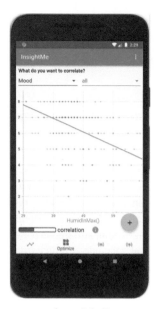

(a) Users can explore data via time-series (b) A scatter plot with linear regression
with a seven-day moving average trend- and Pearson correlation coefficient to vi-
line.   sualize how variables relate to each other.

Fig. 7 Screenshots showing how users can explore data

ration of how variables visually relate to each other. In addition, it draws a linear
regression line and indicates the degree of the linear relationship by visualizing the
correlation coefficient in a bar.

Figure 8 shows a black box with whiskers on a scale, which is the mood estimate
with the 95% prediction interval. This provides the users not only the prediction but
also how much they can rely on the accuracy. Above the prediction in screenshot8, the
authors explain to the user how multiple linear regression calculates the predictions.
Each row represents the contribution of the selected features. The row contains a
red or green bar if the contribution is negative or positive, respectively, and the size
indicates the magnitude of the contribution. The final mood prediction is simply the
sum of all contributions (Fig. 9).

To understand more about the contribution of a feature, the user can tap on one
and see either a bar chart or a regression triangle drawn onto the scatter plot.

The bar chart shows that the contribution (green or red bar) of a feature is the
product of the weight of the feature (deep teal) and the difference from today's value
to its average value (light teal).

The triangle drawn onto the scatter plot shows the same information but with more
context. The triangle's horizontal line represents the difference between the average
and today's value of the feature. The triangle's slope represents the feature's weight,

Fig. 8 Mood prediction: Each row represents the contribution of a feature that was selected by L_1 regularization. The row contains a red or green bar if the contribution is negative or positive, and the size of the bar indicates the magnitude of the contribution. The black box with whiskers is the mood estimate with its 95% prediction interval

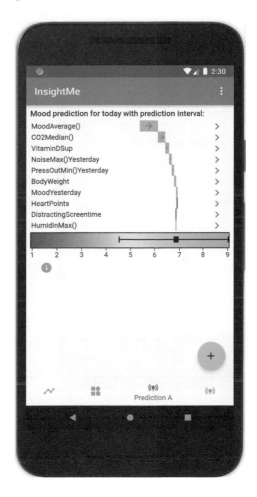

which is a regularized regression line of the two variables. Finally, the vertical length between both lines depicts the contribution.

The bar and triangle chart contain redundant information. Therefore, the authors conducted a small A/B test with 10 participants to determine which chart conveys the information more accurately and faster.

Seven of the participants are male, and three are female. The age ranges from 19 to 32, and all of them have an engineering background. The participants answered four single choice questions concerning the bar chart (A) and four similar but not identical questions about the triangle chart (B). The authors assured proper testing by passively observing them. Each participant worked on both part A and part B; however, 50% of the participants first completed part A and vice versa to control for the order. The authors measured the accuracy and time required to answer the questions of each part and reported their average w.r.t. the number of participants and number of questions. As shown in Table 2, the 'bar chart' results in a slightly

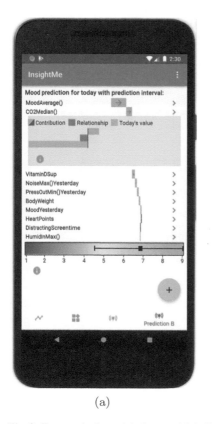 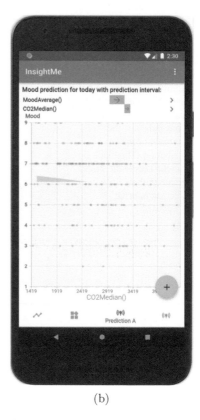

(a)   (b)

Fig. 9 Two methods explain how multiple linear regression computes the contribution of a single feature on the final prediction. The authors call the method **a** 'bar chart' because it contains a chart with three bars and **b** 'triangle chart', because it contains a red or green triangle inside a scatter plot

Table 2 Average required time and accuracy of single choice questions about the bar and triangle chart. The required time is relatively long because it includes the time to read the question

	Bar chart	Triangle chart
Average accuracy (%)	90	85
Average required time (s)	43	45

higher accuracy of 90% and 43 s a marginally faster completion, compared to the 'regression triangle chart' with an accuracy of 85% and 45 s. While these results are not significant, the users also commented that they favor the 'bar chart' as the length of the bar representing the weight is more accessible than the slope of the triangle, especially if the slope or triangle is small.

6 Discussion

Multicollinearity Although 77 variables correlate significantly, there is multicollinearity. The principal component analysis shows in Fig. 10 that only 25 components explain more than 1% of the total variance. Examples of correlated predictors are

- all the fourfold aggregated variables (i.e., the mean, median, fifth-, and 95th percentile of a variable)
- weather indoor and outdoor
- time in bed and time asleep
- walking minutes, heart points, exertion points.

Multiple linear regression versus neural network Multiple linear regression performed better on the test set than the neural network. Neural networks can have an advantage over the linear multiple linear regression method because they can approximate nonlinear relationships. Still, the downside is the need for more training data to optimize additional parameters. In our case, the training set is probably too small, leading to overfitting and difficulty generalizing to new data. The authors expect the neural network will increase its performance with a growing dataset. Besides, an enhanced architecture and improved hyperparameters could lead to better predictions of the neural network. An advantage of multiple linear regression is good explainability, as illustrated in Fig. 9, which is less intuitive for neural networks [33].

Fig. 10 The principal component analysis shows that there are only 25 components that explain more than 1% of the variance, indicating highly dependent features

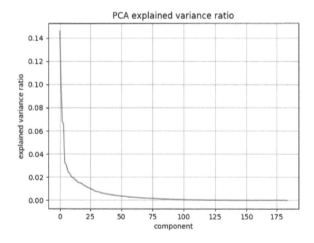

6.1   Limitations

Large unexplained variance Predictions of the multiple linear regression and neural network leave about half the original variance unexplained. Factors limiting performance are

- unmeasured variables which influence a person's mood
- sensor data can be noisy, data imputation of missing values non-optimal
- multiple linear regression assumes a linear relationship between features and mood and neglects nonlinear mechanisms
- the training set might be too small, especially for the neural network.

Assumed linear mood scale Our method of asking the users' mood on an absolute scale from 1 to 9 assumes a linear relationship of these values. However, the genuine relationship might not be linear because there may be a higher degree of change on a comparative scale than on an absolute scale for extreme values [34]. Comparative surveys would reduce these biases. However, the authors decided against them because they require more of the user's time.

Recency and fading affect bias Asking the user's average mood at the end of the day suffers from a potential recency bias [35], where recent events of the evening have a stronger effect than more distant events in the morning. Furthermore, it is prone to the fading affect bias, where negative memories fade faster than positive ones [36]. Asking for a rating multiple times per day would reduce these biases; however, the authors decided against it because it requires more effort and is less sustainable over long periods.

Survey The results of the survey are not conclusive because of the small number of participants. Furthermore, the selection of participants might not represent the actual distribution of the users, w.r.t. age, gender, and background.

Causal inference Features correlating strongly with mood and predictors of multiple linear regression could potentially *causally* affect mood. While this could be the case and interesting to know, correlation does not imply causation, and good predictors are not necessarily causes. For example, the positive correlation between mood and exercise could be caused by exercise causing better mood, or that good mood leads someone to exercise more. Furthermore, both could affect each other, or there could be a confounder like good weather, which causes someone to exercise more and improve mood independently of each other. In some instances, the authors can remove directions from a causal graph. For example, when there is a positive correlation between the variable weekend and good mood, it is unlikely that the good mood causes the day to be a weekend. Unfortunately, this study does not allow for causal inference. Today, randomized controlled trials are still the gold standard for establishing causal conclusions [37].

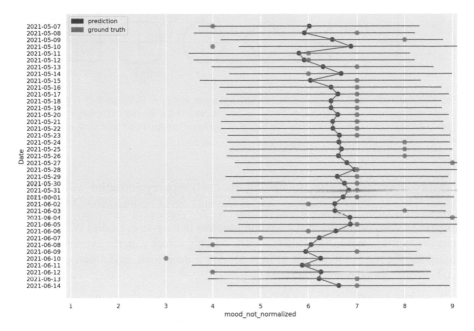

Fig. 11 Sample predictions: Predictions are in red, ground truth values in green. The blue bar represents the 95% prediction interval. It seems that the predictions go in the right direction but are too close to the mean at 6.2

6.2 Future Work

Stronger feature selection and weaker weight penalty Fig. 11 shows sample predictions. It seems that predictions tend to go in the right direction but are too close to the mean at 6.2. The authors hypothesize that the weight penalty is too strong, which pushes predictions to the mean, but feature selection is critical because the overall regularization strength $\alpha \neq 0$. Further evidence is elastic's net optimal L_1-ratio of $\rho = 1$, meaning that performance is best with the lowest allowed weight penalty while maximizing feature selection.

Fortunately, the equation of elastic net regularization, which operates between the L_1 and L_2-norm, can be generalized to the L_p-norm

$$\|x\|_p = \left(\sum_{i=1}^{n} |x_i|^p\right)^{1/p}, \tag{3}$$

with $0 < p < \infty$, allowing an even more extreme feature-selection to weight penalty ratio. Figure 12 shows how feature selection becomes stronger, and the penalty for large parameters weaker as p gets closer to zero. Future work could explore if pre-

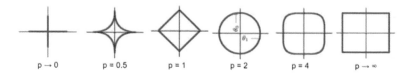

Fig. 12 Visualization of regularization with the L_p-norm. For $p \to 0$, the penalty is proportional to the number of non-zero parameters. For $p \to \infty$, the penalty has the size of the largest parameter. Elastic net regularization works only for $1 \leq p \leq 2$. Figure cropped and reprinted following the publisher's copyright [39]

diction performance improves for $p < 1$ like applying a regularization with the $L_{0.5}$-norm. However, this has the downside of leaving convex optimization [38].

Improved data imputation A common problem in long-term studies is that sensor data are missing. While the authors impute missing values with the average or linear interpolation, the authors plan to impute with a deep multimodal autoencoder to enable better mood prediction [40].

Forecasting tomorrow's mood This project explored predicting the mood of the same day, but the authors also plan to forecast tomorrow's mood. A study shows it is possible with a mean absolute error of 10.8 for workers and 17.8 for students, while the mood's standard deviation is 17.14 [41].

6.3 Conclusion

In this work, the authors presented a meta app that can import data from consumer devices and services and allows for manual tracking. The app allows the user to explore data via plotted time-lines and the relationship between variables through scatter plots and correlation coefficients.

The authors explored conspicuous patterns of certain features through data exploration and processing. They performed a comparative analysis between multiple linear regression and neural networks to find that multiple linear regression performed better. Neural networks can have an advantage over the linear multiple linear regression method because they can approximate nonlinear relationships. Still, the downside is the need for more training data to optimize additional parameters.

Furthermore, the app developed predicts the mood or any other chosen target variable by automatically aggregating the data into daily features and selecting the best ones to predict the user's mood. This project shows that multiple linear regression can explain more than half of the original variance.

We strive for transparency by conveying information about confidence through p-values and prediction intervals and created the app as an open-source project.

We hope the app helps users understand themselves better and improve their wellbeing, health, and physical and cognitive performance.

References

1. Bipolar disorder—Symptoms and causes. https://www.mayoclinic.org/diseases-conditions/bipolar-disorder/symptoms-causes/syc-20355955
2. Fedorikhin, A., & Patrick, V. M. (2010). Positive mood and resistance to temptation: The interfering influence of elevated arousal. *Journal of Consumer Research, 37*(4), 698–711.
3. Brand, S., Reimer, T., & Opwis, K. (2007). How do we learn in a negative mood? Effects of a negative mood on transfer and learning. *Learning and Instruction, 17*(1), 1–16.
4. Vosburg, S. K. (1998). The effects of positive and negative mood on divergent-thinking performance. *Creativity Research Journal, 11*(2), 165–172.
5. Suma, V. (2019). Computer vision for human-machine interaction-review. *Journal of Trends in Computer Science and Smart Technology*
6. Harper, R., & Southern, J. (2020). A Bayesian Deep Learning framework for end-to-end prediction of emotion from heartbeat. *IEEE Transactions on Affective Computing*, 1. https://doi.org/10.1109/TAFFC.2020.2981610. http://arxiv.org/abs/1902.03043. ArXiv: 1902.03043
7. Choudhury, M. D., Gamon, M., Counts, S., & Horvitz, E. (2013). *Predicting depression via social media* (p. 10).
8. Jaques, N., Taylor, S., Azaria, A., Ghandeharioun, A., Sano, A., & Picard, R. (2015). Predicting students' happiness from physiology, phone, mobility, and behavioral data. In *2015 international conference on Affective Computing and Intelligent Interaction (ACII)* (pp. 222–228). https://doi.org/10.1109/ACII.2015.7344575. ISSN: 2156-8111
9. Jaques, N., Taylor, S., Sano, A., & Picard, R. (2015). *Multi-task, Multi-Kernel Learning for estimating individual wellbeing* (p. 7).
10. Jaques, N., Taylor, S., Nosakhare, E., Sano, A., & Picard, R. (2016). *Multi-task learning for predicting health, stress, and happiness* (p. 5).
11. Taylor, S., Jaques, N., Nosakhare, E., Sano, A., & Picard, R. (2020). Personalized multitask learning for predicting tomorrow's mood, stress, and health. *IEEE Transactions on Affective Computing, 11*(2), 200–213. https://doi.org/10.1109/TAFFC.2017.2784832. Conference Name: IEEE Transactions on Affective Computing.
12. Liu, T., Liang, P. P., Muszynski, M., Ishii, R., Brent, D., Auerbach, R., Allen, N., & Morency, L. P. (2020). *Multimodal privacy-preserving mood prediction from mobile data: A preliminary study*. arXiv:2012.02359 [cs, stat]. http://arxiv.org/abs/2012.02359. ArXiv: 2012.02359
13. Daylio—Journal, Diary and Mood Tracker. https://daylio.net/
14. Mood Patterns—Mood tracker & diary with privacy. https://www.moodpatterns.info/
15. Mood Tracker Journal. Mental Health, Depression—Apps on Google Play. https://play.google.com/store/apps/details?id=diary.questions.mood.tracker
16. Moodily—Mood Tracker, Depression Support—Apps on Google Play. https://play.google.com/store/apps/details?id=moodily.rohweller
17. MoodPanda—Your supportive mood diary. https://www.moodpanda.com/
18. MoodPrism Mental health and wellbeing app. https://www.moodprismapp.com
19. AppleInc.: iOS - Health. https://www.apple.com/ios/health/
20. HelloCodeInc.: Exist. https://exist.io
21. Features. https://www.correlate.com/features/
22. Pattern—Correlate, Health Diary, Mood-Tracker—Apps on Google Play. https://play.google.com/store/apps/details?id=com.pattern.health.diary
23. EliteHRV: Best Heart Rate Variability Monitor & App. https://elitehrv.com/
24. Welltory: Welltory—Guide to a life of health and productivity. https://welltory.com/
25. FitBit: Stress Management—Stress Watch & Monitoring | Fitbit. https://www.fitbit.com/global/us/technology/stress
26. Happimeter—Apps on Google Play. https://play.google.com/store/apps/details?id=com.happimeterteam.happimeter&hl=en&gl=US
27. Dürre, A., Fried, R., & Liboschik, T. (2015). Robust estimation of (partial) autocorrelation. *WIREs Computational Statistics, 7*(3), 205–222.

28. Benjamini, Y., & Hochberg, Y. (1995). Controlling the false discovery rate: A practical and powerful approach to multiple testing. *Journal of the Royal Statistical Society: Series B (Methodological), 57*(1), 289–300.
29. Zou, H., & Hastie, T. (2005). Regularization and variable selection via the elastic net. *Journal of the Royal Statistical Society: Series B (Statistical Methodology), 67*(2), 301–320.
30. Xu, B., Wang, N., Chen, T., & Li, M. (2015). *Empirical evaluation of rectified activations in convolutional network.* arXiv:1505.00853 [cs, stat]. http://arxiv.org/abs/1505.00853. ArXiv: 1505.00853
31. Loshchilov, I., & Hutter, F. (2019). *Decoupled weight decay regularization.* arXiv:1711.05101 [cs, math]. http://arxiv.org/abs/1711.05101. ArXiv: 1711.05101
32. Prechelt, L. (1998). Automatic early stopping using cross validation: Quantifying the criteria. *Neural Networks, 11*(4), 761–767.
33. Samek, W., Wiegand, T., & Müller, K. R. (2017). *Explainable artificial intelligence: Understanding, visualizing and interpreting deep learning models.* arXiv:1708.08296 [cs, stat]. http://arxiv.org/abs/1708.08296. ArXiv: 1708.08296
34. Carlsson, A. M. (1983). Assessment of chronic pain. I. Aspects of the reliability and validity of the visual analogue scale. *Pain, 16*(1), 87–101. https://doi.org/10.1016/0304-3959(83)90088-X. http://journals.lww.com/00006396-198305000-00008
35. Cushing, B. E. (1996). *Mitigation of recency bias in audit judgment: The effect of documentation—ProQuest.* https://www.proquest.com/openview/283dcd106582467195de2fee655220f2/1?pq-origsite=gscholar&cbl=31718
36. Skowronski, J. J., Walker, W. R., Henderson, D. X., & Bond, G. D. (2014). Chapter Three—The fading affect bias: Its history, its implications, and its future. In J. M. Olson, M. P. Zanna (Eds.), *Advances in experimental social psychology* (Vol. 49, pp. 163–218). Academic Press. https://doi.org/10.1016/B978-0-12-800052-6.00003-2. https://www.sciencedirect.com/science/article/pii/B9780128000526000032
37. Hariton, E., & Locascio, J. J. (2018). Randomised controlled trials-the gold standard for effectiveness research. *BJOG : An International Journal of Obstetrics and Gynaecology, 125*(13), 1716.
38. Xu, Z., Chang, X., Xu, F., & Zhang, H. (2012). L1/2 regularization: A thresholding representation theory and a fast solver. *IEEE Transactions on Neural Networks and Learning Systems, 23*(7), 1013–1027. https://doi.org/10.1109/TNNLS.2012.2197412. Conference Name: IEEE Transactions on Neural Networks and Learning Systems.
39. Binz, K. (2019). *Intro to regularization.* https://kevinbinz.com/2019/06/09/regularization/
40. Jaques, N., Taylor, S., Sano, A., & Picard, R. (2017). Multimodal autoencoder: A deep learning approach to filling in missing sensor data and enabling better mood prediction. In *2017 seventh international conference on Affective Computing and Intelligent Interaction (ACII)* (pp. 202–208). https://doi.org/10.1109/ACII.2017.8273601. ISSN: 2156-8111.
41. Umematsu, T., Sano, A., Taylor, S., Tsujikawa, M., & Picard, R.W. (2020). Forecasting stress, mood, and health from daytime physiology in office workers and students. In *2020 42nd annual international conference of the IEEE Engineering in Medicine & Biology Society (EMBC)* (pp. 5953–5957). IEEE, Montreal, QC, Canada. https://doi.org/10.1109/EMBC44109.2020.9176706. https://ieeexplore.ieee.org/document/9176706/

3D Realistic Animation of Greek Sign Language's Fingerspelled Signs

Andreas Kener, Dimitris Kouremenos, and Klimis Ntalianis

Abstract Signing avatars make it possible for deaf people to access information in their preferred language. This paper describes the implementation of a new 3D high-realistic avatar, for Greek Sign Language (GSL) fingerspelling and numbering signs. This research study explores open-source 3D platforms such as MakeHuman, Blender, and Unity's game engine as 3D rendering architectures. Creating signing virtual characters is important for training and communicating in virtual environments or further applications. This research looks at how to use the power of C# unity script code and layered motion game technology to achieve programmable transitions of animation segments and to create dynamical, real-time (on the fly), natural, and understandable outputs with controllable playback speed and playback views. Another novelty of the proposed scheme focuses on the precision of the transition time and how it blends between the last sign posture (first sign) and first sign posture animation (second sign).

Keywords Greek Sign Language · Greek letters fingerspelling · Number gesture animation · 3D Signing avatar

1 Introduction

For a significant part of the world population such as people who are deaf or hard of hearing. Sign language is an integral tool of their interactions within social environment in daily basis [1]. For the hearing people been involved in these kinds of interactions respectively, the availability of educational software in this field is crucial for practicing in sign language. This paper looks at the issue of self-teaching sign

A. Kener (✉) · K. Ntalianis
Department of Business Administration, University of West Attica, Agiou Spyridonos Street, 12244 Egaleo, Greece
e-mail: anken73@yahoo.com

D. Kouremenos
National Centre for Scientific Research "Demokritos", Athens, Greece

© The Author(s), under exclusive license to Springer Nature Singapore Pte Ltd. 2023   843
S. Shakya et al. (eds.), *Sentiment Analysis and Deep Learning*,
Advances in Intelligent Systems and Computing 1432,
https://doi.org/10.1007/978-981-19-5443-6_63

languages, starting with the most basic aspect of learning any sign language: finger-spelling. Fingerspelling or dactylology as it is otherwise called, is a necessity for learning signing in GSL. Fingerspelling is also the initial stage in any sign language because the basic hand shapes formed in fingerspelling are the same for most signs [2]. Fingerspelling, on the other hand, is essential for understanding and facilitating this bidirectional communication form. The "manual" alphabets which are hand-shapes—letters from a spoken language's alphabet [3], sequentially form words and undeniably help in cases such as:

- Naming people and places,
- comprehending difficult terminologies,
- signing individual Terms [4],
- For words which they have no corresponding signs,
- And as a an alternative when the word's sign is unknown.

Another great importance's aspect of fingerspelling is dealing the problem that concerns native signers tend having sign names and hearing people don't. Therefore fingerspelling is needful for both participants, and it is used widely in sign languages such as ASL or BSL [5]. Deaf people all over the world use Sign Languages (SL) as their primary mode of communication. They are natural languages with their own syntax and norms that differ significantly from oral languages. An innovative and radical way of improving deaf population's access to information is the adaptation of virtual humans or avatars in sign language. Avatar technologies preserve the anonymity of signers, while at the same enable users to edit, manipulate, and produce the sign language content much easier in relation with video material. Videos suffer from lack of flexibility, and copy/paste functions cannot cope transitions in the context [6]. In Greece, GSL is enforced by law (Act 2817/2002) as the first language of deaf students, and consequently is their primary mean of education. Due to EU standards for access to special education information [7], Greek schools now have unfettered Internet connection, allowing any supportive GSL e-learning platform to be easily applied to real-life classroom routine.

This piece of work is concentrated on, presenting the key points of the GSL finger-spelling alphabet, its undocumented practices and on an animating fingerspelling mechanism using a 3D character. The procedures that are presented, for sequencing and animating all the aspects in GSL fingerspelling are a product of collaboration with native deaf users with several years of experience in GSL computing linguistics.

This document examines a system that converts Greek Unicode text to corre-sponding fingerspelled signs and numbers, including a set of 24 signs related to the modern Greek alphabet, additional non-manual signs, and micromotion move-ments (eyelids, breathing, and standing idle micro-motion). Micro-motion move-ments which are required to represent modifiers in GSL are also extremely helpful for accomplishing the physically realistic human motion of 3D avatar.

The user interface is then displayed, allowing complete control of the finger-spelledword/number to be signed (text/number-to-fingerspelling conversion). The necessary code programming which is done in C Unity Script Embedded Language in combination with Unity's strong game engine [8], enable developers to handle

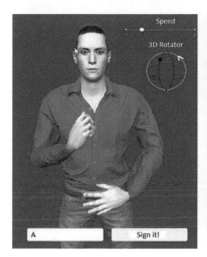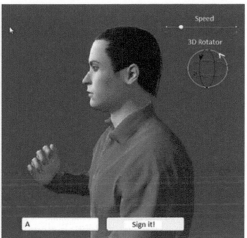

Fig. 1 Speed regulator and 3D rotator giving us a multiview aspect of alphabet sign "A"

animation better than the most open-source 3D Graphical software like Blender [9]. While Blender does support bones, IK rigging, and even motion capture for animation, Unity is powerful enough to take all of these and produce complicated animations. A very essential thing in learning and practicing fingerspelling is the speed of animation. For the amateur observing people fingerspelling at natural speed, the signs are difficult to resolved.

Consequently, they need to practice with sign language at a lower rate until they become capable of resolving words/numbers signed at a natural rate. Equally, the point of view is extremely important for him concerning the blind angles of the hand, thereby our system provides a speed regulator and a 3D view rotator to the end user (Fig. 1).

2 Greek Sign Language—Linguistic Research Background

Thousands of native and non-native signers utilize Greek Sign Language (GSL). GSL was co-shaped by the impact of French sign language and an older sort of Greek sign language, as we now know [10]. The similarity of GSL with most of sign languages, lies mainly in social and linguistic level [11]. It is used extensively and it is estimated that GSL users are about 40,600 [12]. Also, another large segment of its users, are the families of deaf people and hearing students of GSL.

Because of the recent increase in mainstreamed deaf students in education, as well as the population of deaf students spread in other institutions, local town units for the deaf, and private tuition, the total number of secondary and potential sign language users of course needs to be changed. GSL is officially utilized at 11 deaf clubs around Greece, as well as 14 deaf educational institutions at all levels [13].

3   Related Work

The deaf community expresses words that do not have a conversational sign, such as a name, using fingerspelling as a letter-by-letter technique. Sinhala Sign Language (SSL) employs a phonetic pronunciation method to decipher such words due to the presence of one or more modifiers after a consonant [14].

The numerals have a similar notation and are divided down into components before being interpreted in sign motions. Variations are used in the Sinhala project to make the 3D avatar-based interpretation system resemble a human interpreter's genuine fingerspelled SSL.

The Blender animation software is used to complete the task of creating a 3D avatar animation based on phonetic English. The Sinhala conversion system is composed by 101 SSL signs-fingerspelling letters and numbers. Any word written is possible to be transposed with the modern Sinhala alphabet including the numbers which can reach up to billions. Teaching of SSL fingerspelling to deaf children with this helpful tool became quite easy [14].

The transition from desktop to mobile technologies in recent years, exacerbated the problem that already concerned hearing people, about the difficulty of familiarization with fingerspelling-nevertheless constituting one of the hardest aspects of sign language due to the lack of practice tools.

The study "Enabling Real-Time 3D Display of Lifelike Fingerspelling in a Web App" explores a web-enabled system that enables real-time fingerspelling on a 3D character to address these challenges. Furthermore, strengthens the realistic motion and in contrast with video-based efforts provides potentials for more interaction and customizing, enhance more self-practice tools for fingerspelling reception [3].

An interesting Greek Sign Language Fingerspelling attempt occurs in work [13]. The educational platform being created as part of the SYNENNOESE project is described in this document. It's a 3D animation software platform, taking advantage of GSL electronic linguistic resources.

4   System Description

The aim of visualizing the data which have been obtained from the proof of concept in demonstration mode (builder mode), is the main concern of the GSL fingerspelling system—implemented in Unity platform. Accuracy and realism in animation of a signer avatar is the mail goal and the proposed synthesis aims to build new fingerspelling and number signs in GSL. Finally, it constructs compound sentences by:

- retrieving motion chunks, matching to isolated fingerspelling signs and numbering motion clips and
- building continuous fingerspelling or number sign motion, through addition of transitions between those chunks.

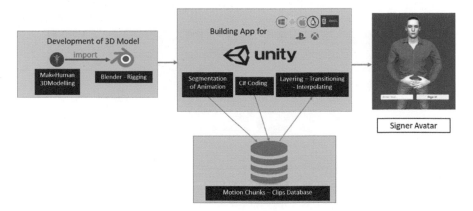

Fig. 2 Architecture presentation of the signer avatar system

The synthesis system is based on the Unity platform using C coding. We created a 3D rigged model with the MakeHuman open-source (AGPL3) tool, and the animation motion clips are produced with the Blender animation software [9]. The suggested scheme's overall architecture is described in Fig. 2.

5 Modeling the 3D Human—A Realistic Hand for Natural Animation

The 3D model of the MakeHuman open source (AGPL3) tool simplifies the creation of virtual humans, using a graphical user interface. The ultimate goal is to be able to quickly create a realistic virtual human with just a few mouse clicks and render or export it for usage in Blender. The avatar includes a full skeleton of 163 bones (Fig. 3), in order to maximize the potential for detailed rigging.

6 Animation (Blender)

Rigging the model is the procedure in Blender that creates a bone hierarchical skeleton, and is needed for animation and physics simulations. It enables you to animate a model, with even a more complex model like a human body (Fig. 4). Rigging is the process of giving a solid model articulation at specific points and in specific ways, usually to replicate a biological skeleton. The bone is assigned to the proper vertex groups and vertex weights for each group in order to have proportionate control.

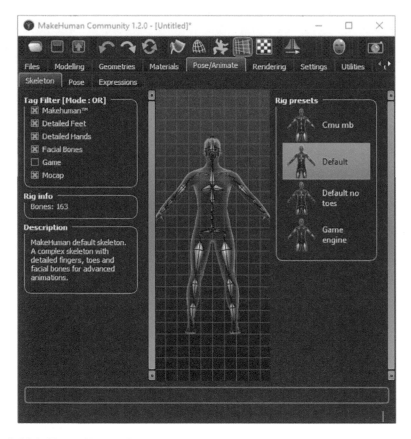

Fig. 3 MakeHuman 3D modeling

In our project, the hand gestures-signs are mainly implemented by inverse kinematics (IK) and the facial expressions by shape keys. Blender's IK algorithm streamlines the animation process, allowing for more complex animations to be created with less work. IK allows you to put the final bone in a bone chain while the others are automatically positioned.

Facial expressions are obligatory for some signs (e.g., RELIEVED, SHOCK, and BEG) where the facial expressions reflect the emotions associated with the meaning [15] and also are used as modifiers for a specific sign. The main concern of the GSL finger spelling system implemented in Unity platform is visualising the data obtained from the proof of concept in demonstration mode (builder mode). Utilizing Blender shape keys is very important to create basic facial expressions for a character, and how to create corrective shape keys to fix deformation issues (Fig. 5).

Also, natural embedded small movements (micro-motion) like eyelids, breathing, and standing idle micro-motion reinforce in the naturalness of the 3D signer [16].

The outcome of the Blender engagement was a.fbx file which included all the motions (hand gestures and facial expressions) in a single timeline.

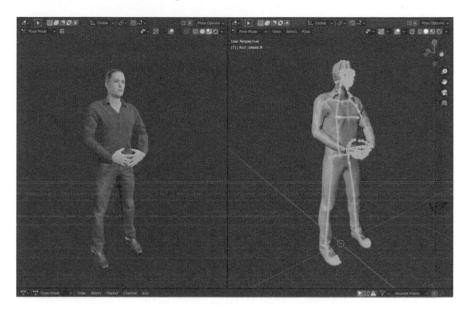

Fig. 4 Blender skeleton consisting of a bone hierarchy

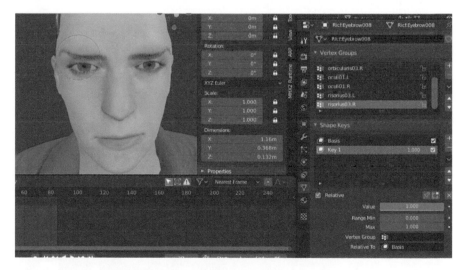

Fig. 5 Blender shape keys tool

7 Unity WebGL2

"Mecanim"—a visual environment in which the entire operation for the humanoid avatar will take place, powered by Unity's powerful state machine animation system. All models with a humanoid (human-like) rig can use the animation operation. It's simple to change the mapping from one character to another. The easy implementation of motion interpolation through scripting, offers a smoother and more realistic movement. Importing.fbx files in Unity also, is another asset for developing realistic motion playbacks.

Unity, a cross-platform engine, provides developers the ability to switch any time the platform according to their demands. The WebGL2 Unity tool (Fig. 6) provides to user, a more responsive interface that renders and animates fingerspelling on the client's device, reducing concurrently the mobile bandwidth usage. Each product of this system can be a package for Desktop PCs (with a web browser) or Android TV or Apple TV are examples of such devices.

When the blender '.fbx' file is imported to Unity, it provides a corpus of animation that has to be segmented in order to create a database of motion chunks (Fig. 7), regarding the hand gesture—sign motion, the facials, and reflected body movements.

An '. anim' file is produced, every time a new animation clip is created in Unity. It contains all the data that the Animator needs to play a clip, such as object positions, names, rotations etc. (Figs. 8 and 9).

Actually, these files are YAML's type1, and can be simply edited with any text editor. The purpose of this segmentation concerns the multi-layered synthesis of the fingerspelling perspective of the project.

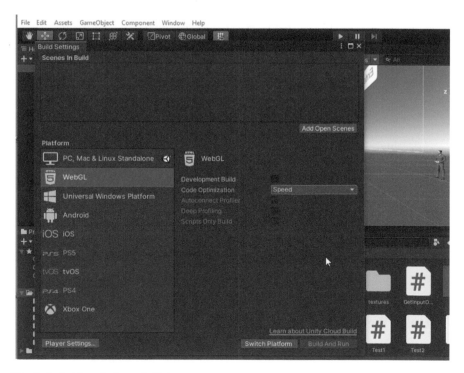

Fig. 6 Switching platforms in Unity

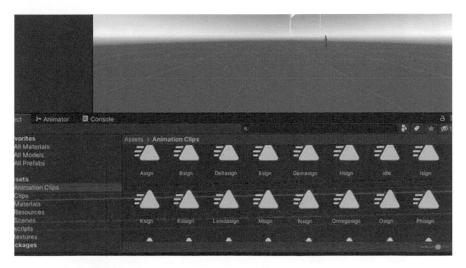

Fig. 7 Motion chunks database

Facial and non-manual sign expressions are an integral part of Greek Sign Language. The capability of facial and non-manual synchronization in the module should be tested, for better sign comprehension.

Interpolation smooths down the effect of performing patches at a constant frame rate, removing the problems of jittery viewing movements due to physics and visuals not being properly synced. At this point, it is important to note that the animation clips were segmented using the Naert et al. and Kita et al. form: Preparation-Meaning-Retraction [6, 17]. This form has to do with the Preparation—substance of the sign and reinstatement in the previous position. More specifically, when hand (or arm) prepares to move toward the starting position/configuration of the hand for the upcoming sign, is called preparation of the sign. For the same reason, before making the next sign or because there is only one letter-character to be signified, the hand/arm first retracts to follow the next sign. This stage is named retraction [6, 17] and concerns the frames before and after the meaning, which is contained in the designation of the .anim file and it constitutes the denotation that the system needs to identify. The three phases together shape the animation clip length (ACL).

In order to achieve a smooth transition from one sign-gesture animation clip to another, it's critical to calculate the exact exit time of the meaning phase as shown above in Fig. 10. In unity, from all the transition settings which can be modified through C coding or its GUI the key setting is exit time (ET), in our case. The exit time gives you the capability of adjusting the transition between animations. This value embodies the exact time of the transition taking effect and represents normalized time. In unity, for example, an exit time of 0.75 is the 75% of animation's full length. The ET' Eq. (1) we devised to deal with this problem of converting seconds subdivisions into normalized time is as follows:

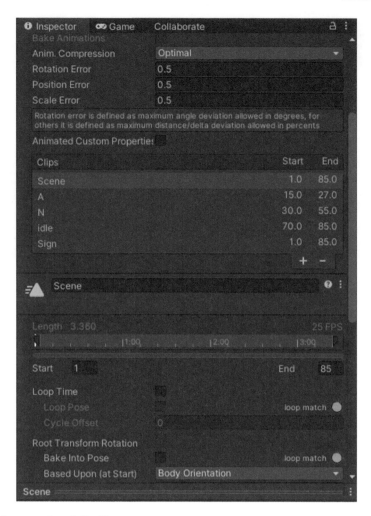

Fig. 8 Segmentation of .fbx file

$$ET = \frac{\cdot LF\,FPS}{\cdot ACL} \tag{1}$$

where FPS is the frames per second, and LF (last frame) is the desired frame we want. Given an animation e.g. that is 1.5 s long that plays at 25FPS, say we'd like to know how long it takes to get to the 27th frame, we use: $(27/25)/1.5 = 0.72$, in such wise, the Transition Duration (TD) is the rest of the normalized time: $1-0.72 = 0.28$ and can be calculated from Eq. (2).

$$TD(\sec) = (1 - ET) \cdot ACL \tag{2}$$

```
%YAML 1.1
%TAG !u! tag:unity3d.com,2011:
--- !u!74 &7400000
AnimationClip:
  m_ObjectHideFlags: 0
  m_CorrespondingSourceObject: {fileID: 0}
  m_PrefabInstance: {fileID: 0}
  m_PrefabAsset: {fileID: 0}
  m_Name: Isign
  serializedVersion: 6
  m_Legacy: 0
  m_Compressed: 0
  m_UseHighQualityCurve: 1
  m_RotationCurves:
  - curve:
      serializedVersion: 2
      m_Curve:
      - serializedVersion: 3
        time: 0
        value: {x: 0.7071068, y: 0, z: -0, w: 0.7071067}
        inSlope: {x: 0, y: 0, z: 0, w: 0}
        outSlope: {x: 0, y: 0, z: 0, w: 0}
```

Fig. 9 ". anim" file content

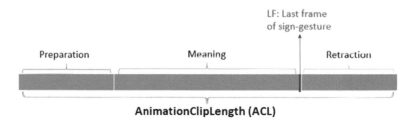

Fig. 10 Preparation-Meaning-Retraction form

For the better understanding of the reader, we illustrated a transition graph in Fig. 11.

In contrast to Naert's methodology, which computes the TD value based on the average distances (Geodesic Euclidean) between the last posture of the previous sign and the first posture of the next sign, we focus on the precision of the exit time from one sign into another because fingerspelling is very bounded spatially (Listing 1).

On the other hand, the delimitation of the animation clips into phases (Preparation-Meaning-Retraction), was held by a professional sign linguist who authored and signified meticulously the exact time limits.

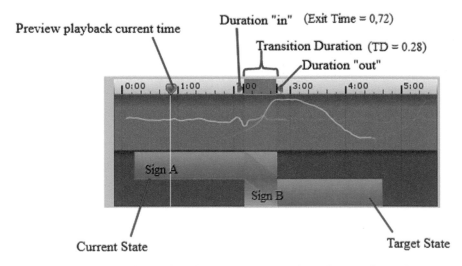

Preview playback current time

Duration "in"   (Exit Time = 0,72)

Transition Duration (TD = 0.28)

Duration "out"

Sign A

Sign B

Current State

Target State

Fig. 11 The transition settings and graph, e.g., two sequential Greek Signs ("A" + "B") as shown in the Inspector

```
foreach (ChildAnimatorState childAnimatorState in newState.
    states)
{AnimatorState state = childAnimatorState.state;
if (state.name == "idle")
{var trans = state.AddTransition(stateA1);
trans.hasExitTime = true;
trans.duration = 0.28;
trans.exitTime = 0.72;
trans.AddCondition(UnityEditor.Animations.
    AnimatorConditionMode.If, 0, "a");}}
```

Listing 1 Exit time-Transition Duration C# code

8 Implementation

The synthesis system visualizes the motion clip data that was imported from the Blender platform using an animated 3D model. The signer avatar is animated through the implemented system with outstanding accuracy and realism, in order to improve the understanding of signs. This is due to the reproduction of imported handcraft signed motion clips, focusing on synchronization of facial and hand gestures, as well as their graph topologies [16].

Greek Unicode text is converted to GSL signs using Unity C# coding (Fig. 13) (Listing 2) (Listing 3), which is fed with Blender animation clips [9]. The whole concept is based on a central script managing the connection of sign gesture clips composing the written word or a number.

Each Greek letter corresponds from the central script to another script, which synthesizes motion chunks that make up the completed sign character—utterance while at the same time enhancing them with transition settings for the next letter that

Fig. 13 Interconnected Scripts

```
.....................................................
{case "A*":
output = output + GameObject.Find("MH2").AddComponent<Script
    Name** >();
break;
case "B":
output = output +..........
.....................................................
*Greek letter
**The script that, calls the corresponding motion chunk, makes
    Layering, Transition and Interpolation
```

Listing 2 Converting letters of alphabet to animation C# code

```
sign1 = sign1 + GameObject.Find("MH2").AddComponent<Script
    Name*>();
break;
default:
isDone = true;
break;
} if (!isDone)
{ if (Number.Substring(0, pos) != "0" && Number.Substring(pos) !=
    "0")
{ try
{sign = ConvertNumber(Number.Substring(0, pos)) + sign1 +
    ConvertNumber(Number.Substring(pos)); }
catch { }
} else
{sign = ConvertNumber(Number.Substring(0, pos)) +
    ConvertNumber(Number.Substring(pos)); }
sign = " and " + sign.Trim();   }
if (sign.Trim().Equals(sign1.Trim())) sign = "";  } }
catch { }
return sign.Trim();  }
.....................................................
Script name* The script that, calls the corresponding motion chunk,
    makes Layering, Transition and Blending
```

Listing 3 Converting numeric values to animation—algorithm C\# code

```
..........................
AnimatorStateMachine sm = ac.layers[1].stateMachine;
AnimationClip newClip = AssetDatabase.LoadAssetAtPath<
    AnimationClip>("Assets/Clips/Alphalips.anim");
AnimatorState stateA = sm.AddState("stateA1");
stateA.speed = 0.6f;
stateA.motion = newClip;
ac.layers = layers;
..........................
```

Listing 4 Layering C# code

the central script is corresponding. The mathematically calculation of transitioning between postures is modeled for smooth alternations.

Typically, animations are created in Unity projects using third-party applications such as Blender. Since Blender once imported all kinds of animation into Unity, it is necessary to set up the Animation Clips. Therefore, each sign of the alphabet has to have consisted of a sign clip (the fingerspelled letter), a facial clip (the appropriate for the fingerspelled letter), and a reflected body movements clip (according to the movement of the hand that formed the fingerspelled letter), which are combined through corresponding animation layers. In order synthesize these layers, it was definitely indispensable to make-the-most-of pliability (future requirements), to leave aside the Unity GUI (shake off the Unity game logic) and start building C coding scripts that manipulate the avatar and broadly cope with the complexity of the project (Listing 4).

Another vast challenge, that the scripts answer, is the naturalness and plasticity of the transition from one fingerspelled letter to the other because our synthesis system aims to build new fingerspelling words or numbers in GSL.

Using the Unity layering technology [18], we can add reflected motion chunks and interact with different features and one another instead of the rigging methodology [19]. The reflected motion chunks add natural and high-realism sign language animation. Furthermore, the whole unity project application is developed by coding (C Unity script API) instead of using Unity's graphical user interface. The Unity script API gives the developer flexibility to handle complicated motion chunks.

9 Conclusions and Future Work

The new 3D development architecture (Unity, MakeHuman, Blender) [8, 9] which was proposed in this paper, builds high realism and quality fingerspelling animations through layered motion, using the unity game engine. Unity's layering motion technology delivers greater spontaneity in fingerspelling for practice and allows more flexibility, exploring fingerspelling such as different point of view.

This project is about (a) a new technological step for fingerspelling receptive practice tools, which is originated from the input that native signers' community kindly offered and (b) a significant technological breakthrough for the purpose of creating a fingerspelling educational platform using the WebGL2 Unity tool. The cross-platform output of this tool achieves to have a wide range of application like desktops, mobile devices, Android and Apple TV, game consoles (XBOX, Nintendo, PlayStation), etc.

With the existing designing collaboration with GSL trainers and teachers, the educational gamification platform for learning GSL fingerspelling. When the new 3D gamification fingerspelling tutor is published, it will be further tested with sign language students to see how the extra interaction affects the students' learning experience.[17, 20, 21, 22]

References

1. World Federation of the Deaf. (2021). Who are we. http://wfdeaf.org/our-work/. [Accessed 21 November 2021].
2. Flodin, M. (2004). *Signing illustrated: The complete learning guide*. Penguin.
3. Montgomery, J., McDonald, J., Gong, F., Baowidan, S., & Wolfe, R. (2020). Enabling real-time 3D display of lifelike fingerspelling in a web app. In *International conference on computers helping people with special needs* (pp. 38–44).
4. Battison, R. (1978). *Lexical borrowing in American sign language*. Linstok Press.
5. Sutton-Spence, R. (1994). *The role of the manual alphabet and fingerspelling in British sign language* [Unpublished Ph.D. dissertation University of Bristol].
6. Naert, L., Larboulette, C., & Gibet, S. (2017). Coarticulation analysis for sign language synthesis. In *International conference on universal access in human-computer interaction* (pp. 55–75).
7. Publication Office of the European Union. (2021). *Communication from the commission to the council, the European parliament, the economic and social committee and the committee of the regions—Towards a barrier free Europe for people with disabilities*. https://op.europa.eu/en/publication-detail/-/publication/d5aae2cb-c4eb-40f2-97aa-d235916556e3/language-en. [Accessed 10 September 2021].
8. Unity Technologies. Cross-platform game engine. (2021). https://unity.com. [Accessed 10 September 2021].
9. Blender, F. (2021). *The free and open source 3D creation suite*. https://www.blender.org. [Accessed 2 June 2021].
10. Lampropoulou, V., Panteliadou, S., & Gibet, S. (2000). Special education in Greece-Critical view. In *Proceedings of the special education conference* (pp. 12–14).
11. Kyle, J., Woll, B., & Ackerman, J. (1989). *Gesture to sign and speech*. Final Report. No. C 00: 23: 2327.
12. Efthimiou, E., & Fotinea, S. (2007). *An environment for deaf accessibility to educational content*. Greek national project DIANOEMA (p. 62), (GSRT, M3. 3, id 35). ICTA 2007, Hammamet, Tunisia.
13. Sapountzaki, G., Efthimiou, E., Karpouzis, K., & Kourbetis, V. (2007). Open ended resources in Greek sign language: Development of an e-learning platform. In *4th International conference on language resources and evaluation* (pp. 13–19). Lisbon, Portugal.
14. Punchimudiyanse, M., & Meegama, R. G. N. (2015). 3D signing avatar for Sinhala sign language. In *2015 IEEE 10th international conference on industrial and information systems (ICIIS)* (pp. 290–295).
15. Sutton-Spence, R., Woll, B. (1999). *The linguistics of British sign language: An introduction*. Cambridge University Press.

16. Naert, L., Larboulette, C., & Gibet, S. (2020). A survey on the animation of signing avatars: From sign representation to utterance synthesis. *Computers Graphics, 92,* 76–98.
17. Kita, S., Van Gijn, I., Van der Hulst, H., Dimitropoulos, K., Daras, P. (1998). Movement phases in signs and co-speech gestures, and their transcription by human coders. In *Gesture and sign language in human-computer interaction: International gesture workshop Bielefeld* (pp. 23–35), Germany, September 1719, 1997: Proceedings, Vol 1371. Springer, Berlin.
18. Adamo-Villani, N. (2008). 3D rendering of American sign language fingerspelling: A comparative study of two animation technique. *International Journal of Computer and Information Engineering, 2,* 2676–2681.
19. Stefanidis, K., Konstantinidis, D., Kalvourtzis, A., Dimitropoulos, K., Daras, P. (2020). 3D technologies and applications in sign language. In *Recent advances in 3D imaging, modeling, and reconstruction* (pp. 50–78). IGI Global.
20. Kouremenos, D., Ntalianis, K., & Kollias, S. (2018). A novel rule based machine translation scheme from Greek to Greek sign language: Production of different types of large corpora and language models evaluation. *Computer Speech & Language, 51,* 110–135.
21. Kouremenos, D., Ntalianis, K., Siolas, G., & Stafylopatis, A. (2018). Statistical machine translation for Greek to Greek sign language using parallel corpora produced via rule-based machine translation. In *CIMA@ICTAI* (pp. 28–42).
22. Ntalianis, K., Kener, A., & Otterbacher, J. (2019). Feelings rating and detection of similar locations, based on volunteered crowdsensing and crowdsourcing. *IEEE Access, 7,* 90215–90229.

Architecture, Applications and Data Analytics Tools for Smart Cities: A Technical Perspective

Jalpesh Vasa, Hemant Yadav, Bimal Patel, and Ravi Patel

Abstract Rapid urbanization in recent years has imposed new challenges to the sustainable development of Cities. Governments and industries are focused on developing smart cities which face the challenge of efficient integration and dissemination of information. Moving toward smart cities reality depends on automation of resource allocation and utilization, which can be achieved via smart sensors and RFID. Sensor grids, generally known as the Internet of Things or IoT nowadays, disseminate large amounts of data continuously. Identification of areas to automate and tools for automating such areas are necessary for a successful and efficient deployment. This paper identifies five important areas for automation using concepts of IoT. Comparative analysis of tools based on various characteristics like collection, storage, and summarization of information is also discussed. Finally, Smart Utilities with their application area in smart cities along with its data analytics tools that will help us to live desired living by making better decisions is explored.

Keywords Internet of Things (IoT) · Radio frequency identification (RFID) · Smart sensors · Smart utilities · Smart city · Analytics tools

J. Vasa (✉) · H. Yadav · B. Patel
Smt. Kundanben Dinsha Patel Department of Information Technology, Faculty of Technology and Engineering (FTE), Chandubhai S Patel Institute of Technology (CSPIT), Charotar University of Science and Technology (CHARUSAT), Changa, Gujarat 388421, India
e-mail: jalpeshvasa.it@charusat.ac.in

H. Yadav
e-mail: hemantyadav.it@charusat.ac.in

B. Patel
e-mail: bimalpatel.it@charusat.ac.in

R. Patel
Vadodara, Gujarat, India

S. Shakya et al. (eds.), *Sentiment Analysis and Deep Learning*,
Advances in Intelligent Systems and Computing 1432,
https://doi.org/10.1007/978-981-19-5443-6_64

1   Introduction

There has been a growing trend in recent years of significant numbers of people shifting to urban living. The concept of smart city development is one of the responses to tackle the challenge of the rising population in the world, specifically in urban areas. Smart Cities operate in a complex urban context that includes a wide range of infrastructural systems, human activity, technology, social and political processes, and the economy. Transportation, parking, residential and commercial structures, and the environment are all managed rationally in a smart city. By concept smart cities are data-driven. Big data technology and analytics play a significant part in city government, with many cities also having a Chief Data Officer. Big data analysis and smart city technologies converge to help cities enhance management in critical segments.

One of the enabling technologies in smart cities is the Internet of Things, which primarily provides data from numerous linked components. IoT is already being deployed for data collection and informed decision making in several industrial and consumer applications such as vehicle and traffic tracking, smart home management, water utilization and distribution and environment monitoring.

"According to the latest United Nations Population, more than half of the world's population now lives in urban areas. By 2050, it is expected that around 70% of the world's population will be living in cities" [1–3]. Such a large rise will place undue strain on the climate, energy supply, environment, and living conditions. Smart city development necessitates the creation of an integrated information infrastructure for the creation of cooperative urban utilities that gather and analyze data in order to maximize resource allocation efficiency for the benefit of inhabitants [4].

The paper is organized as follows. Section 2 describes Smart city architecture. Section 3 addresses the IoT Communication stack for smart cities and its applications. Section 4 addresses the smart application required for a more urban world to mitigate the current issues. Section 5 gives an idea about how to integrate all smart applications as a single system utility. Section 6 discusses IoT data analytics tools and its elements, along with modern technologies. Finally, Sect. 7 concludes the paper.

2   Information System Architecture for Smart City

A growing number of cities around the world have begun developing their own smart strategies to mitigate the challenge of rapid urbanization to improve the well-being of citizens. China, being the world's most populated country, has more than 200 smart city initiatives in the works [2]. Various cooperative applications, such as smart urban Mobility, smart governance, smart health care, smart environments, smart buildings and smart living, can benefit citizens by embedding billions of devices in a city's infrastructure.

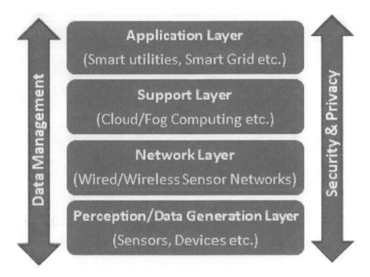

Fig. 1 IoT-based smart city architecture

Figure 1 shows general information system architecture for IoT-based smart cities. It is divided in four layers. The first layer is about data collecting, which is aided by sensors and monitoring equipment. Layer-1 items collaborate via a network layer, which collects data in the cloud data center. Data centers in the cloud are essential parts for hosting applications and storing data. Security and privacy remain one of core requirements at each layer [5, 6].

3   IoT Communication Stack in Smart City

This section discusses proposed protocols, standards, technologies and applications in each layer of the IoT communication stack discussed in Sect. 2. The IoT technology stack leads to the seamless integration of items and components in a system via the Internet. Figure 2 shows layer-wise details of technologies suitable for specific layers.

(1) **Physical Layer**: It is the first layer which connects IoT systems with actual devices and sensors. The PHY layer design of IoT systems should have certain considerations for wireless systems regarding the practical constraints of Quality of Service, cost-effectiveness, energy efficiency and ease of use [7]. Physical IoT objects are connected to upper layers via IoT Protocols like 3GPP, IEEE802.15.4, RFID/NFC, Bluetooth 4.0, and Wi-Fi.

- Protocols designed by 3GPP such as GSM, UMTS, 5G and LTE [8] are used when it is needed to communicate in licensed spectrum, specifically in large range communication.

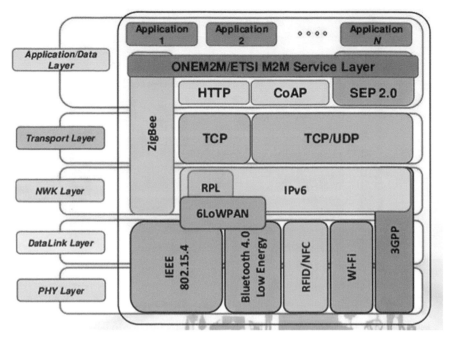

Fig. 2 IoT technology stack [13]

- IEEE802.15.4 is applied specifically in small range communication such as Zigbee [9].
- Bluetooth is generally applied as a communication with low energy devices [10].
- RFID/NFC is specially applied in designing passive communication devices such as toll tickets [11].
- Wi-Fi in IoT used for group communication between devices as it is based on radio frequency and has a range frequency and has a range of 20–100 m [12].

(2) **Data Link Layer:** Data collected via the physical layer are being communicated to upper layers via data link layer. Data link layer standards are generally incorporated into physical layer protocols such as Wi-Fi and Bluetooth. Also, it can be classified in low power and Internet enabled protocols [14].

(3) **Network Layer:** Data coming from data link layers are processed for routing and security protocols like Zigbee, IPv6, and RPL can be applied.

(4) **Transport Layer:** Here this layer provides a reliable connection facility so that retransmission and packet loss can be reduced and better securities can be achieved. Protocols like TCP/UDP or smart TCP/UDP can be used.

(5) **Application layer:** Protocols like https/CoAP/SEP2.0 are generally preferred for developing applications.

In general, there are two common approaches to offering access to IoT data. The first is the utilization of multi-hop mesh networks in the unlicensed spectrum with short-range communication between network nodes. The second is the use of long-range cellular technology in the licensed frequency band. A new way of communication technology to provide alternative IoT connectivity called Low-Power Wide Area Network (LPWAN) is being introduced in [15, 16]. As the name suggests, using star topology, this communication technology can provide low-speed, long-range transmission within unlicensed frequency bands. For certain smart city applications these features can be very useful.

IoT devices include transducers such as sensors and actuators and myriad items sometimes referred to as "smart," "intelligent" or plain old "wired" (smart light bulbs, wired valves and pumps, smart meters, connected vehicles, smart or intelligent building parts, smart home appliances, etc.

4 IoT Data Analytics Tools/Frameworks for Smart City

In recent years, the terms "real-time data", "fast data", "streaming data", and "fog computing" have become increasingly popular in the big data sector [17]. Smart city projects include vast quantities of similar data in these categories. Environment monitoring applications, for example, can make use of continuously created data such as temperature and humidity measurements. To identify any forecasts or data-driven judgments, analytics on this climate data can employ both historical data stored and real-time data collected. Smart applications such as traffic control and public safety management systems use video surveillance cameras generating live videos. These apps' streaming data must be used for further analysis on a continuous basis. In many V's of big data description, if speed of data generation is rapidly increasing then data is viewed as fast data [10]. Fast data is a system that uses data such as sensor data, click-stream data, log-aggregation, and so on. In this case, the volume of real-time data produced every time period is a crucial component in further analytics.

Real-time Big Data Analytics as in Table 1 differs from traditional analytics in that it considers real-time updates and quick analysis in order to take suitable measures based on early predictions and timely decisions [14]. In a smart city, wireless sensor networks generate a significant amount of data [18]. Wireless sensor networks have recently been implemented in Smart City infrastructures, resulting in massive volumes of data being created every day across a variety of domains, with applications ranging from environmental monitoring to health care monitoring to transportation monitoring. New methods and techniques for successful data management and analysis are necessary to provide information that can help intelligently and dynamically control resource utilization to take advantage of the enormous amounts of data. To analyze this big real-time data, currently the following tools are being used on the market.

Table 1 Various types of data analytics

Data analytics types			
Descriptive	Diagnostic	Predictive	Prescriptive
What happened is explained	Why something happened	What is going to happen	Prescribe what steps should be taken to avoid a future problem
Business Analytics and Data Mining	Analytic Dashboard	Forecasting	Simulation and optimization
Juggles raw data from different data sources to provide valuable historical insights	Provides in-depth analysis of a certain problems using BI tools	Predict the future data based on Machine Learning and Deep Learning models	Suggest all favorable outcomes to a specified course of action and vice-versa

4.1 Streaming Frameworks for IoT-Stream Data Analysis

Apache Kafka [19]: Apache Kafka is a massive event store and stream processing platform that is community-driven and distributed. Kafka is a distributed commit log-based system that solves the challenge of spreading massive amounts of event data to multiple subscribers. Kafka is used in real-time streaming data infrastructures to provide real-time analytics. Kafka is a fast, scalable, robust, and fault-tolerant publish-subscribe messaging system. It is crucial, then, to be able to feed this high-volume activity data to both our front-end and back-end systems in real-time. Kafka is a real-time data stream processing system that aims to deliver a unified, high-performance, low-latency platform. Kafka has a binary TCP-based protocol that is geared for reliability, as well as a "message collection" abstraction that blends messages to reduce network round trip overhead.

Apache Spark streaming [20, 21]: Spark Streaming is a Spark API core extension that provides for scalable, high-throughput, and fault-tolerant live data stream processing. DStreams is a Spark Streaming abstraction that represents a continuous stream of data, just as RDDs. Internally, each DStream is expressed as a sequence of RDDs that arrive at each time step. Spark Streaming is based on a "micro-batch" architecture, which considers streaming computing to be a continuous series of batch calculations on small batches of data. Spark Streaming collects data from a variety of sources and breaks it into manageable chunks. New batches are generated at regular intervals. A fresh batch is formed at the start of each time period. The batch has finished increasing at the end of the time period. A parameter called the batch interval is used to calculate the size of the time interval.

Apache Storm [21]: Apache Storm is a distributed real-time stream processing computer framework developed in the Clojure programming language that is free and open source. After being acquired by Twitter, the product, which was originally built by Nathan Marz and his colleagues at Back Type, was made open source. Apache Storm makes it simple to efficiently process unbound data streams, similar to how Hadoop handles batch processing in real-time. Apache Storm is easy to use, reliable, and compatible with a variety of programming languages.

Table 2 Comparison table of streaming frameworks

Characteristics	Samza	Storm	Spark	Flink
Processing type	Stream	Stream	Hybrid	Hybrid
Message delivery	At-least-once	At-least-once	Exactly-once	Exactly-once
Throughput	High	Low	High	High
Latency	Milliseconds	Milliseconds	Seconds	Milliseconds
Auto scaling	Yes	Yes	Yes	No
Inflight modification	No	Yes	No	No

Apache Flink [21, 22]: The Apache Software Foundation built Apache Flink, an open source stream processing application. Apache Flink is a distributed streaming data-flow engine based on Java and Scala. Flink performs any data-flow program in a pipelined and parallel manner. Apache Flink is a platform for stream and batch processing that is simple to use. Due to pipelined data transfers across parallel tasks, including pipelined shuffles, Flink's runtime supports both domains natively. From the time tasks are created until the time they are received, records are sent out quickly (after being collected in a network transfer buffer). Blocking data transfers may be used to conduct batch jobs.

Samza [23, 24]: The Apache Software Foundation's Apache Samza is an open source, near-real-time, asynchronous computing framework for stream processing written in Scala and Java. It was created in collaboration with Apache Kafka. Users can use Samza to create stateful apps that process data in real-time from a variety of sources, including Apache Kafka. Fault tolerance, isolation, and stateful processing are all features of Samza. It enables continuous processing and output, as opposed to batch systems like Apache Hadoop or Apache Spark, resulting in millisecond reaction times. The comparison table on the above survey on streaming framework in shown in Table 2.

5 Smart City Applications for Sustainable Development

In this article, we look at how Big Data analytics is used in real-time in some of the most well-known smart city applications. Real-time analytics assists the government in monitoring and managing traffic, vehicle carbon footprints, lowering energy consumption through intensity-based street light monitoring, managing cleanliness through timely waste collection via smart waste monitoring, and automating toll booths to eliminate vehicle lineups and payment with RFID to reduce human interference [25].

This smart city infrastructure contains various intelligent gadgets, such as auto-rotating solar panels to boost energy output by allowing more efficient use of solar panels, which the government deploys in locations such as solar farms, to keep our city clean garbage control and collection system used for urban garbage management that will alert the administrator of picking up public dustbins when its full. Smart parking allows visitors to secure a parking place before arriving at their destination in congested areas. Automatic street lights that detect the movement of an automobile or a walker and turn on and off after a certain period of time; Real-time traffic monitoring to efficiently manage traffic in different sections of the city utilizing image processing/video surveillance, RFID-based toll booth management will eliminate human intervention for cash collection, making the system automated, fast, reliable, and temper-proof, resulting in reduced carbon footprints generated at toll booth.

One goal of smart city development is to assist citizens in a variety of areas that are closely related to their living standards, such as power, climate, business, living, and services as well as safety from events like fire [26]. We explain some emerging smart city technologies and describe them in the following section. One important aspect of data collected via such a sensor needs a mechanism for reliable storage and accessibility in a decentralized manner as in [27].

5.1 Auto Rotating Solar Panel [28]

This system consists of an Arduino ATmega328p microcontroller, two servo motors for x-axis and y-axis rotations, a sun direction sensor consisting of LDRs to determine the actual and precise direction of the sun, and several monitoring devices that can measure the power produced by the system. Because it is based on IoT, it is connected via Wi-Fi and sends data to a server at regular intervals so that the data can be used in the future and mining can be done, and we may come to various solar power generating conclusions [8]. Here is the prototype for the Auto Rotating Solar Panel which is shown in Fig. 3.

5.2 Smart Street Lights [1]

This technology is extremely efficient in terms of energy conservation and is most effective around midnight, when most highways are deserted. This device conserves energy over time, resulting in a more productive system. This device is driven by IR sensors and is controlled by a microprocessor. Any movement detected by the IR sensor instructs that the lights be turned on [1]. Here is the prototype for Automated Street Lights as shown in Fig. 4.

5.3 RFID-Based Toll Booth Manager [29]

Managing several toll booths at same time seems to be a complicated task. So, to make an ease in management of toll booths, here is the proposed RFID-based toll

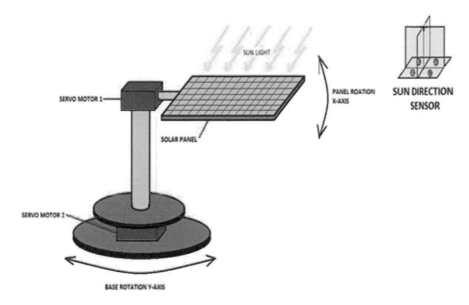

Fig. 3 Prototype for Auto Rotating Solar Panel

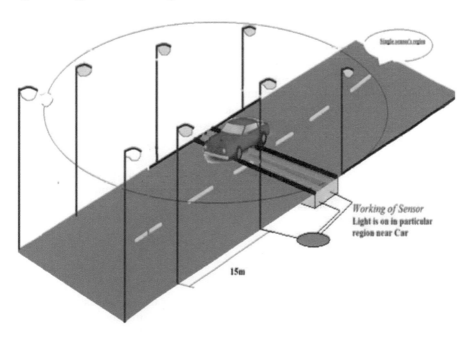

Fig. 4 Prototype for Automated Street Lights

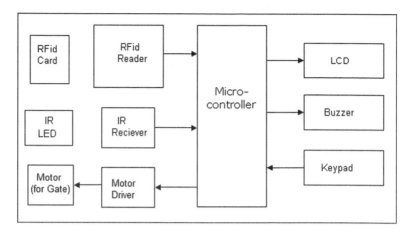

Fig. 5 Block diagram of RFID-based toll booth manager

booth management system that ensures all toll booths are monitored from a single location using IOT. Firstly, the system has a centralized internet server that has all details of users and their present balance. To identify the vehicles of owners, vehicles have a RFID card with account number stored in it. So, when the vehicle arrives at any toll booth, the RFID card is scanned and the system checks whether the card is valid or not on the Internet server and gets the details of user and present balance. If the system finds sufficient balance to be deducted then, the system deducts the fixed amount of vehicle and sends a signal to the scanner stating that the user account is billed followed by the system sending the signal to a motor to open the toll gate for the vehicle. Wi-Fi connection is used for interaction of the system over the Internet with the web server of the verification process. Moreover, the system is capable of storing data over the Internet of all passed vehicles with their time intervals for surveillance and other purposes. Thus, this system automates the management and monitoring of all toll booth collection from a single location with ease using RFID-based Toll booth management system using IOT [7]. Here is the block diagram of RFID-based toll booth manager as shown in Fig. 5.

5.4 Smart Parking System [30]

Illegal parking of vehicles on the street is worsening traffic congestion in metro cities. Multi-story parking lots are designed to reduce traffic congestion, but they take longer to find empty spaces. Because we visit the parking location frequently, the parking spot/space is usually taken. The solution we propose here comprises two processing steps to solve the problem of finding free parking slots in multi-story parking systems. The user goes to the parking area's website/app on the first level. In this case, the client must reserve the vacant space for a particular time period. The

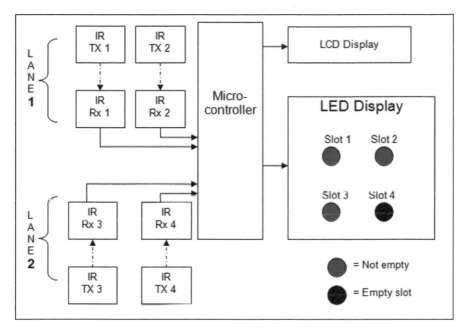

Fig. 6 Block diagram of smart parking system

input from the website/app is shown on a digital IoT display at the second level, and the reserved and unreserved slots are marked with LEDs using this data. The charges should be paid based on the length of the reserved slot [9]. Here is the block diagram of the smart parking system as shown in Fig. 6.

5.5 Smart Garbage Monitoring System [31]

As the cities grow in size, there is also a high amount of waste they produce. In cities, large public dustbins are usually found overflowing and garbage spread here and there, and they disrupt the area's cleanliness. This device is controlled by a microprocessor and sensors that compute the depth of the bin, the level of bin filling, and notify the firm that the dustbin is about to full and that the empty one should be replaced. We can figure out where to put extra garbage bins and clean up the area by using knowledge discovery. We can have clean cities due to this smart system [2]. Figure 7 shows the block diagram of the Smart Garbage monitoring system.

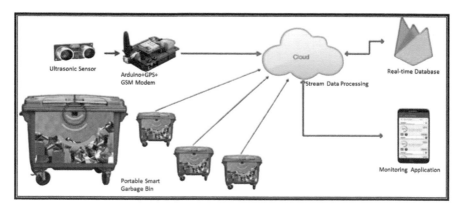

Fig. 7 Block diagram of smart garbage monitoring system

Fig. 8 Block diagram of real-time traffic monitoring system

5.6 Real-Time Traffic Monitoring [32]

People are more likely to break rules such as disobeying traffic signals, traveling without a helmet, riding without a seatbelt, overloaded passengers, and so on, thus cities have a big traffic problem that cannot be solved manually all of the time [33]. Image processing was applied to detect offender for breaking traffic regulations, based on their registration plate recovered from multiple CCTV footages installed in smart city [10]. Figure 8 shows the process flow for traffic monitors.

6 All Applications Under Single Umbrella

City gets a name "Smart City" by combining all the above applications such as Auto Rotating Solar Panel, Automated Street Lights, and RFID-based toll booth manager, Smart Parking System, Garbage monitoring system and traffic monitors into a single system or a cluster and can be controlled from a single controlling and monitoring location of the city. After getting data from all IoT devices installed in the city, we can also apply some data mining techniques to IoT data and get some useful insights as mentioned in [4]. Also, these all systems need an automated electrical device monitoring system which can take care of its functioning without human visiting to

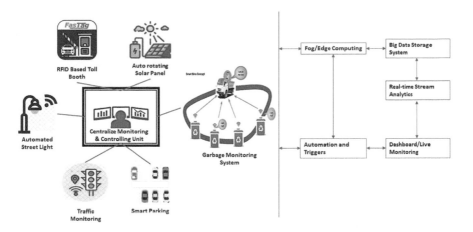

Fig. 9 Smart city control and monitoring system

check devices which is discussed in [34, 35]. Figure 9 shown overall combination of smart applications which make cities smarter and are connected to a single controlling system.

7 Conclusion and Future Work

Innovation is an attempt toward a life with ease. Smart city reshapes the task processed in a regular city by improvising it with modern day innovation and scientific application to make an easier, simple and eco-friendly living. Here in this paper, we demonstrated the need for smart city and urban planning along with its architecture. Also, from the communication stack we can devise that recently LoWPAN protocol is to communicate between the smart devices. Also, some applications are discussed here to reduce the carbon footprint and efficiently manage the city with such small smart applications. Among all Smart Infrastructure, Smart Traffic Monitoring, Smart Parking, and Smart Waste Management are the top-5 applications nowadays. Real-time data/stream and its analysis to get better decision and help urban government in a greater manner to better manage the city. Smart Cities will utilize analytics to give efficient solutions which would result in energy saving cost up to 30%. Also, some real-time processing tools and underlying technologies are discussed and finally it is concluded that Samza is a good platform for real-time analysis purposes. We can compare other tools and applications with urbanization as a future work.

References

1. Srivastava, S. (2013). Automatic street lights. *Advance in Electronic and Electric Engineering, 3*, 539–542.
2. https://www.un.org/development/desa/en/news/population/2018-revision-of-world-urbanizat ion-prospects.html
3. *Some facts on urbanization-2014.* https://www.un.org/en/ecosoc/integration/pdf/fact_sheet.pdf
4. Chovatiya, F., Prajapati, P., Vasa, J., & Patel, J. (2017). A research direction on data mining with IoT. In *International conference on information and communication technology for intelligent systems* (pp. 183–190). Springer, Cham.
5. Patel, B. H., Shah, P. D., Jethva, H. B., & Chavda, N. (2013). Issues and imperatives of Adhoc networks. *International Journal of Computer Applications, 62*(13).
6. Shakya, S. (2021). Collaboration of smart city services with appropriate resource management and privacy protection. *Journal of Ubiquitous Computing and Communication Technologies (UCCT), 3*(01), 43–51.
7. Sharma, S. K., Tadilo, E. B., Symeon, C., Xianbin, W., & Long, B. L. (2016). Physical layer aspects of wireless IoT. In *2016 International symposium on wireless communication systems (ISWCS)* (pp. 304–308). IEEE.
8. 3GPP. https://en.wikipedia.org/wiki/3GPP
9. IEEE-802.15.4. https://en.wikipedia.org/wiki/IEEE_802.15.4
10. IEEE802.15. https://en.wikipedia.org/wiki/IEEE_802.15#IEEE_802.15.1:_WPAN_/_Blu etooth
11. NFC. https://en.wikipedia.org/wiki/Near-field_communication
12. IEEE-802.11. https://en.wikipedia.org/wiki/IEEE_802.11
13. *Introduction to IoT Architecture and Protocols.* https://www.slideshare.net/AbdullahAlfadhly/ introduction-to-iot-architectures-and-protocols
14. Palattella, M. R., Accettura, N., Vilajosana, X., Watteyne, T., Grieco, L.-A., Boggia, G., & Dohler, M. (2013). Standardized protocol stack for the internet of (important) things. *IEEE Communications Surveys & Tutorials, 15*, 1389–1406.
15. Song, Y., Lin, J., Tang, M., & Dong, S. (2017). An internet of energy things based on wireless LPWAN. *Engineering, 3*(4), 460–466.
16. Centenaro, M., Vangelista, L., Zanella, A., & Zorzi, M. (2016). Long-range communications in unlicensed bands: The rising stars in the IoT and smart city scenarios. *IEEE Wireless Communications, 23*(5), 60–67.
17. Lai, K. L., Chen, J. I. Z., & Zong, J. I. (2021). Development of smart cities with fog computing and internet of things. *Journal of Ubiquitous Computing and Communication Technologies (UCCT), 3*(01), 52–60.
18. Sivaganesan, D. (2021). A data driven trust mechanism based on blockchain in IoT sensor networks for detection and mitigation of attacks. *Journal of trends in Computer Science and Smart technology (TCSST), 3*(01), 59–69.
19. Wang, G., Koshy, J., Subramanian, S., Paramasivam, K., Zadeh, M., Narkhede, N., Rao, J., Kreps, J., & Stein, J. (2015). Building a replicated logging system with Apache Kafka. *Proceedings of the VLDB Endowment, 8*(12), 1654–1655.
20. Zaharia, M., Xin, R. S., Wendell, P., Das, T., Armbrust, M., Dave, A., Meng, X., Rosen, J., Venkataraman, S., Franklin, M. J., & Ghodsi, A. (2016). Apache spark: A unified engine for big data processing. *Communications of the ACM, 59*(11), 56–65.
21. Sakr, S., & Zomaya, A. Y. (2019). *Encyclopedia of big data technologies.* Springer International Publishing.
22. Katsifodimos, A., & Schelter, S. (2016). Apache flink: Stream analytics at scale. In *IEEE international conference on cloud engineering workshop (IC2EW)* (pp. 193–193). IEEE press.
23. Noghabi, S. A., Paramasivam, K., Pan, Y., Ramesh, N., Bringhurst, J., Gupta, I., & Campbell, R. H. (2017). Samza: Stateful scalable stream processing at LinkedIn. *Proceedings of the VLDB Endowment, 10*(12), 1634–1645.

24. Zhuang, Z., Feng, T., Pan, Y., Ramachandra, H., & Sridharan, B. (2016). Effective multi-stream joining in apache samza framework. In *IEEE international congress on big data (BigData Congress)* (pp. 267–274). IEEE press.
25. Annappa, B. (2018). Real time big data analytics in smart city applications. In *International Conference on Communication, Computing and Internet of Things (IC3IoT)* (pp. 279–284). IEEE Press.
26. Sungheetha, A., & Sharma, R. (2020). Real time monitoring and fire detection using internet of things and cloud based drones. *Journal of Soft Computing Paradigm (JSCP), 2*(03), 168–174.
27. Sivaganesan, D. (2021). Performance estimation of sustainable smart farming with blockchain technology. *IRO Journal on Sustainable Wireless Systems, 3*(2), 97–106.
28. Watane, N. D., & Dafde, R. A. (2013). Automatic solar tracker system. *International Journal of Scientific and Engineering Research, 4*(6)
29. Win, A., & MyatNwe, C. (2014). RFID based automated toll plaza system. *International Journal of Scientific and Research Publications, 4*(6)
30. Basavaraju, S. R. (2015). Automatic smart parking system using Internet of Things (IOT). *International Journal of Scientific and Research Publications, 5*(12), 629–632.
31. Gupta, S., & Kumar, P. (2015). Real time solid waste monitoring and management system: A case study of Kanpur city. *International Journal of Science, Environment and Technology, 4*(2), 514–517.
32. Cao, J. (2016). Research on urban intelligent traffic monitoring system based on video image processing. *International Journal of Signal Processing, Image Processing and Pattern Recognition, 9*(6), 393–406.
33. Patil, P. J., Zalke, R. V., Tumasare, K. R., Shiwankar, B. A., Singh, S. R., & Sakhare, S. (2021). IOT protocol for accident spotting with Medical facility. *Journal of artificial intelligence, 3*(02), 140–150.
34. Karuppusamy, P. (2020). A sensor based IoT monitoring system for electrical devices using Blynk framework. *Journal of Electronics and Informatics, 2*(3), 182–187.
35. Patel, B., & Shah, P. (2021). Operating system support, protocol stack with key concerns and testbed facilities for IoT: A case study perspective. *Journal of King Saud University-Computer and Information Sciences.*

Evaluation of Software Engineering Virtual Laboratory in Determining Undergraduate Students' Conceptual Understanding: A Blended Learning Model Using Collaborative-Creative Virtual Learning Environment Employing Critical Pedagogy

Ashraf Alam and Atasi Mohanty

Abstract The most measured learning outcomes, according to the extant literature, are students' grades. However, such grades may not be the most accurate reflection of pupils' learning efficacy. Due to the inherent unpredictability of the grading process, grades may not be an effective instrument for the measurement of learning and for assessing students' laboratory performance. In this situation, a self-designed and customized tool might be best suited to consistently help assess students' learning gains. With support from the Ministry of Human Resource Development (MHRD), Government of India, a Software Engineering Virtual Lab (SE VLab) (http://vlabs.iitkgp.ernet.in/se/) was built for engineering students at the Indian Institute of Technology Kharagpur, India. This lab introduces students to a variety of essential concepts in software engineering. Existing literature states that the evaluation procedure in a standard traditional laboratory setting is fundamentally flawed, subjective, and vulnerable to injustice, with the possibility of prejudice. Researchers in this investigation has explored a unique technique with a novel approach for measuring learning gains and educational outcomes. Here, a tool is developed to assess the efficacy of virtual learning. The findings are statistically confirmed. Using the developed tool, researchers also tested the SE VLab in several pedagogical scenarios and in diverse technical setups. In terms of learning gains, the results suggest that SE VLab is more efficient than identical traditional SE laboratories.

Keywords Critical pedagogy · Evaluation · Computer science engineering education · Virtual laboratory · Software engineering · Curriculum · Learning outcomes

A. Alam (✉) · A. Mohanty
Rekhi Centre of Excellence for the Science of Happiness, Indian Institute of Technology Kharagpur, Kharagpur, India
e-mail: ashraf_alam@kgpian.iitkgp.ac.in

© The Author(s), under exclusive license to Springer Nature Singapore Pte Ltd. 2023   875
S. Shakya et al. (eds.), *Sentiment Analysis and Deep Learning*,
Advances in Intelligent Systems and Computing 1432,
https://doi.org/10.1007/978-981-19-5443-6_65

1   Introduction

One of the critical features of computer science and engineering (CSE) education is learning by doing. The rapid burst in the use of Internet along with the lockdown across the globe because of Covid-19 pandemic has drawn attention to the importance of online laboratory (lab)-based learning in CSE education [1]. Research says,—"You only understand something when you know how to do it" [2]. However, bringing such experiences online is very tricky, but learning technology researchers have been making substantial progress in enabling it over the past decade [1, 3].

Contextually, online virtual lab-based learning is a modern teaching mode, which attracts wide attention of educational institutes, industries, societies, and governments [4]. This is also elaborated in the Washington Accord [5] and in the National Accreditation Board (NAB) document [3] one of the important program outcomes maintaining the skills of a graduate engineer. It states—"Graduate will demonstrate an ability to design and conduct experiments, analyze and interpret data" [2, 3, 5]. So, online laboratory-based learning is an emerging area of research [3, 6]. However, in the online laboratory learning environment, the instructor has significantly reduced role, and students take more responsibility for their learning [7]. This shows that in online lab-based learning, the involvement of students is greater [8].

Traditional classroom-based lab (TLab) teaching does not seem to be effective [9–12]. So, the necessity to engage students in self-learning is imminent [4, 6, 9]. In such cases, the students also get an opportunity to perform their lab experiments outside the classrooms, such as at home and during vacation [13].

This paper has been organized as follows. In Sect. 1, we present the motivation toward evaluation of students' performance using VLab and in Sect. 2, we present the evaluation process in TLab. In Sect. 3, we discuss the development of CP-VLLM tool and in Sect. 4, we present our research hypotheses and questions. In Sect. 5, we describe in details the methods for our data collection and in Sect. 6, we present the evaluation of students' performance using SE VLab. In Sect. 7, we present the students' perceptions from CP-VLLM tool and finally conclude the paper in Sect. 8.

2   Motivation Toward Evaluation of Students' Performance Using VLab

As there is no standard measurement tool for evaluating the effectiveness of virtual lab-based learning, we developed CP-VLLM tool and evaluated the SE VLab using the same. We have based our work on the critical pedagogy (CP) theory. Reason for it is: "it addresses many important educational issues such as instructional design delivery, students learning outcomes, assessments, students' empowerment, social presence, critical thinking skills, and alignment" [7, 14].

3 Evaluation Process in TLab

The daily lab activities in the TLab environments are typically recorded through lab log books. The evaluation process is based on the grading in assignments and oral examinations [4, 12]. Sometimes students are not properly evaluated by teachers [10, 15] and the teachers assign marks based on the set of questions in the lab session. Research results indicate that both the physical observation of students and lab assignment records are used for students' performance evaluation [3, 16]. It is also observed that final lab activity grades (or marks) in lab exams, and regularity in maintaining lab records are also factors for performance assessment in the TLab environments [4, 14]. However, in the TLab environment, the evaluation process is inherently deficient, subjective, and unfair with possibilities of bias [3, 9]. However, the evaluation of lab is challenging, specifically for large class sizes [6, 17]. Sometimes evaluation of lab work is based on the end product, where the interim progress of the lab work between the lab session and students' competencies is neglected [4, 13]. The Accreditation Board of Engineering and Technology (ABET) [2, 8, 15] also notes that the lab evaluation practices should evolve continuously.

4 Development of CP-VLLM Tool

The development of CP-VLLM tool was based on the following research question and hypotheses:

> Are there correlations between the parameters of evaluation used in the CP-VLLM measurement tool?

In this work, the parameters of evaluations were Instructional Design Delivery (IDD), Students Learning Outcomes (LO), Assessments (ASS), Students Empowerment (EM), Social Presence (SP), Critical Thinking (CT) and Alignment (AL). Also, it was expected that students' perceptions of responses would tend to be correlated and the following null hypothesis (H0) and alternate hypothesis (H1) were considered.

- (H0): There is no correlation in a given pair of parameters for learning assessment.
- (H1): There is correlation within a given pair of parameters for learning assessment.

In H1, a pair of parameters is said to have a correlation when the two variables vary, where the correlation may be either negative or positive. Paired correlations are assessed based on targeted opinions, where the answer from the respondents is close to Strongly Agree. Seven parameters were considered to evaluate the SE VLab learning. Each parameter was paired with one of the other parameters, resulting in twenty-one paired comparisons. In order to check the existence of correlation among students' responses across the seven parameters, the overall opinion for each lab experiment was computed using the Pearson Product Moment Coefficient (PPMC) that considers all of the possible opinions scores. In this manner, the combined score was ranked.

4.1 Reliability and Validity

To ensure the reliability and validity of CP-VLLM, the experimental error rate was considered and was set to 0.1 and Type-I error rate for each was $0.1/21 = 0.0047619$.

In this case, if the p-value was found to be less than or equal to 0.0047619, then pair correlation was significant and if the p-value was found to be greater than 0.0047619, then the correlation was significant.

Table 1 shows the comparison of pair correlation using PPMC. The results depicted in Table 1 indicate that all correlations are positive. It was also observed that the correlation coefficient values are greater than 0.5, which indicates a very strong positive correlation and the p-values are less than 0.0047619. This shows that the error rate is 0.1, and that the results are statistically significant.

The above-mentioned positive correlation indicates that by rejecting the null hypothesis (H0), we approach the alternative hypothesis (H1), that is, the responses from students indicate that there is a correlation within a given pair of parameters for learning assessment.

5 Research Hypotheses and Questions

Evaluation questions were related to the overall positive LO. Participation in SE VLab sessions, in which the students undertake the evaluation process, is intended to increase SE knowledge, as compared to hands-on lab sessions. In this context, the hypothesis is:

H1: Students get more knowledge after the SE VLab session than before using it.

As the experiments are executed online, rather than in a conventional lab, the students are, therefore, unlikely to learn more by using the VLab than the TLab. Better success is achieved in terms of students who are interested for online learning and who decline to learn in a traditional environment. So, the hypothesis is:

H2: Students executing experiments using SE VLab learn more as compared to students executing them using TLab.

Though self-assessment, subject knowledge test, and LO can be measured, an assessment using the CP theory also shows positive average correlations of performance. This raises the following research question:

Table 1 Pair correlation comparisons using PPMC

Pair-parameter comparisons	Correlation coefficient	p-value
IDD versus LO	0.815748	0.000000144*
IDD versus ASS	0.821827	0.000001009*
IDD versus EM	0.687841	0.0000567*
IDD versus SP	0.797884	0.000000407*
IDD versus CT	0.670106	0.0000983*
IDD versus AL	0.817397	0.00000023*
LO versus ASS	0.938091	0.00000000000052*
LO versus EM	0.726263	0.0000134*
LO versus SP	0.829669	0.0000000546*
LO versus CT	0.829668	0.0000006*
LO versus AL	0.877493	0.00000000124*
ASS versus EM	0.737592	0.00000781*
ASS versus SP	0.85412	0.00000000913*
ASS versus CT	0.844494	0.0000000196*
ASS versus AL	0.94416	0.000000000000147*
EM versus SP	0.606952	0.00074*
EM versus CT	0.683416	0.0000672*
EM versus AL	0.693269	0.0000537*
SP versus CT	0.822527	0.0000000912*
SP versus AL	0.801582	0.000000328*
CT versus AL	0.858291	0.00000000671*

* Significant Correlation p-value is < 0.0047619

Q1: Can students assess their learning performance, so that they do not need any knowledge test for future evaluation?

Along with the knowledge test, other factors such as IDD, LO, ASS, EM, SP, CT, and AL of SE VLab should also be assessed. In this context, the following research question arises.

Q2: Do students find the following important advantages toward online lab delivery?

- Instructional Design Delivery
- Students Learning Outcomes
- Assessments
- Students Empowerment
- Social Presence
- Critical Thinking and
- Alignment

5.1 Research Design

This research employed a pre-test/post-test design to examine the learning gains applied across 4 pedagogical contexts. The participants were assigned group-wise to one of the following four experimental conditions:

- *Group-I*: Traditional Lab (TLab). In this group, we included students who already completed the SE theory classes as well as completed their TLab experiments.
- *Group-II*: Virtual Lab (VLab). In this group, we included students who already completed the SE theory classes, and did not complete their TLab experiments. They were allowed to access the SE VLab and they had completed their experiments using the VLab.
- *Group-III*: TLab and Virtual Lab (TVLab). In this group, we included students who had already completed the SE theory classes and also completed their TLab. Simultaneously, they were allowed to work on VLab environment.
- *Group-IV*: SE Virtual Lab (SE VLab). In this group, we included students who had not completed SE theory classes. They were new to the SE subject. They were allowed to access the SE VLab and had completed all theories available in SE VLab. Simultaneously, they were allowed to work on the VLab environment.

A pre-test was administered to assess SE knowledge test for Groups-I, II, and III, immediately after completion of theory classes, i.e., prior to the start of lab sessions. But for Group-IV, a pre-test was administered immediately after accessing theory sessions from SE VLab environments, i.e., prior to start of the SE VLab experiments. After completion of the lab sessions, a post-test was administered to assess the learning gains from each environment.

6 Method

This section describes the settings, participants, data collection tools, procedures for data collection, and data analysis.

6.1 Setting and Participants

The evaluation procedure was comprised two types of study—a pilot study and a main study. The pilot study covered 106 UG and 18 postgraduate (PG) students from IIT Kharagpur. In this work, we present the results from only the main study. In the main study, there were 239 student participants from UG engineering colleges in Chennai. Chennai is the capital of the Indian state of Tamil Nadu. All the participating engineering colleges were affiliated to Anna University and were AICTE approved. In our study among 239 students, there were 107 girls and 132 boys with an average

age of approximately 21.7 years. They had English as the medium of instruction in college, and were belonging to similar demographic background.

6.2 Data Collection Tools

In this work, a combination of quantitative (pre-test and post-test) and qualitative (observations, interviews) instruments were used to foster a grounded understanding of the learning gains and LO of SE VLab. Pre-test assessment tools were administered on all participants after the completion of theory classes and before accessing the lab sessions. After completion of lab experiments, the post-test assessments were administered to evaluate students' performance in terms of learning gains and LO of SE VLab. All statistical analyses were performed using the IBM SPSS Statistics version 21.0 in the Ubuntu 13.10 operating system environment.

Pre-test Tool. Pre-test we administered using a quantitative test containing 15 SE lab subject related multiple choice questions (MCQ) (choosing one answer out of four options) covering the standard syllabus approved by AICTE, IIT Kharagpur and Anna University in the UG level. This questionnaire also included basic demographic questions such as gender and ethnicity. This test measured the students' SE knowledge after the completion of the SE theory classes. The maximum possible score in this questionnaire was 15 and the minimum was zero. As this questionnaire is SE content related, content validity was reviewed for clarity and consistency. Questions were scrutinized and all questions that weakened the measurement tool validity were omitted. Modifications were made based on recommendations of five experts (faculty members) in the field of SE subject from IIT Kharagpur. We also checked the reliability of the questionnaires using the α coefficient in Cronbach's Alpha (α) test method [10, 18, 19] and all assessments were found to be reliable ($\alpha > 0.8$) in our pilot study.

Post-test Tools. The post-test was a mix-mode approach containing a combination of both quantitative and qualitative questions. There were four types of questionnaires, one for pre-test, a SE knowledge test after completing the lab experiments (though the post SE knowledge test questions cover the whole syllabus the post-test question contents were different from the pre-test questions. Covering the whole theory and practical syllabus, we prepared 30 questions, and randomly 15 questions were put in the pre-test tool and 15 questions in the post-test tool. This was done so that students do not feel any ambiguity in question settings) and the other was CP theory-based questionnaire. The CP-VLLM tool was based on students' perceptions.

The other two questions were open-ended and interview-oriented questions. As all post SE knowledge test, CP-VLLM tools, open-ended, and interview-oriented questionnaires were content related. So, content validity was evaluated using five SE subject experts from IIT Kharagpur. The CP-VLLM tool evaluated students' perceptions related to IDD, LO, ASS, EM, SP, CT, and AL.

6.3   Procedure

With prior permission from the heads of the institutions and with the help of the departmental heads, we met the SE subject teachers and requested them to allow us for data collection from students. During the first week of the semester under study and before the commencement of the teaching of SE theory course, we met the students and informed them about the detailed process of data collection. We also informed the students that the participation was voluntary and their responses would be kept confidential and would only be used for academic/research purpose in an aggregated manner. They were also informed that their responses will not affect their semester grades/marks. As there were four groups for data collection, there were different instructions for each group of students. For Groups-I, II, and III, a pre-test on SE knowledge test was administered immediately after the completion of the theory classes. Group-IV, did not have the necessity of attending SE theory class. We allowed them to access SE VLab and they could study the Theory part of the SE VLab. After the collection of pre-test data, Group-I was allowed to use SE TLab, Group-II was allowed to access SE VLab experiments, Group-III was allowed to use both TLab and VLab, and Group-IV was allowed to continue with the SE VLab experiments. As both the SE TLab and VLab consisted of 10 experiments, we met them after two months interval for post-test data collection.

7   Evaluation of Students' Performance Using SE VLab

This Section comprises of different comparisons aimed at addressing students' performance using the SE VLab. In this section, we first give a description of comparison between each group on prior SE knowledge test scores, then a comparison between each group on post SE knowledge test scores, and lastly a comparison of learning gains for each group.

7.1   Comparison Between Each Group on Prior SE
Knowledge Test Scores

To test the significant differences, if any, in the post-lab test of four different lab environments, four groups were first compared in terms of the mean score achieved in the prior SE knowledge test. The prior SE knowledge test contains 15 SE subject questions to measure prior SE knowledge covering the standard syllabus. The test contains objective type questions. The maximum score for each item was one, and therefore, the maximum total score was 15. The scores were calculated in percentage and the detailed descriptive statistics for each of the four groups is given in Table 2.

Table 2 Groupwise descriptive statistics for prior SE knowledge test scores

Different pedagogical context	N	Mean	Standard deviation	Standard error	95% Confidence interval for mean	
					Lower bound	Upper bound
TLab	59	27.52	17.10	2.24	20.74	30.10
VLab	60	31.41	19.21	2.51	24.32	34.36
TVLab	59	34.71	20.41	2.53	29.63	39.71
SE VLab	61	29.32	18.23	2.33	22.61	31.84
Total	239	30.49	19.10	1.42	27.10	32.41

Initial differences across the four groups in prior SE knowledge test scores were explored using one-way ANOVA test. Before exploring this, we examined whether the data met the assumptions for one-way ANOVA. For examining this, we had made 4 assumptions, which are as follows: (i) whether distributions within the groups are normally distributed, (ii) whether there is homogeneity of variance, (iii) whether observations are independent, and (iv) dependent variable is measured on at least an interval scale [13, 20, 21]. From these four assumptions, we found that assumption (iii) is met, as there were different participants in each group. Assumption (iv) is met as the dependent variable that is prior SE knowledge test score is on a continuous scale. In order to check whether the distribution of scores within each group is normally distributed that is assumption (i) is satisfied, we considered the values of skewness and kurtosis (K-S) in conjunction with Quantile-Quantile (Q-Q) plots. This was done, as in the case of large samples, K-S are likely to be significant even when they are not too different from normal [3, 8, 17]. The results form skewness, kurtosis, and Q-Q plots indicate that the normality assumption is tenable. The Levene's test for homogeneity of variance was used to check assumption (ii). This test examines the null hypothesis, which examines whether the variances in different groups are not significantly different [10, 16, 22]. For prior SE knowledge test scores, homogeneity of variances across groups was assumed as the Levene's statistic was found to be non-significant (F (3238) = 0.513, ns).

The results of one-way ANOVA on prior SE knowledge test scores is given in Table 3. The results indicate that the group mean scores did not differ significantly (F (3238) = 2.309, ns). So, there was no significant pre-treatment differences across the four groups in terms of prior SE subject knowledge.

Table 3 One-way ANOVA on prior SE knowledge test scores

Groups	Sum of squares	df	Mean square	F	Sig.
Between groups	2326.63	3	776.76	2.309	0.082
Within groups	77,701.72	236	339.41		

7.2 Comparison Between Each Group on Post SE Knowledge Test Scores

In post SE knowledge test score, it was found that Group-III (TV Lab) achieved an average highest overall score of 73.62% and Group-IV (SE VLab) with the second highest score of 51.42%. However, the performance of the other two groups, Group-I with an average score of 43.21%, and Group-II, with average score of 48.63%, was similar. So, at an inferential level, one-way ANOVA was used to test for significant difference in aggregate LO across different groups. In order to do this, we first examined the data for the assumption of one-way ANOVA. The test for K-S indicates slight deviations from normality [3, 13, 17]. When group sizes are equal, the F-statistic (also known as fixation indices) is robust to violations of normality [10, 23]. Variances were found to differ significantly in different groups, as indicated by the Levene's test, F $(3236) = 3.107, p < 0.05$. So, the Welch test, which is known to be appropriate under violation of the homogeneity assumption, was used [19, 24]. The results for Welch test are given in Table 4.

The results of one-way ANOVA show that overall there are significant differences between the groups in terms of LO achieved, F $(3213.421) = 32.207, p < 0.01$. It was also found that there is an overall significant effect of different pedagogical groups on LO, as measured by the total post SE knowledge test score. Furthermore, analysis was conducted to test which groups differ significantly in their post SE knowledge test scores. As we did not have a specific hypothesis prior to the experiment regarding this in their sum of the post SE knowledge test scores, post-hoc tests were used for pairwise comparison of all different combinations of the experimental groups. Since group variances were not homogeneous, as indicated by the Levene's test, the Games Howell test, which is mostly used under violations of the homogeneity assumption, was performed [25]. Further, this test is considered to be accurate if the sample sizes are unequal. The post-hoc analysis results are given in Table 5. The results in Table 5 indicate that Group-III and Group-IV perform significantly better than the other two groups on post SE knowledge test scores. It was found that there was significant mean difference when the score of the Group-III (TVLab) was compared with each of the other three groups, when the mean value was compared with all other three groups, and it was found that $p < 0.05$. The comparison of mean value with Group-I and Group-IV gave the value $p < 0.05$, which was significant. So, we arrived at the conclusion that the performance using the SE VLab is better than that using the TLab.

Table 4 Robust tests of equality of means for post SE knowledge test scores

Tests	Statistics	df1	df2	Sig.
Welch test	32.207	3	213.421	0.000
Brown-Forsythe test	29.109	3	328.764	0.000

Table 5 Post-hoc analysis of means for post SE knowledge test scores

Different groups (p)	Compared groups (q)	Mean difference (p-q)	Standard error	Sig.
TLab	VLab	−5.42	2.88	0.697
	TVLab	−30.41	3.08	0.000
	SE VLab	−8.21	2.94	0.363
VLab	TLab	5.42	2.88	0.697
	TVLab	−24.99	2.84	0.000
	SE VLab	−2.79	3.66	0.893
TVLab	TLab	30.41	3.08	0.000
	VLab	24.99	2.84	0.000
	SE VLab	22.2	2.61	0.000
SE VLab	TLab	8.21	2.94	0.363
	VLab	2.79	3.66	0.893
	TVLab	−22.2	2.61	0.000

7.3 Comparison of Learning Gains for Each Group

In this Section, we compare the learning gains from both the pre-test and post-test SE knowledge test scores. In order to examine the learning gains for each group, paired-sample t-test was used to compare pre-test and post-test performance. Each group reported a significantly higher mean score in the post-test, as compared to the pre-test scores. The groupwise results are reported here. On an average, the participants in Group-III (TVLab) achieved significantly higher scores in the post-test ($M = 73.62$, $SE = 2.17$), as compared to the pre-test ($M = 34.71$, $SE = 2.53$), $t(59) = −26.421$, $p < 0.01$. The mean score for Group-I (TLab) was significantly higher in the post-test ($M = 43.21$, $SE = 2.09$), as compared to the pre-test ($M = 27.52$, $SE = 2.24$), $t(59) = −8.963$, $p < 0.01$. On an average, participants in Group-II (VLab) achieved significantly higher scores in the post-test ($M = 48.63$, $SE = 2.24$) as compared to the pre-test ($M = 31.41$, $SE = 2.51$), $t(60) = −10.368$, $p < 0.01$. Similar results were reported for Group-IV (SE VLab), where participants reported significantly higher scores on the post-test ($M = 51.42$, $SE = 2.21$), as compared to the pre-test ($M = 29.32$, $SE = 2.33$), $t(61) = -14.631$, $p < 0.01$. Groupwise comparison showed that learning gains, calculated as the mean difference between the pre-test and the post-test scores, was highest for Group-III (TVLab). The details of paired samples test for comparison of the pre-test and the post-test scores are shown in Table 6.

Table 6 Paired samples test for comparison of pre-test and post-test scores

Different groups	Mean	SD	SE mean	t	df	Sig. (2-tailed)
TLab	−15.69	10.81	1.93	−8.963	59	0.000
VLab	−17.22	14.17	1.67	−10.368	60	0.000
TVLab	−38.91	8.73	1.09	−26.421	59	0.000
SE VLab	−22.1	11.62	1.29	−14.631	61	0.000

8 Students' Perceptions from CP-VLLM Tool

A survey was conducted using the CP-VLLM tool from Group-IV after the completion of SE VLab experiments. The participants were asked to respond using the CP-VLLM tool. A summary of survey outcomes from students is given in Table 7. The minimum score considered was one (1) and maximum was five (5). A five-point Likert scale: 5-strongly agree to 1-strongly disagree was administered [6]. The assessment results had Cronbach's α reliability coefficient > 0.8 and p-value < 0.05 in t-test for it to be statistically significant [10]. It also indicates that there is a significant difference between students' mean value responses (M) and the response option, "Neither agree nor disagree" in the measurement tool. Figure 1 shows a comparison of the pre-test and the post-test scores of SE knowledge test for each group.

Table 7 shows that the mean Likert scale scores (M) of the questions in IDD is 2.74 and the students responses for questions on IDD are shown in Fig. 2. The figure indicates that 54% of students strongly agree and 37% moderately agree to IDD1, and similarly, 17% of students strongly agree, and 50% moderately agree to IDD2. Students' response for IDD3 was 22% for strongly agreed and 37% for moderately agreed.

Positive responses were also received regarding the students' LO (Likert scale scores $M = 2.86$). Figure 3 shows the results for responses to questions on LO. Figure 3 shows that 57% of students strongly agreed and 33% moderately agreed to the question LO1 and 32% of students strongly agreed and 56% moderately agreed to question LO2.

Similarly, for LO3, it was found that 27% of students strongly agreed and 55% moderately agreed. In conclusion, almost all of the users found SE VLab to be satisfactory.

Similar results were obtained for ASS (Likert scale scores $M = 2.79$ in Table 7). The results from questions on ASS is shown in Fig. 4. It shows that 18% of students strongly agreed and 65% moderately agreed to the question ASS1 and 6% of students strongly agreed and 56% moderately agreed to the question ASS2.

Table 7 Students' perception results using CP-VLLM tool

Different parameters	No. of questions	N	α	M	SD	P
Instructional design delivery	3	61	0.84	2.74	1.23	0.0042
Students learning outcomes	3	61	0.87	2.86	1.03	0.0031
Assessments	3	61	0.94	2.79	1.19	0.0007
Students empowerment	3	61	0.83	2.82	0.87	0.0017
Social presence	3	61	0.91	2.89	1.12	0.0047
Critical thinking	3	61	0.98	2.91	0.91	0.0028
Alignment	3	61	0.94	2.93	0.96	0.0023

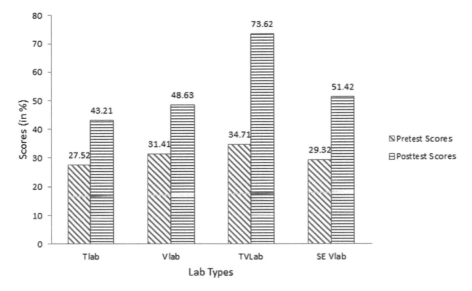

Fig. 1 Comparison of pre-test and post-test scores for each group

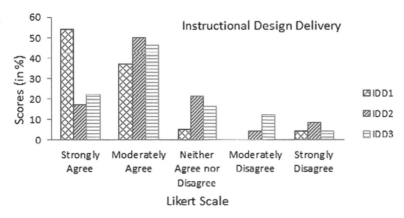

Fig. 2 Results from questions on IDD

Similarly, for ASS3, it was found that 21% of students strongly agreed and 46% moderately agreed.

The results for students' responses on their empowerment was excellent (Likert scale scores $M = 2.82$). Figure 5 shows the responses to questions on EM. From Fig. 5 we can conclude that 12% students strongly agreed and 69% moderately agreed to question EM1, and similarly, 28% strongly agreed and 56% moderately agreed to EM2. In EM3, the response was 9% for strongly agree and 71% for moderately agree.

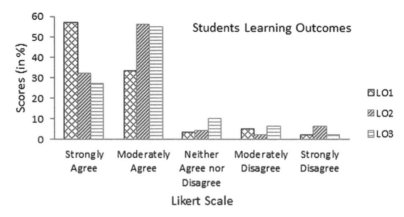

Fig. 3 Results from questions on LO

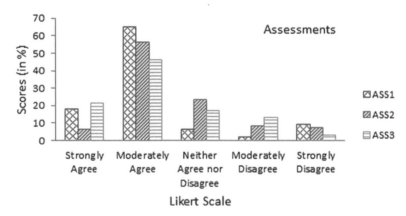

Fig. 4 Results from questions on ASS

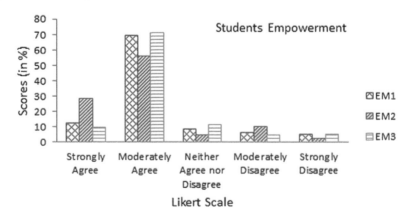

Fig. 5 Results from questions on EM

A very good response on SP (Likert scale scores $M = 2.89$) was also found. Figure 6 shows the responses of students to questions on SP. Figure 6 indicates that 16% of students strongly agreed and 59% moderately agreed to SP1 and 11% of students strongly agreed and 63% moderately agreed to SP2. For SP3, the responses were 13% for strongly agree and 76% for moderately agree.

We also found a very good response on students' CT skills (Likert scale scores $M = 2.91$ in Table 7). The results for responses to questions (refer to Appendix) on CT skills are shown in Fig. 7. The results in Fig. 7 indicate that 36% of students strongly agreed and 50% moderately agreed on CT1, 29% strongly agreed and 59% moderately agreed on CT2, and 21% of students strongly agreed and 69% moderately agreed on CT3.

The responses on AL was also found satisfactory (Likert scale scores $M = 2.93$ in Table 7) and the response results to questions on AL (refer Appendix) is shown in Fig. 8. The results in Fig. 8 indicate that 17% students strongly agreed and 68%

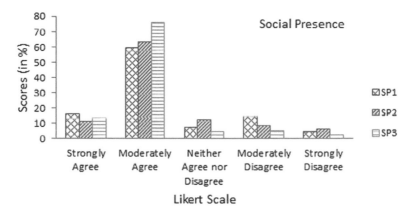

Fig. 6 Results from questions on SP

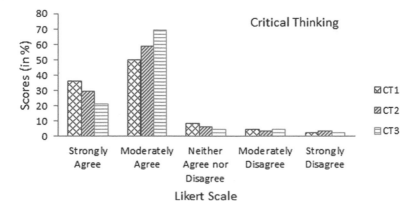

Fig. 7 Results from questions on CT

moderately agreed on AL1, 24% strongly agreed and 59% moderately agreed on AL2, and 19% strongly agreed and 64% moderately agreed on AL3.

Figure 9 indicate the students' overall acceptance toward SE VLab, as per CP contexts. We found that students preferred IDD and LO more compared to the other parameters in CP.

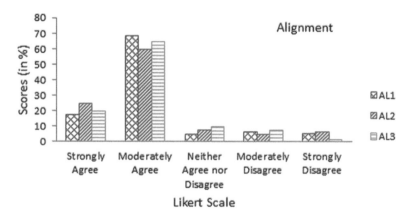

Fig. 8 Results from questions on AL

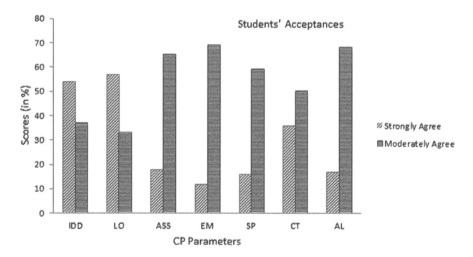

Fig. 9 Students' overall acceptance

9   Concluding Remarks

In this section, we presented the results of evaluation of engineering students' performance using a virtual laboratory, named as SE VLab, on the SE subject. The SE VLab can help in overcoming certain limitations of SE Tlab. We also developed a novel virtual learning measurement tool and evaluated students' learning performance using the same. Pre-test and post-test were administered for performance evaluation. Overall, the output of the research demonstrates a positive response toward the use of SE VLab.

9.1   Discussion

We now discuss the results obtained from students' perception using the CP-VLLM tool. The findings indicate that virtual lab-based learning is likely to be influenced by the CP contexts, within which the virtual learning activity is positioned.

Internet plays an important role in our daily life. It provides a link to connect to every part of the world. In the last few years, Internet media is being applied to the educational area. In India, very little attention has yet been devoted to Web-based courses. This paucity of research is surprising and is partly due to the relative novelty of Internet technology use in education. In this research, students' feedback indicates that they like this innovative working environment and feel encouraged to learn more effectively. We found that the VLab could complement a real lab.

9.2   Major Findings

We present below discussion of our findings in the context of contemporary research on technology enhanced virtual learning. An extensive review on the critical factors contributing toward enhancing the pedagogical effectiveness of online courses [3] reveals that some important concepts such as students' interaction and collaboration, active learning or students' initiated learning, prompt feedback, user-friendly technology access, teacher as a facilitator/mentor, learning communities, learning objectives, social presence, diverse learning styles, authenticity of the content are found to be very significant in determining the effectiveness of an online/virtual learning environment from students' perspectives. Using CP-VLLM tool, we also tried to incorporate these components in our study. As indicated by the results, the students viewed the IDD and LO parameters to be highly effective in our SE VLab.

In the context of Web-based virtual learning environment [8], researchers have also confirmed better learning performance, as well as higher levels of learning climate, self-efficacy and learners' satisfaction in comparison to traditional learning environment. In our work, when we have evaluated learners' satisfaction in both virtual and traditional learning environment, it is found to be better in virtual lab and best in the mixed (Tlab + VLab) scenario. This has also been supported by empirical findings. Students' learning outcomes are found to be better in blended learning environment as this allowed them to face better the real problems of the current technological world in a better way [9]. Keeping in view the diversity of the student population, Dalgarno et al. [7] suggested that using the virtual laboratory, students should be provided with wide range of learning opportunities from where they can choose various activities and resources to satisfy their own learning needs. In this context, in our SE VLab, we have also incorporated many URLs, and references for catering to the students' needs. In order to enhance the students' motivation and engagement, we have introduced some pedagogical agents as well.

Specifically, animated and interactive features embedded in the SE VLab experiments have induced positive outcomes in mediating the students' conceptualization of highly complex and abstract theoretical notions of SE [11]. The performance assessment tools adopted here in this study are found to be effective in improving the learning gains of students even with lower level of prior-subject knowledge.

In our research work, we tried to analyze the data both quantitatively as well as qualitatively, as promoted by (similar study undertaken by) Wang [11]. As it is well proven that working with real data instead of simulated ones may induce higher levels of students' motivation, it has been suggested by the researchers that creating a virtual learning environment through careful design, computer mediated technologies and proper coordination of group interaction and individual activities has been proved to be beneficial. Even in case of a highly abstract subject like Software Engineering [12], we have also tried to synchronize all these components for better learning outcomes. In this study, we introduced problem-based assignments in the "Exercises" section, which proved to be successful in enhancing the learning outcomes, such as higher post-test scores in content knowledge. This technique had also been adopted by other researchers in Web-based education and found to be effective [7].

Even though in this study we have found multiple advantages of virtual lab learning, we do not suggest that these VLabs would be the substitution of traditional labs. Rather the SE VLab can complement the traditional classroom or lab teaching–learning process for boosting the students' learning and motivation. Previous research indicates that the LO achieved using lab learning depends upon the instructional structure within which the lab experiments are implemented [9]. Findings of the present study reaffirms that students get more benefits from virtual lab learning environment, as compared to only TLab. The LO of students using the TVLab (those who have gained experience from both traditional and virtual lab sessions) are found to be more effective, as compared to the other considered groups. With regard to students' engagement in virtual classrooms there is substantial body of research indicating that learners get multiple benefits from lab learning through virtual mode [1, 4, 25]. In this study, we also found the participants to be actively engaged in virtual learning environment. In comparison to only TLab (Group-I) learning, students were found to be more active and involved in the SE VLab learning (Group-IV) environment.

Moreover, in this quasi-experimental study, we have also adopted a few open-ended questionnaires and interview sessions through which we collected some qualitative data.

9.3 Students' Responses Through Open-Ended Questions

From all of the 180 (Groups-II, III, and IV) SE VLab users' (students') responses to the open-ended questionnaires and interview sessions, most of them were found to indicate a positive attitude toward the use of SE VLab. They also reported that SE Vlab motivated and encouraged them to learn better. Most of the students (89%) found that SE Vlab engages them to achieve LO.

Feedback from students revealed a positive as well as constructive response toward the use of the SE VLab and some of these responses are given as follows.

> I like this SE VLab, I can do it at home in my free time. I can repeat experiments many times. I can interact with my virtual teacher. I do not feel bored, rather I found many interesting references books, research papers, and URLs. After using this lab, my confidence towards the SE subject increased. I feel that I can now independently solve my lab exercises.

However, based on a student's previous knowledge from SE theory classes and knowledge from conventional lab, a student stated that:

> I sometimes feel the absence of real-time contact with my friends and teacher. I need audio-visual connection in this SE VLab.

Some students, who had experienced difficulties in the TLab session found the SE VLab to be very useful for learning the lab experiments. One student mentions—

> I feel, doing the lab in virtual environment is more convenient way to perform better result. As SE VLab contains many hyper-links to connect to many reference sites, we can get more information when needed.

Students were found to be more encouraged in learning in a VLab environment, as VLab provides a user-friendly interface and a simple platform to learn and practice each experiment.

> I can repeat experiments and understand many concepts much easily than by reading a book. I can evaluate myself immediately after completing each experiment by using quiz sessions.

Around 92% of SE VLab users felt the need for video conferencing system. A user writes—

> I need more audio-video facilities to avoid the feelings of absence of my teacher. VLab is not the substitute of real lab. It is only for the additional hands-on practice purpose.

However, some students who had difficulties in evaluating themselves indicated that they found the VLab to be more interesting. One student writes—

The self-evaluation section is interesting. I can evaluate myself after using this SE VLab through self-evaluation section. If I am poor in the current experiment, I can reuse the experiment and re-evaluate my knowledge.

Virtual learning environment helps to increase teacher/student interaction and to make learning more student-centered. However, some students opined negatively as well. One student writes—

I feel isolation, frustration, confusion, and anxiety. Sometimes my interest in the lab experiment gets extremely reduced.

9.4 Educational Implications

The present study has the following educational implications:

1. An effective teaching–learning tool for distance learning and mass-education.
2. Can boost personalized instruction self-learning through quality education.
3. As self-learners, the students will develop skills to own-up their learning responsibilities.
4. Knowledge construction, dissemination, and management would be enhanced across the subject topics.
5. Students' motivation and creativity (through problem solving) be improved by promoting learning communities through both synchronous and asynchronous communication networks.

9.5 Limitations

The limitations that we come across during our study are listed as follow:

1. In this work, we did not consider the responses from teachers. As our sample was not very large, the drop rate toward the VLab users was almost zero. However, as per the existing research literature, drop rates tend to be higher in technology-mediated learning systems [5].
2. In this work, the SE VLab is based on only text and animations, and it does not include video and audio facilities.
3. As there are no other publicly available online virtual labs on software engineering other than this one, all the results reported in this study are specific to this SE VLab.
4. The data collected for this research is only from a demographically limited region, namely, the Indian states of West Bengal and Tamil Nadu. However, cross-cultural and trans-national evaluations in responses have not been performed in this study.

9.6 Conclusion and Directions for Future Work

Here we present a consolidated view of our research results reported in the previous sections. This section further contains our contributions, summary of research findings, and possible directions for future research.

In the modern society, software is developed for use in both offices and home. This introduces challenge of training large number of software engineers in the undergraduate and postgraduate programs. As the real-world problems are evolving rapidly, software engineers of the twenty-first century face new challenges. In order to cope with the situation, the introducing of laboratory-based learning in Software Engineering education is indispensable. However, large number of academic institutions, in developing countries currently involved in engineering education do not have necessary faculty and infrastructure that can ensure imparting quality hands-on training necessary for SE education. In order to address this lacuna, the SE VLab developed at IIT Kharagpur was evaluated.

CP-VLLM: Development of a Virtual Learning Measurement Tool. In an undergraduate course, practical activities are of different types, such as experimental investigation, control assignments, and project works. The control assignments constitute a major part of the laboratory activities. Often it is very difficult to assess lab activities, as there is no such standard measurement tool/instrument available online/offline. In view of this, we developed a learning measurement tool, named CP-VLLM, based on the theory of CP. We statistically tested the reliability and validity of the developed tool. This tool can be used for measuring students' learning outcomes and performance in online/virtual learning environments.

Evaluation of Students' Learning Performance in a Virtual Laboratory Learning Environment. As SE VLab is designed and implemented, it is important to evaluate this laboratory by considering students' feedback. In this context, we considered the CP-VLLM measurement tool, along with two SE subject knowledge test tools, and administered them for both pre-test and post-test data collection.

Directions for Future Research. There exist several unexplored points and open issues related to the research reported here that warrants further investigation. A tighter coupling with the social networking platforms (e.g., Facebook, Twitter, and Instagram) is envisioned to ensure a greater user participation, as well as a global audience. Such a feature would also enable group-activities and interval discussion among the participants. More case-study, new experiments (as per different syllabus), and exercises can be added in future.

In this research, we only considered critical pedagogy theory and developed tool following the said theory. In future, it is possible to develop other virtual learning measurement tools by considering Bloom's taxonomy [9], Kolb's model, Peter Honey, Alan Mumford's model, Anthony Gregorc's model, Neil Fleming's VAK/VARK model, and other cognitive approach to learning styles [5, 7, 11, 12].

The research work we conducted considers the evaluation process based on the data collected from IIT Kharagpur and few other engineering colleges from Tamil

Nadu, India. It is possible to collect data from other geographical regions of India. It is also possible to collect data from other countries. As this lab is available online on Web, students can use this lab and send their feedback online.

In the future, it can be extended to change or add some new experiments as per the current syllabus. As mentioned earlier, currently there is no publicly available SE VLab. In the future, when other SE VLabs are developed, a comparison can be made between them. However, for extending the evaluation of this SE VLab in the future, one can attempt to find similarities in user experiences between this VLab and virtual labs developed in other areas of computer science and engineering.

Appendix

Table 8 shows the details about CP-VLLM tool. This tool contains 5-point Likert scale: 1-"Strongly agree" to 5-"Strongly disagree".

Table 8 CP-VLLM tool

Parameters	Items	Questionnaires (5-point Likert scale) (5-*strongly agree* to 1-*strongly disagree*)
Instructional Design and Delivery (IDD)	IDD1 IDD2 IDD3	"The lab course materials are well organized" "The lab course with textual and visual contents improve students' learning" "The lab course contents are useful and latest"
Students' Learning Outcomes (LO)	LO1 LO2 LO3	"The lab course can be easily understood, comprehended, and remembered" "The lab course can be transferred or applied to other similar context to solve problems" "The components of this lab course can be properly analyzed and concept can be synthesized"
Assessments (ASS)	ASS1 ASS2 ASS3	"How students will be evaluated in the lab session is clearly explained" "Lab assignments with proper levels of difficulty are provided" "Feedback on lab assignments is provided"

(continued)

Table 8 (continued)

Parameters	Items	Questionnaires (5-point Likert scale) (5-*strongly agree* to 1-*strongly disagree*)
Students' Empowerment (EM)	EM1 EM2 EM3	"I can design a new software project" "I have opportunities to think how to solve a problem" "I can self-evaluate my learning achievements"
Social Presence (SP)	SP1 SP2 SP3	"I am getting encouraged to post questions for my clarifications" "I am getting opportunities to interact with other students I am getting encouraged to post my self-introduction"
Critical Thinking Skills (CT)	CT1 CT2 CT3	"The course content intimate to think in-depth about a subject" "The course content helps in solving problems in software designing" "The course content help in analyzing, interpreting, and synthesizing information"
Alignment (AL)	AL1 AL2 AL3	"Lab assignments provides students interests and abilities" "Lab assignments are based on the lab course content and learning objectives" "Lab learning outcomes are based on the requirements of the current syllabus"

References

1. Gunawan, G., Nisrina, N., Suranti, N. M. Y., Herayanti, L., & Rahmatiah, R. (2018). Virtual laboratory to improve students' conceptual understanding in physics learning. *Journal of Physics: Conference Series, 1108*(1), 012049.
2. Alam, A. (2022). Social robots in education for long-term human-robot interaction: Socially supportive behaviour of robotic tutor for creating robo-tangible learning environment in a guided discovery learning interaction. *ECS Transactions, 107*(1), 12389. https://doi.org/10.1149/10701.12389ecst
3. Kanij, T., & Grundy, J. (2020, November). Adapting teaching of a software engineering service course due to COVID-19. In *2020 IEEE 32nd conference on software engineering education and training (CSEE&T)* (pp. 1–6). IEEE.
4. Alam, A. (2020). Challenges and possibilities in teaching and learning of calculus: A case study of India. *Journal for the Education of Gifted Young Scientists, 8*(1), 407–433. https://doi.org/10.17478/jegys.660201
5. De Vries, L. E., & May, M. (2019). Virtual laboratory simulation in the education of laboratory technicians–motivation and study intensity. *Biochemistry and Molecular Biology Education, 47*(3), 257–262.
6. Alam, A. (2022, April). Educational robotics and computer programming in early childhood education: A conceptual framework for assessing elementary school students' computational

thinking for designing powerful educational scenarios. In *2022 International conference on smart technologies and systems for next generation computing (ICSTSN)* (pp. 1–7). IEEE. Retrieved from https://doi.org/10.1109/ICSTSN53084.2022.9761354

7. Schnieder, M., Williams, S., & Ghosh, S. (2022). Comparison of in-person and virtual labs/tutorials for engineering students using blended learning principles. *Education Sciences, 12*(3), 153.

8. Alam, A. (2020b). Pedagogy of calculus in India: An empirical investigation. *Periódico Tchê Química, 17*(34), 164–180. https://doi.org/10.52571/PTQ.v17.n34.2020.181_P34_pgs_164_180.pdf

9. Estriegana, R., Medina-Merodio, J. A., & Barchino, R. (2019). Student acceptance of virtual laboratory and practical work: An extension of the technology acceptance model. *Computers & Education, 135*, 1–14.

10. Alam, A. (2020c). Possibilities and challenges of compounding artificial intelligence in India's educational landscape. *International Journal of Advanced Science and Technology, 29*(5), 5077–5094. http://sersc.org/journals/index.php/IJAST/article/view/13910

11. Gamage, K. A., Wijesuriya, D. I., Ekanayake, S. Y., Rennie, A. E., Lambert, C. G., & Gunawardhana, N. (2020). Online delivery of teaching and laboratory practices: Continuity of university programmes during COVID-19 pandemic. *Education Sciences, 10*(10), 291.

12. Alam, A. (2020d). Test of knowledge of elementary vectors concepts (TKEVC) among first-semester bachelor of engineering and technology students. *Periódico Tchê Química, 17*(35), 477–494. https://doi.org/10.52571/PTQ.v17.n35.2020.41_ALAM_pgs_477_494.pdf

13. Faulconer, E. K., & Gruss, A. B. (2018). A review to weigh the pros and cons of online, remote, and distance science laboratory experiences. *International Review of Research in Open and Distributed Learning, 19*(2).

14. Alam, A. (2022). Positive psychology goes to school: conceptualizing students' happiness in 21st century schools while 'minding the mind!' Are we there yet? Evidence-backed, school-based positive psychology interventions. *ECS Transactions, 107*(1), 11199. https://doi.org/10.1149/10701.11199ecst

15. Jamshidi, R., & Milanovic, I. (2022). Building virtual laboratory with simulations. *Computer Applications in Engineering Education, 30*(2), 483–489.

16. Giray, G. (2021). An assessment of student satisfaction with e-learning: An empirical study with computer and software engineering undergraduate students in Turkey under pandemic conditions. *Education and Information Technologies, 26*(6), 6651–6673.

17. Alam, A. (2021, November). Possibilities and apprehensions in the landscape of artificial intelligence in education. In *2021 International conference on computational intelligence and computing applications (ICCICA)* (pp. 1–8). IEEE. Retrieved from https://doi.org/10.1109/ICCICA52458.2021.9697272

18. Mamani, N. M., García-Peñalvo, F. J., Conde, M. Á., & Gonçalves, J. (2021, September). A systematic mapping about simulators and remote laboratories using hardware in the loop and robotic: Developing STEM/STEAM skills in pre-university education. In *2021 International symposium on computers in education (SIIE)* (pp. 1–6). IEEE.

19. Singh, G., Mantri, A., Sharma, O., & Kaur, R. (2021). Virtual reality learning environment for enhancing electronics engineering laboratory experience. *Computer Applications in Engineering Education, 29*(1), 229–243.

20. Triejunita, C. N., Putri, A., & Rosmansyah, Y. (2021, November). A systematic literature review on virtual laboratory for learning. In *2021 International conference on data and software engineering (ICoDSE)* (pp. 1–6). IEEE.

21. Ramírez, J., Soto, D., López, S., Akroyd, J., Nurkowski, D., Botero, M. L., Bianco, N., Brownbridge, G., & Molina, A. (2020). A virtual laboratory to support chemical reaction engineering courses using real-life problems and industrial software. *Education for Chemical Engineers, 33*, 36–44.

22. Hartikainen, S., Rintala, H., Pylväs, L., & Nokelainen, P. (2019). The concept of active learning and the measurement of learning outcomes: A review of research in engineering higher education. *Education Sciences, 9*(4), 276.

23. Alam, A. (2021, December). Should robots replace teachers? Mobilisation of AI and learning analytics in education. In *2021 International conference on advances in computing, communication, and control (ICAC3)* (pp. 1–12). IEEE. Retrieved from https://doi.org/10.1109/ICAC353642.2021.9697300

24. Alam, A. (2022). Psychological, sociocultural, and biological elucidations for gender gap in STEM education: A call for translation of research into evidence-based interventions. In *Proceedings of the 2nd international conference on sustainability and equity (ICSE-2021)*. Atlantis Highlights in Social Sciences, Education and Humanities. ISSN: 2667-128X. Retrieved from https://doi.org/10.2991/ahsseh.k.220105.012

25. Tobarra, L., Robles-Gomez, A., Pastor, R., Hernandez, R., Duque, A., & Cano, J. (2020). Students' acceptance and tracking of a new container-based virtual laboratory. *Applied Sciences, 10*(3), 1091.

26. Glassey, J., & Magalhães, F. D. (2020). Virtual labs—love them or hate them, they are likely to be used more in the future. *Education for Chemical Engineers, 33*, 76.

27. Alam, A. (2022, April). A digital game based learning approach for effective curriculum transaction for teaching-learning of artificial intelligence and machine learning. In *2022 International conference on sustainable computing and data communication systems (ICSCDS)* (pp. 69–74). IEEE. Retrieved from https://doi.org/10.1109/ICSCDS53736.2022.9760932

A Customer-Focused System Structure for Cyber Security Operations Center

Taslina Akter, Kuraish Bin Quader Chowdhury, Md Tamimur Rahman, and Shatabdee Bala

Abstract In most cases, a Security Information and Event Management (SIEM) architecture is constructed to regularly guarantee events via several avoidance measures with ensigns' warnings to ensure an endeavor's network security. The alarms are assessed by experts in the security operation center (SOC) to determine whether they are malicious. Regardless, the volume of notices is usually huge, in the company of the weight of themselves deceiving upsides along with the SOC's ability to play with every alarm being exceeded. As a result, harmful assaults and infected hosts may go undetected. Machine learning is a feasible option for lowering false-positive rates and increasing SOC analyst productivity. In this evaluation, everyone encourages by a Customer-Basic AI build for the digital protection activity emphasis in a real-world organization scenario. We'll go over the most well-known details of history in SOC, how they perform, and make use of to managing those data sets which help AI thrive.

Keywords Machine learning system · Cyber security operation center · Danger detection

1 Introduction

Organizational security will have an impact on the relationship. In order to detect suspicious activity, associations or nations use the Security Information and Event Management (SIEM) architecture. Endpoints, fire dividers, Intrusion Detection and Prevention Systems (IDS/IPS), Data Loss Prevention (DLP), Domain Name System (DNS), Dynamic Host Configuration Protocol (DHCP), Windows/Unix

T. Akter (✉) · K. B. Quader Chowdhury
Department of Electrical and Electronic Engineering, Independent University, Dhaka, Bangladesh
e-mail: 2222796@iub.edu.bd

M. T. Rahman · S. Bala
Department of Computer Science and Engineering, Gono Bishwabidyalay, Dhaka, Bangladesh

© The Author(s), under exclusive license to Springer Nature Singapore Pte Ltd. 2023   901
S. Shakya et al. (eds.), *Sentiment Analysis and Deep Learning*,
Advances in Intelligent Systems and Computing 1432,
https://doi.org/10.1007/978-981-19-5443-6_66

health stages, VPN branches, and other event logs are all guaranteed to be available. There are a few different types of safety situations. Each day, the records contain massive amounts of data, measured in gigabytes [1]. The Security Operation Center (SOC) group creates such use cases from the security event logs based on the investigators' interactions. They are frequently decision-based frameworks that link at the minimum sign via several branches. The indicated criteria would be based on network or length. When a pre-defined application case is activated, the SIEM device will send a real-time notice. The notifications will then be investigated by SOC analysts to determine if the user associated with the warning is dangerous (a true helpful side) or not (untrue-helpful side). SOC specialists will create Open Source Ticket Request System (OTRS) tokens if this investigation reveals certain cautions are suspect. Following the underlying request, some OTRS tickets will be elevated to level 2 examination frameworks (e.g., Co3 Process) as important security events that the Emergency Backup Squad will investigate and target. SIEM, on the other hand, often creates a massive number of warnings, with quite significant wrong-helpful charges. The statistics of warnings received every period might reach a huge number, far beyond the capabilities of the SOC to investigate them all [2].

As a result, SOC may decide to analyze just high-severity warnings or suppress all alerts of the same type. This might result in some serious attacks being missed. As a result, to detect dangerous users, a more sophisticated and automated approach is necessary. The AI computation is at the heart of the SOC's work process, combining various event logs, SIEM alerts, and SOC examination findings to produce a holistic client risk rating for the security war room. SOC professionals might use AI risk ratings to focus their examinations, starting with individuals who pose the greatest risk, rather than sifting through many SIEM alarms in search of a pile [3].

This will essentially increase their productivity, improve their work line for leaders, and aid the organization's security in the long term. Our strategy, in particular, constructs a framework of buyer-based AI techniques to distinguish customer risk assessment based on cautious data. This method can provide security investigators with a comprehensive gambling evaluation of a client, allowing them to focus on those who have high gambling ratings. To the authors' knowledge, no prior study has been conducted on developing a full system solution for this application [4–7]. This paper's key contribution is about as follows:

- To assess user hazards, an innovative user-centric deep learning approach is presented and tested using real-world industrial data. The solution may significantly minimize the resources required to manually assess alarms while also improving company security.
- To construct includes and convey names for AI calculations, an ideal information designing approach is introduced that combines ready signs, security logs, and SOC expert examination notes.

2 Data in Raw Format and Error Checking

The verifiable facts come via Symantec's information gathering records. SIEM stage warnings, notes from logical testing, and logs from various sources such as firewalls, IDS/IPS, HTTP/FTP/DNS display, DHCP, susceptibility separation, Windows safety events, VPNs, and so more are all fully covered. Substance aim is critical for data traffic otherwise other data roots with active IP addresses since, in large companies, customers are gradually assigned different inner IP addresses, causing their IP addresses to shift over time [8]. Without precise unique IP to client planning, correlating movement across several organization records would be difficult and incorrect. To organize traffic data, IP to client planning is used to ensure that the client ID is connected as the primary key for dataset creation, data coordination, and model evaluation. Finally, examination study reports are frequently saved in text format in such a labeling structure. The following information is frequently remembered for the notes: why an alert was issued, specific data from the framework logs and other assets (like VirusTotal and IPVoid), and the examination ending in case an alarm is an sure as an alternative misleading positive [9].

3 Label Generation and User Feature Engineering

3.1 Page Numbering and Running Heads

Because our primary purpose is to estimate user risk, the characteristics are built at the individual user level. Over 100 characteristics have been developed to define a user's activity. Sum together attributes derived from the insights list (number of warnings per day), worldly elements derived from relapse analysis (movement appearance recurrence), and connected aspects derived from social data analysis (client centrality from client occasion chart, and so on) [10].

3.2 Page Numbering and Running Heads

Since all of the characteristics have been created, we must assign a benchmark or grade to our machine learning algorithms. Mining analysts' inquiry notes provide the primary labels. To derive the user's real situation from the remarks, Text/data mining algorithms like as important phrase/word extraction and sentiment classification are applied [11]. After text mining, only a few users with annotations (2% in general) will be classified as dangerous. If we just employ these customers for machine learning algorithms, we have two concerns:

Table 1 A dataset for final modeling as an example

User ID	Summary feature 1	Indicator feature 2	Temporal feature 3	Relational feature 4	Label
User 1	13	1	0.65	5.17	1 (risky Initial)
User 2	25	0	2.74	9.34	1 (risky Derived)
User 3	4	0	1.33	3.52	0 (normal)

- The majority of consumers who do not have annotations are excluded from the model, yet they may have helpful info.
- Many ML algorithms/models do not perform well when faced with a severely imbalanced categorization task.

Label propagation strategies are required to generate more labels in order to address these two difficulties. The fundamental notion here is that if we know about specific problematic users, we may identify other users who exhibit similar behaviors as risky. We employed two label propagation techniques: matrix plunging-based clustering as well as controlled PU learning [12–14].

Finally, analyzing text and tab transmission as objectives for our ML algorithms to mix the tabs. Table 1 is an example of the final analytic dataset here:

4 Algorithms and Implementations in ML

4.1 Algorithms for ML

An expert system algorithm is a method through which an AI system achieves its goal, which is to anticipate output values from given input data. Machine learning algorithms can be classed as either supervised or unsupervised. Unsupervised algorithms work with data that has not been classified or labeled, whereas supervised learning algorithms have both input and output data labeled for them. In our framework [15], Multi-Layer Neural Networks (MLN) with two secret layers, Random Forests with Gin-split trees, Support Vector Machines with large workpieces, and Logistic Regression were among the organizations used. The results of our experiments largely point to Multi-layer Neural Networks and Random Forests performing admirably in our case. The approval results for these models will be released later.

4.2 Contact Author Information

Showing data should be partitioned into getting ready and testing sets with no evident end objective in mind, as is a standard approach, and multiple models should be examined on test holdout data [16]. In addition to the Area under the Bend, we define two model quality measurements in Eqs. (1) and (2) [17, 18].

Rate of model detection

$$= \left(\frac{\text{Number of the risky host}}{\text{in certain estimations}} \right) \Big/ \left(\frac{\text{Total number}}{\text{of risky hosts}} \right) * 100\% \tag{1}$$

Lifted model

$$= \left(\frac{\text{Proportion of risky host}}{\text{in certain predictions}} \right) \Big/ \left(\frac{\text{Total proportion}}{\text{of risky host}} \right) * 100\% \tag{2}$$

Detection rate and lift, in contrast to the area under the graph, which evaluates the model on the entire test data, demonstrate how successful the model is at detecting problematic users among distinct segments of predictions. To generate these two measures, the data is first sorted in descending order by the model scores. The proportion of results obtained with and without a characterization model is referred to as the model's identification pace. Assume there are 60 potentially hazardous clients in the test data, and the model identifies 30 of them from the top 10% of forecasts; the recognition rate is equal to 80%. Lift is a metric that determines how often using a model is preferable than not using one. If the experimental data covers 5000 customers, the lift is equivalent to $(30/500)/(60/5000) = 5$. On specific expectations, a vehicle with a higher lift will perform better.

4.3 Model Evaluation Techniques

We examine one month of model execution data and establish execution metrics to access the feasibility of our AI platform. We randomly divided the data into two categories: preparation (75% of the cases) and testing (25% of the examples) (the leftover 25%). The AUC is calculated using text data. The multi-facet Neural Network and Random Forest provide acceptable exactness with AUGs greater than 0.80. Table 2 displays the data as follows.

Table 3 displays the area charges over a lengthy period in light of the top 5–20% estimates separately. With only 20% of the top conjectures, Irregular Forest can uncover 80% of certified dangerous examples, which is empowering.

The lift achieved by the multi-facet Neural Network for the top 5% gauges is shown in Table 4, which is 6.82, which is more than multiple times better than the current rule-based technique. The Multi-layer Neural Network has the best average

Table 2 Model AUC on test data

	MNN	RF	SVM	LR
Mean	**0.807**	**0.829**	0.775	0.754
Standard Error	0.006	0.004	0.016	0.008

Table 3 Model detection rates on top 5–20% predictions

Top % predictions (%)	MNN (%)	RF (%)	SVM (%)	LR (%)
5	32	24.50	21.20	32.77
10	59	42.93	47.57	51.20
15	71	71.20	71.00	69.23
20	79	**79.50**	79.20	75.57

Table 4 Model lifts on top 5–20% predictions

Top % of predictions (%)	MNN	RF	SVM	LR
5	**6.82**	5.30	4.19	6.82
10	6.25	4.55	4.92	5.30
15	4.92	4.92	4.92	4.80
20	4.09	4.19	4.19	4.00
Average	**5.52**	4.74	4.56	5.23

lift of over 5.5 when we look at the regular lifts of the top 5–20% values, as shown in Table 4's last half. That's especially exciting.

4.4 What Will Be Done with Your Paper

The AI framework is now being used in a verifiable creation environment. The properties and names are updated regularly based on previous data. The AI model is then updated and sent to the scoring motor regularly to ensure that it receives the most up-to-date data patterns. When fresh alarms are triggered, the risk assessments are done in stages, allowing SOC examiners to respond quickly to high-risk individuals. Finally, the comments will be compiled and integrated into current data in order to continue working on the model. The entire interaction has been mechanized, from data combining to score generation. Furthermore, the framework efficiently promotes innovative ideas from research.

5 Conclusion

We propose a client-driven AI framework in this review that recognizes trouble-some customers by combining data from various security logs, prepared information, and master experiences. For organization security exercises, this solution provides a complete structure and response for hazardous client disclosure. As part of this work, we will talk about creating marks from SOC test notes, utilizing IP, host, and clients to design client-driven ascribes, evaluating AI computations, and implementing such a process in a SOC advancement setting. We also show how, even with simple AI algorithms, the resulting structure can distinguish additional pieces of information from data with severely unbalanced and constrained markings. The top 20% projection level for the multi cerebrum network model is often a few times higher than the present rule-based approach. The entire AI framework is entirely robotized in a creation setting, drastically improving and upgrading business risk ID and the board, from information gathering to everyday model reviving to constant scoring. In terms of future developments, we'll look into further learning techniques to improve location precision even more.

References

1. Khan, N. R., Rabbi, M., Al Zabir, K., Dewri, K., Sultana, S. A., & Lippert, K. J. (2022). Internet of things-based educational paradigm for best learning outcomes. In *2022 International conference on advances in computing, communication and applied informatics (ACCAI)* (pp. 1–8). https://doi.org/10.1109/ACCAI53970.2022.9752569
2. Rahman Khan, M. N., Yesmin, S., Aktar, M., Quader Chowdhury, K. B., Labeeb, K., & Abedin, M. Z. (2021). Techniques for multi-omics data incorporating machine learning and system genomics. In *2021 6th International conference on communication and electronics systems (ICCES)* (pp. 1524–1528). https://doi.org/10.1109/ICCES51350.2021.9489222
3. Khan, M. N. R., & Lippert, K. J. (2022). A framework for a virtual reality-based medical support system. In C. K. K. Reddy (Ed.), *The Intelligent systems and machine learning for industry: Advancements, challenges, and applications* (pp. 8–28). CRC Press, Taylor & Francis Group.
4. Khan, M. N. R., & Lippert, K. J. (2022). Immersive technologies in healthcare education. In C. K. K. Reddy (Ed.), *The Intelligent systems and machine learning for industry: Advancements, challenges, and applications* (pp. 111–132). CRC Press, Taylor & Francis Group.
5. Khan, M. N. R., Mashuk, A. K. E. H., Durdana, W. F., Alam, M., Roy, R., & Razzak, M. A. (2019). Doctor who?'—A customizable android application for integrated health care. In *2019 10th International conference on computing, communication and networking technologies (ICCCNT)* (pp. 1–6). https://doi.org/10.1109/ICCCNT45670.2019.8944501
6. Khan, M. N. R., Shahin, F. B., Sunny, F. I., Khan, M. R., Haque Mashuk, A. K. E., & Al Mamun, K. A. (2019). An innovative and augmentative android application for enhancing mediated communication of verbally disabled people. In *2019 10th International conference on computing, communication and networking technologies (ICCCNT)* (pp. 1–5). https://doi.org/10.1109/ICCCNT45670.2019.8944655
7. Khan, M. N. R., Durdana, W. F., Roy, R., Poddar, G., Ferdous, S., & Mashuk, A. K. E. H. (2018). Health guardian—A subsidiary android application for maintaining sound health. In

2018 International conference on recent innovations in electrical, electronics & communication engineering (ICRIEECE) (pp. 1406–1409). https://doi.org/10.1109/ICRIEECE44171. 2018.9008485

8. Khan, M. N. R., Sonet, H. H., Yasmin, F., Yesmin, S., Sarker, F., & Mamun, K. A. (2017). 'Bolte Chai'—An android application for verbally challenged children. In *2017 4th International conference on advances in electrical engineering (ICAEE)* (pp. 541–545). https://doi.org/10. 1109/ICAEE.2017.8255415

9. Khan, M. N. R., Pias, M. N. H., Habib, K., Hossain, M., Sarker, F., & Mamun, K. A. (2016). Bolte Chai: An augmentative and alternative communication device for enhancing communication for nonverbal children. In *2016 International conference on medical engineering, health informatics and technology (MediTec)* (pp. 1–4). https://doi.org/10.1109/MEDITEC.2016.783 5391

10. Khan, M. N. R., Iqbal, M. M. T., Yesmin, S., Mashuk, A. K. E. H., Shahin, F. B., & Razzak, M. A. (2018). Development of an automatic detection of pressure distortion and alarm system of endotracheal tube. In *2018 IEEE-EMBS conference on biomedical engineering and sciences (IECBES)* (pp. 476–479). https://doi.org/10.1109/IECBES.2018.8626676

11. Haque, H., Labeeb, K., Riha, R. B., & Khan, M. N. R. (2021). IoT Based water quality monitoring system by using zigbee protocol. In *2021 International conference on emerging smart computing and informatics (ESCI)* (pp. 619–622). https://doi.org/10.1109/ESCI50559.2021. 9397031

12. Labeeb, K., Chowdhury, K. B. Q., Riha, R. B., Abedin, M. Z., Yesmin, S., & Khan, M. N. R. (2020). Pre-processing data in weather monitoring application by using big data quality framework. In *2020 IEEE international women in engineering (WIE) conference on electrical and computer engineering (WIECON-ECE)* (pp. 284–287). https://doi.org/10.1109/WIECON-ECE52138.2020.9397990

13. Lippert, K., Khan, M.N. R., Rabbi, M. M., Dutta, A., & Cloutier, R. (2021). A framework of metaverse for systems engineering. In *2021 IEEE international conference on signal processing, information, communication and systems*, 3–4 December.

14. Khan, M. N. R., Shakir, A. K., Nadi, S. S., & Abedin, M. Z. (2022). An android application for university-based academic solution for crisis situation. In: S. Shakya, V. E. Balas, S. Kamolphiwong, & K. L. Du (Eds.), *Sentimental analysis and deep learning. Advances in intelligent systems and computing*, Vol. 1408. Springer. https://doi.org/10.1007/978-981-16-5157-1_51

15. Khan, M. N. R., Ara, J., Yesmin, S., & Abedin, M. Z. (2022). Machine learning approaches in cybersecurity. In: I. J. Jacob, S. Kolandapalayam Shanmugam, & R. Bestak (Eds.), *Data intelligence and cognitive informatics. Algorithms for intelligent systems*. Springer. https://doi. org/10.1007/978-981-16-6460-1_26

16. Khan, M. N. R., Saha, H., Yesmin, S., & Abedin, M. Z. (2022). An innovative framework by using metaheuristic algorithms for detecting fake news on social media. In: I. J. Jacob, S. Kolandapalayam Shanmugam, & R. Bestak (Eds.), *Data intelligence and cognitive informatics. Algorithms for intelligent systems*. Springer. https://doi.org/10.1007/978-981-16-6460-1_27

17. Khan, M. N. R., Saha, H., Yesmin, S., & Abedin, M. Z. (2022). Review of city pricing system analysis based on big data. In: I. J. Jacob, S. Kolandapalayam Shanmugam, & R. Bestak (Eds.), *Data intelligence and cognitive informatics. Algorithms for intelligent systems*. Springer. https://doi.org/10.1007/978-981-16-6460-1_25

18. Khan, M. N. R., Bala, S., Yesmin, S., & Abedin, M. Z. (2022). Bioinformatics: The importance of data mining techniques. In: S. Shakya, V. E. Balas, S. Kamolphiwong, K. L. Du (Eds.), *Sentimental analysis and deep learning. Advances in intelligent systems and computing*, Vol 1408. Springer. https://doi.org/10.1007/978-981-16-5157-1_32

Credit Card Defaulters Prediction Using Unsupervised Features

Thomaskutty Reji, Andrea Rodrigues, and Jossy P. George

Abstract Banks take multiple risks while providing credit cards and loans to their customers. A customer becomes a credit card defaulter when they fail to pay their credit amount within the due time. Risk minimization is possible when banks are able to predict the probability of whether a customer becomes a defaulter or not. This paper proposes an enhanced machine learning model for predicting default customers, using unsupervised features from the Gaussian mixture model and K-means clustering. The method has been tested against various classification algorithms like gradient boosting, AdaBoost, SVM, linear discriminant analysis, and logistic regression. Experimental results showed that the performance of unsupervised features models had been significantly improved by the proposed methodology. Comparing the performance measures indicated that the gradient boosting machine outperformed other models with accuracy and recall of 88% and 89%, respectively.

Keywords Unsupervised learning · Clustering · Machine learning · Credit card defaulter · Risk minimization · Boosting

1 Introduction

People spend ahead of time and borrow credit to enjoy certain goods sooner than later. While the need for credit has increased, the ability to pay back has not, which has led to a significant increase in the number of defaults in the credit industry. Credit card default has an impact not only on the status of the bank's credit card holders but also on the standing of other banks in general. As a result of card default, customers

---

T. Reji (✉) · A. Rodrigues · J. P. George
Department of Data Science, Christ (Deemed to be University), Lavasa, Pune, Maharashtra, India
e-mail: thomaskutty.reji@science.christuniversity.in

A. Rodrigues
e-mail: andrea.rodrigues@christuniversity.in

J. P. George
e-mail: frjossy@christuniversity.in

© The Author(s), under exclusive license to Springer Nature Singapore Pte Ltd. 2023 909
S. Shakya et al. (eds.), *Sentiment Analysis and Deep Learning*,
Advances in Intelligent Systems and Computing 1432,
https://doi.org/10.1007/978-981-19-5443-6_67

will face credit bureau block listing, a stricter criterion for loan approval, and other issues in credit-based services.

There are rules and regulations to be followed according to the bank's credit policies. One of the most important conditions is the time limit for repaying borrowed funds. It is recommended that one does not skip a monthly credit card charge after this specified period. However, even if customers miss such dates, they will be given additional time to complete the payments. The account will default if the amount is not accommodated during this grace period [1]. There are multiple reasons one could be a credit defaulter: excessive spending, delaying tactics, disruptions in the mail, and ignoring the due date. It would always help find those customers to minimize the credit risk. In short, the customer becomes a defaulter when they fail to repay the amount within the allowed period.

Banking systems are using various risk minimization models to solve the issues. However, since this is a prediction problem, machine learning models can also be used to predict card default probability. Models which can predict the likelihood of potential defaulters can be adapted by banking systems to minimize the credit risk. There are classification algorithms that can perform binary classification. Machine learning is better suited to solve these kinds of problems. If a model can predict the card default probability, the bank may be able to prevent loss and provide the customer with other options like forbearance and debt consolidation [2].

In the risk modeling problem, the type II error is more serious. An intelligent model should not miss any customer who is likely to default. To minimize the type II error, the model's false positive rate should be low. Hence, the recall metric would be a highly efficient model selection criteria for credit card defaulter's prediction. Since the target class is usually unbalanced in banking use cases, the main challenge of this work would be getting a good recall score.

In this work, machine learning algorithms are used to predict the probabilities of credit card default. Five different models will be trained, namely gradient boosting classifier, AdaBoost classifier, linear discriminant analysis, support vector machines, logistic regression, and two unsupervised learning models such as Gaussian mixture model and K-means clustering.

1.1 Contribution

Credit card defaulters problem has been approached by many researchers with machine learning and deep learning techniques. Only a few works have been concerned with the unsupervised learning techniques, especially in the feature engineering process. Existing works focused more on getting a good overall accuracy. This paper proposes unsupervised features for predicting the probability of credit card default. Different machine learning classification algorithms are trained and evaluated in separate phases, with and without the unsupervised features. Model evaluation is done based on the recall metric rather than the overall accuracy.

1.2 Organization

The paper is organized in the following manner. Section 2 contains the literature review, Sect. 3 covers the proposed methodology and the model workflow, Sect. 4 includes the experimental results and comparative analysis, and conclusions of the work are presented in Sect. 5.

2 Literature Review

Researchers have tried to predict the probability of credit card default with the credit card default dataset in the UCI repository. They have used various machine learning classification algorithms to solve the problem. Deep learning works have also been conducted in credit risk modeling, including artificial neural networks, perceptrons, and self-organizing maps.

Shivanna & Agarwal [3] used a deep support vector machine, decision trees, and perceptron to predict the card defaulters. Among the various models, deep support vector machines performed better. They used the UCI credit card dataset of 30,000 customer information and 25 features. The proposed support vector machines achieved an accuracy of 82.2% and an AUC of 0.74.

Keramati & Yousefi [4] conducted a comprehensive literature survey related to the various data mining methods in the credit card industry. They studied machine learning methods like linear discriminant analysis (LDA) and K-nearest neighbors, CART algorithm, and used AUC score as the evaluation metric. They have investigated how ensemble models help in credit risk minimization. They have also introduced some hybrid models for credit score modeling.

Agarwal et al. [5] have completed a comparative study and proposed enhancing classification techniques using principal component analysis for credit card predictions. They evaluated various classification algorithms like logistic regression, decision tree, K-nearest neighbor, and Naïve Bayes. The existing credit card defaulter dataset from the UCI repository is used for the analysis. The objective was to predict the likelihood of potential defaulters. The principal component analysis is applied to evaluate its impact on the performance of the machine learning algorithms. They found that logistic regression outperforms all other models.

There are various unsupervised machine learning methods for classification tasks. Rai & Dwivedi [6] proposed a neural network-based unsupervised learning for detecting frauds in the credit card dataset. The proposed method outperformed the existing autoencoder, K-means clustering, isolation forest and achieved an accuracy of 99.87%. Panchal & Verma [7] used self-organizing maps (SOMs) to identify potential credit card defaulters in future. Self-organizing maps are helpful in anomaly detection and dimensionality reduction.

Alam et al. [8] conducted an exhaustive study on credit card defaulter prediction on an imbalanced dataset. Multiple credit card datasets, namely Taiwan clients credit dataset, South German clients credit dataset, Belgium clients credit dataset, are used for the model evaluation. They have developed a model after applying various resampling techniques and data normalization. Experimental results show that the gradient boosted decision tree method outperforms other models.

Zhu et al. [9] used gradient boosting models to predict credit users' late repayment rate. They used the receiver operator curve (ROC) and other measures for the model evaluation. The entire study is executed using P2P company data, consisting of 30,000 customer details.

Soui et al. [10] conducted an exploratory analysis of the Taiwanese consumer payment dataset using the genetic programming algorithm to increase the model's accuracy. The findings showed that the genetic programming algorithm could create an effective assessment model based on IF-THEN criteria. The model can attain an average of more than 86% precision, recall, and accuracy.

Vishwakarma & Hajela [11] did a comparative study of the performance of various machine learning algorithms in predicting credit card default. Different classification techniques, including boosting, have tested and compared its confusion matrix and area under the curve (AUC).

3 Proposed Methodology

The proposed methodology includes three phases. In the first phase, baseline models are trained with the necessary transformation of the data. In the second phase, baselines models are again re-trained by applying the SMOTE technique for handling imbalanced data. Finally, the models are re-trained by adding the features generated by the unsupervised learning techniques such as Gaussian mixture models and K-means clustering. Figure 1 shows the entire workflow from data gathering to the final model of the proposed system.

After data gathering, the system splits the source dataset into training and testing sets with an 80:20 split ratio. The second step consists of preprocessing, including feature generation from the unsupervised algorithms, feature selection, and data transformation. Both training data and testing data are subjected to the preprocessing pipeline. Model training is done in three phases. For all the phases, the workflow is almost the same. Several classification models are trained using the training data, and the final model is selected after evaluating the recall and accuracy measures.

3.1 Dataset

The source dataset is taken from UCI machine learning repository, and it consists of 24 columns including credit limit, sex, education, marital status, age and history of past payment, amount of bill statements, and amount of the previous payments for the past six months. 1001 customer instances are available in the dataset. Limit

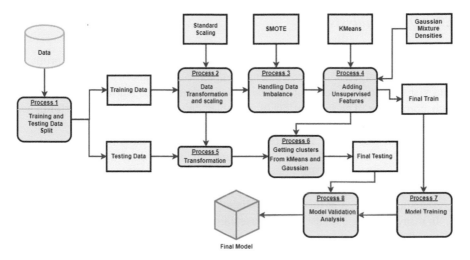

Fig. 1 Proposed model architecture

balance feature is the amount of the given credit for a particular customer. The bill amount and pay amount features are the bill statements and the previous payments, respectively, in different months. The response variable is binary numerical where one represents a default customer, and zero represents the non-default. Among the customer instances, 787 are non-default customers and 214 are defaulters. Table 1 gives a detailed view of feature descriptions.

Table 1 Dataset feature descriptions

Feature	Description
Limit balance	Amount of the given credit in dollars
Sex	Gender of the customer (1 = male; 2 = female)
Education	Education (1 = graduate, 2 = university, 3 = high school, 4 = others)
Marriage	Marital status (1 = married, 2 = single, 3 = others)
Age	Age of the customer
Pay_	History of repayment status in different months
Bill amount	Amount of bill statements in different months
Pay amount	Amount of the previous payments in different months
Target	Binary (1 = default customer, 0 = non-default customers)

3.2   Data Analysis

Jupyter Notebooks and major Python packages like Pandas and NumPy are primarily used for data analysis and preprocessing. Matplotlib and Seaborn libraries are used for data visualization. Correlation analysis showed that the sequential variables exhibit high correlation. The correlation of variables can affect the model performance. Figure 2 represents the correlation heatmap of the data.

As represented in Fig. 1, two blocks of the correlated variables are evident. The correlation blocks show that pay columns and bill amount columns are highly correlated among themselves. By considering these correlation blocks, two additional features have been generated from the pay columns and bill amount columns by aggregating the mean of the feature values.

In many banking use cases, the dataset target class ratios show a significant difference in the ratio. Target ratio analysis showed that the distribution of classes is highly imbalanced in the credit card defaulter dataset as expected. The final model should be intelligent enough to capture all the defaulters in the testing dataset. For handling target imbalance, the SMOTE oversampling technique is used. Synthetic minority over sampling technique (SMOTE) is a resampling technique that creates new data points for the minority class based on the nearest neighbors algorithm [12]. Figure 3 shows the target class distribution before and after applying smote.

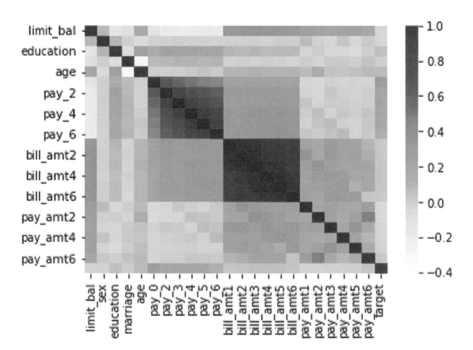

Fig. 2 Correlation analysis heatmap

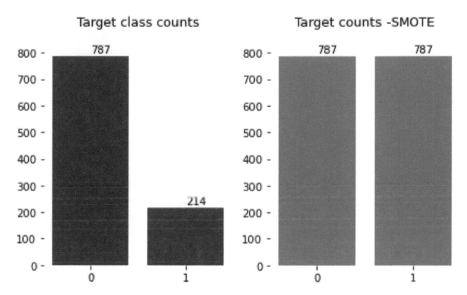

Fig. 3 Target class distribution

The significant difference between the counts in the first subgraph can severely impact the evaluation metric, especially the recall value. In the second subgraph, the value counts for both zero and one are equal (zero labels are oversampled). The Python class of SMOTE is imported from the Imblearn package. Since the dataset contains no missing values, all the further imputation processes are skipped.

3.3 Model Training

Model training happens in three phases. In the first phase, baseline models are trained without handling the data imbalance. A custom function is created for transforming the limit balance columns. The function takes the input value, and if it is negative, then it will be reassigned as zero, else it applies the NumPy log function. Since the dataset contains no null values, missing value imputation steps are skipped. After storing the final training and final validation data, baseline models are trained. The validation results are then collected in a data frame.

In the second phase, all the baseline models are further improved by handling the data imbalance using the SMOTE technique, feature standardization, and addition of two aggregate features from the bill amount and pay amount explanatory variables. In the third phase, the Gaussian mixture model and K-means clustering models are used for additional cluster features. The Sklearn library has a class for the K-means and the Gaussian mixture models. Those classes are imported into the Jupyter to create the cluster features. Along with unsupervised features, the two features which are generated in the second phase have also been used with SMOTE in the phase three models. The recall metric is taken as the primary model evaluation criteria to decrease the number of false-negative rates.

3.4 Algorithms

The primary machine learning algorithm in this study includes gradient boosting classifier, AdaBoost classifier, linear discriminant analysis, support vector machines, logistic regression, K-means clustering, and Gaussian mixtures.

Gradient boosting is one of the ensemble-boosting algorithms used for classification and regression. In general, boosting algorithms are trying to update the model by minimizing past errors sequentially. First, the algorithm initializes the model with a constant value. Each iteration computes the pseudo residuals, fits the base learner to the errors, and finally updates the model. This process continues till it reaches the given threshold [13].

AdaBoost is another boosting algorithm, which unlike gradient boost initializes the training with a decision tree stump as the base learner. The algorithm starts giving equal weights to each data point and updates the weights in each iteration so that the misclassified data points get large weights. Decision tree stumps are constructed continuously with the updated weights [14].

Linear discriminant analysis (LDA) is a classification and dimensionality reduction methodology. The algorithm reduces the feature space from a higher to a lower dimension by maximizing the variability between the classes. LDA is used for multi-class classification, and it performs better than logistic regression. LDA might not perform well if the number of observations exceeds the number of features. LDA linear decision boundaries may not be flexible. The algorithm assumes that the features are typically distributed [15].

Another supervised machine learning algorithm that can be used for both classification and regression tasks is support vector machines. The algorithm finds an optimal hyperplane that maximizes the separation between the target classes. If the data is lower-dimensional, the SVM kernel trick will protect the data into higher dimensions to find a good separation line. SVM uses hinge loss for optimization. The algorithm is more computationally expensive than other classification methods [16].

Logistic regression is a classic algorithm that tries to fit a logistic curve to predict the probability of the default class. The algorithm uses the maximum likelihood principle for finding the best fit using the log loss as the loss function. Logistic regression uses the sigmoid function to find the probability, unlike the linear sum product in the linear regression. The method performs well if the data is linearly separable. However, there are many statistical transformations available for making the data linear. Logistic regression can also be used for multi-class classification [17].

This paper uses features generated from two unsupervised learning algorithms: K-means clustering and Gaussian mixture models. The K-means algorithm starts by placing centroids randomly in the training dataset. Then, the algorithm finds the distance between the centroid and all the data points. It will then assign a centroid label for each of the training data. This step is known as cluster assignment. In the second step, the mean of the cluster points is found first, then the mean is assigned as the new cluster. These two steps will continue till it reaches the threshold. In hard clustering, each data point will get only one cluster. So the K-means algorithm performs hard clustering. The primary distance metric used for the cluster assignment is the Euclidean distance metric. There are other varieties of K-means, namely K-means ++ and mini-batch K-means. One of the major difficulties while training K-means clustering is the selection of cluster counts. The elbow method helps select the optimal number of clusters. In the elbow method, the sum of squared distance is computed between the centroid and each training data point (WCSS) with different k values. WCSS will decrease when the number of clusters increases. The k value corresponding to a rapid change in the graph will be the optimal number of clusters [18]. The second unsupervised algorithm used for feature generation was the Gaussian mixture model. Mixture models consist of several Gaussian distributions, and each such Gaussian represents a cluster. Each cluster has two parameters, namely the mean and covariance. These parameters are estimated by fitting the data point to a Gaussian distribution. The expectation-maximization algorithm performs the estimation process. Once the fitting is done, the model can be used for density estimation and clustering tasks [19].

4 Experimental Results and Analysis

Five different machine learning models were trained, namely gradient boosting classifier, AdaBoost classifier, linear discriminant analysis, support vector machines, and logistic regression. Grid search CV is applied for optimizing the model hyperparameters for all the models. Since the primary model evaluation criteria are recall scores, all the recall scores and accuracy are recorded after each model training phase. Table 2 shows the model scores for the first phase.

All the baseline models have achieved around 80% accuracy, but they failed to give even 50% recall. Both AdaBoost and gradient boost obtained slightly better recall scores than other models. The graphical representation of the evaluation scores for phase 1 is given in Fig. 4.

Table 2 Phase 1 model results

Algorithm	Accuracy	Recall
Gradient boosting model	79.1	35.13
AdaBoost classifier	80.01	29.7
Linear discriminant	80.01	16.21
Support vector classifier	81.59	18.2
Logistic regression	79.6	13.5

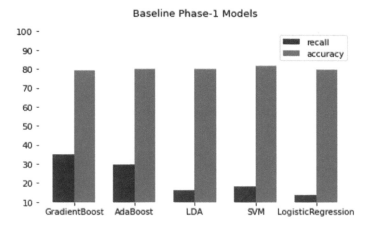

Fig. 4 Baseline phase 1-model performance

For all the baseline models, the overall accuracy is high and almost equal, but the recall value is significantly low due to the target distribution imbalance. Among the models, the boosting ensembles have slightly better recall measures of above 30%. So, in short, the baseline model performance is not satisfactory in terms of recall measure. Since, prediction of defaulters aims to capture all the ones in the target, the false-negative predictions need to be minimized, and the recall measure should be the first criteria for selecting models.

In the second phase, the SMOTE technique is applied along with the addition of two aggregate features to improve the baseline model's performance. Table 3 shows the model evaluation result after handling the data imbalance.

The SMOTE technique improved the recall scores by 40%. Among the models, gradient boost still balances accuracy score and recall value. The accuracy value of the support vector model and logistic regression has been reduced by around 20%. The graphical representation of the results is shown in Fig. 5.

For all the models, except SVM, the results are almost balanced. For SVM, the recall value is high compared to the accuracy score. After handling the target imbalance using SMOTE, the models gave a much better recall score. Still, there are methods to reduce all the false-negative predictions.

Table 3 Phase 2 model results

Algorithm	Accuracy	Recall
Gradient boosting	82.2	81.8
AdaBoost classifier	76.8	77.9
Linear discriminant	73.3	71.1
Support vector classifier	52.06	68.4
Logistic regression	58.7	55.7

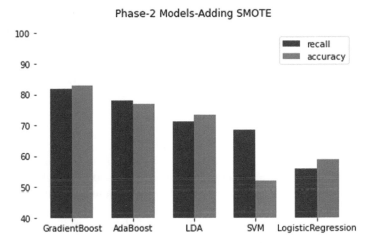

Fig. 5 Baseline phase 2-model performance

It has been observed a significant improvement in the models recall measures by comparing the phase 1 and phase 2 results. In the final phase, unsupervised features from the Gaussian mixtures and K-means clusters are added to the training feature space. The elbow method is implemented in Python to find the optimal number of clusters. The cluster count can be found from the elbow point in the curve. Figure 6 shows the value of the within-cluster sum of squares plotted against the number of clusters.

In the above elbow curve, the changes can be found with six and ten on the x-axis. Six and ten can be taken as the number of clusters. However, the target class count is

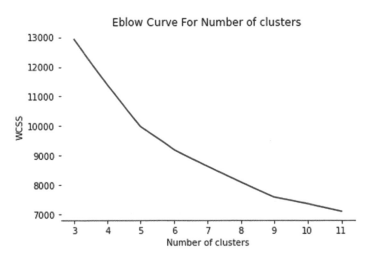

Fig. 6 Elbow curve for optimal clusters

Table 4 Phase 3 model results

Algorithm	Accuracy	Recall
Gradient boosting classifier	88.8	89.4
AdaBoost classifier	77.4	80.5
Linear discriminant analysis	73.3	76.5
Support vector classifier	56.19	69.23
Logistic regression	60.0	54.4

two, and the number of clusters is taken as six. In case of the K-means and Gaussian mixture models, the number of clusters is assigned as six. Table 4 represents the model scores after adding the unsupervised features from K-means and Gaussian mixture models.

The evaluation scores have improved significantly. The gradient boosting algorithm achieved a high recall of 89.4 and an accuracy of 88.8. Both boosting models achieved high recall scores. Figure 7 shows the graphical comparison of the results in the third phases.

After comparing Figs. 5 and 7, it was found that gradient boosting accuracy and recall have increased by around 6%. For the AdaBoost model, the recall has increased by around 3%. Other models also exhibit improvements in their scores. Unlike other models, logistic regression has a lower recall value than the accuracy score.

Five different models were trained in three separate phases along with other data exploration. Analysis showed that two blocks of variables are correlated, and the data is imbalanced. The additional aggregated features are added to the training data, and the SMOTE technique is applied for handling the imbalance. The baseline model results indicate that the recall measures for all the models are too low. Second phase results have significantly improved recall scores by around 40%. In the final

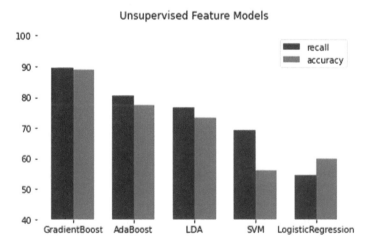

Fig. 7 Unsupervised feature model performance

phase, both accuracy and recall have been improved by around 6% for the gradient boosting model. Other models also exhibit improvements in the recall score. After comparing the obtained results, it can be concluded that the proposed methodology of unsupervised features improved the model accuracy scores significantly. The banking systems can adopt these methodologies to minimize credit risk.

5 Conclusion

In this paper, unsupervised features are used to improve the prediction of credit card default. The proposed methodology is efficient for predictive banking systems. Various machine learning algorithms have been trained with the credit card defaulter dataset, and the results showed that the performance of training algorithms has been improved significantly by the addition of unsupervised features. On comparing the performance measures like recall and accuracy, gradient boosting classifier outperformed and achieved the highest accuracy and recall of 88% and 89%, respectively. Future works will involve artificial neural networks for model training.

References

1. Faraj, A. A., Mahmud, D. A., & Rashid, N. B. (2021). Comparison of different ensemble methods in credit card default prediction. *UHD Journal of Science and Technology.*
2. Nautiyal, H., Jyala, A., & Bhandari, D. (2021). A review on credit card default modelling using data science. In Nautiyal2021ARO.
3. Shivanna, A., & Agarwal, D. P. (2020). Prediction of defaulters using machine learning on Azure ML. In *2020 11th IEEE annual information technology, electronics and mobile communication conference (IEMCON)* (pp. 0320–0325).
4. Keramati, A., & Yousefi, N. (2011). A proposed classification of data mining techniques in credit scoring.
5. Agarwal, A., Rana, A. K., Gupta, K., & Verma, N. (2020). A comparative study and enhancement of classification techniques using principal component analysis for credit card dataset. In *2020 International conference on intelligent engineering and management (ICIEM)* (pp. 443–448).
6. Rai, A. K., & Dwivedi, R. K. (2020). Fraud detection in credit card data using unsupervised machine learning based scheme. In *2020 International conference on electronics and sustainable communication systems (ICESC)* (pp. 421–426).
7. Panchal, U. K., & Verma, S. (2019). Identification of potential future credit card defaulters from non defaulters using self organizing maps. In *2019 10th International Conference on computing, communication and networking technologies (ICCCNT)* (pp. 1–5).
8. Alam, T. M., Shaukat, K., Hameed, I. A., Luo, S., Sarwar, M. U., Shabbir, S., Li, J., & Khushi, M. (2020). An investigation of credit card default prediction in the imbalanced datasets. *IEEE Access, 8*, 201173–201198.
9. Zhu, X., & Chen, J. (2021). Risk prediction of P2P credit loans overdue based on gradient boosting machine model. In *2021 IEEE international conference on power, intelligent computing and systems (ICPICS)* (pp. 212–216).
10. Soui, M., Smiti, S., Bribech, S., & Gasmi, I. (2018). Credit card default prediction as a classification problem. In Soui2018CreditCD.

11. Vishwakarma, S. K., Rasool, A., & Hajela, G. (2021). Machine learning algorithms for prediction of credit card defaulters—A comparative study. In Vishwakarma2021MachineLA.
12. Chawla, N., Bowyer, K., Hall, L., & Kegelmeyer, P. (2002). SMOTE: Synthetic minority over-sampling technique. *The Journal of Artificial Intelligence Research, 16*, 321–357.
13. Konstantinov, A. V., & Utkin, L. V. (2021). Interpretable machine learning with an ensemble of gradient boosting machines. In Konstantinov2021InterpretableML.
14. Randhawa, K., Loo, C. K., Seera, M., Lim, C. P., & Nandi, A. K. (2018). Credit card fraud detection using Adaboost and majority voting. *IEEE Access, 6*, 14277–14284.
15. Tharwat, A., Gaber, T., Ibrahim, A., & Hassanien, A. E. (2017). Linear discriminant analysis: A detailed tutorial. *AI Communications, 30*, 169–190.
16. Christianini, N., & Taylor, J. S. (2000). An introduction to support vector machines and other kernel-based learning methods. In Cristianini2000AnIT.
17. Wan, X. (2021). Research on predicting credit card customers' service using logistic regression and Bp neural network. In Wan2021ResearchOP.
18. Zahra, S., Ghazanfar, M. A., Khalid, A., Azam, M. A., Naeem, U., & Bennett, A. (2015). Novel centroid selection approaches for KMeans-clustering based recommender systems. *Information Sciences, 320*, 156–189.
19. Chakravarti, P., Balakrishnan, S., & Wasserman, L. A. (2019). Gaussian mixture clustering using relative tests of fit. arXiv: Methodology.

Social Media Mining to Detect Mental Health Disorders Using Machine Learning

Udaipurwala Rashida and K. Suresh Kumar

Abstract The growth of social media has caused people to post a significant content online. They find it easier to communicate in the virtual world rather than to talk to a person and share their problems. This has given rise to a pertinent question—can we use social media to help these youngsters to understand the state of their mental health? If so, how can we do it without manual intervention so that they do not feel uncomfortable to share their problems? The main purpose of this research was to find out a way in which we can detect mental health disorders without manual intervention. Machine learning can be the right solution to this problem. We used social media sites to analyse the mental state of the users. Machine learning models are trained to analyse the mental health of the people. Machine learning algorithms are employed to analyse the mental health disorders based on social media posts. Sentiment analysis can be used to detect the tone of the user from the message. This tone is useful in identifying whether the tweets are positive or negative. The proposed system uses the popular social media platform—Twitter to detect depression in the users. The tweets collected from Twitter users are used to train the machine learning model to detect whether the tweet is positive or negative. We have used logistic regression to train the model. This results in a model that is able to accurately predict whether the Twitter user is depressed or not. This can be used to detect the depression in the users and link to health workers or helplines which help the depressed person cope with their mental illness.

Keywords Machine learning · Artificial intelligence · Sentiment analysis · Twitter · Natural language processing · Machine learning algorithms

U. Rashida (✉) · K. Suresh Kumar
Department of Information Technology, Saveetha Engineering College, Chennai 602105, India
e-mail: u.rashida@gmail.com

K. Suresh Kumar
e-mail: sureshkumar@saveetha.ac.in

© The Author(s), under exclusive license to Springer Nature Singapore Pte Ltd. 2023   923
S. Shakya et al. (eds.), *Sentiment Analysis and Deep Learning*,
Advances in Intelligent Systems and Computing 1432,
https://doi.org/10.1007/978-981-19-5443-6_68

1   Introduction

With the increasing use of the Internet, the number of online transactions is increasing. On social media sites, a lot of people are posting their messages, and every day, a huge volume of data is being generated. The data are being collected from various sources, and the features are extracted, and they are being mined.

Using various natural language processing techniques, the data are being cleaned and the essential features are extracted. After extracting the features, it is given to the classifier module to classify the mental state of the people [1].

Predicting the results of popular political elections and polls is also an emerging application to sentiment analysis. One such study was conducted by Tumasjan et al. [2] in Germany for predicting the outcome of federal elections which concluded that Twitter is a good reflection of offline sentiment.

2   Ethical and Social Issues

The complex interaction between individuals on social media causes a lot of challenges in social media mining. Some of them are the nature of the social media users, inaccessibility of social media platforms, difficulty in obtaining meaningful data from the data posted online, and an unconventional method of approaching the data. The major ethical challenges are that there is no clear distinction between private and public data. Preserving the privacy of individuals is a very important ethical challenge. The intellectual property of individuals is not protected and consent to use the social media data is not clearly sought.

A major concern with using machine learning as a detection tool for depression is the incorrect diagnosis of the disease based on the social media posts. This could create more harm to the person than do them any good. Some tweets can be confusing and might be misinterpreted by the algorithms. This would waste the time of not only the users but also the intervention provided to them. However, filtering out such requests requires a very intelligent algorithm. In case the algorithm does not detect correctly, the person might feel social stigma due to the other people knowing about the diagnosis.

Users post their thoughts on social media without realizing that their data might be used by other programmes for various diagnostics. Processing of such data raises the question of getting consent from the users before any action is taken on the data posted by them. Identification of individuals is a challenge as the usernames and credentials used by them might not be their real identity. Using various resources to trace and find out the actual users is a time-consuming process that might not yield the quick results required for taking timely action for the treatment.

3 Impact of the Solution

Depression is a disease that affects millions of people and remains undetected most of the time. Early identification is the most important step towards its treatment. With the availability of datasets for depression detection, it has become easier to develop algorithms to detect depression using social media data. As social media users share their thoughts freely, the people working in the health sector get the opportunity to know what goes on in the minds of people without meeting them in person.

Machine learning provides techniques that can find out unique hidden patterns in online communication, thereby determining the mental state of the users. This project aims to reach individuals who are unwilling to come forward and get therapy. Healthcare workers will be more aware of how people with depression react to certain situations and provide them with the help required.

Automated depression detection systems are crucial for depression detection and treatment. Sentiment analysis provides an objective assessment and monitoring of psychological disorders. Effective health monitoring systems make the health worker's job easier and lower the healthcare expenses of the patients. Although these machine learning techniques cannot replace psychologists, however, they could be a powerful aid to health workers.

4 Related Work

Researchers have been continuously working to harness the power of social media for numerous analysis purposes. Social media has been used in various fields such as news media, crime intelligence, education, production and supply chain, health care, detection of cyber-bullying, drug usage, rumour detection and many more such applications.

Table 1 shows the various literature and the strategies for mining data on social media that was studied as a reference to our study.

5 System Architecture

Twitter is a micro-blogging website where members interact with each other using a short string of messages known as "tweets". The proposed system architecture is shown in Fig. 1. This website provides a platform to users where they can express their opinions and views freely. These tweets serve as a good source for researches and market analyst to perform sentiment analysis on these tweets. Furthermore, with the advent of machine learning algorithms, sentiment analysis has become more accurate and is able to predict and provide insights on the mental state of the users [18].

Table 1 Literature review of articles for machine learning algorithms for social media

Author	Topic	Strategy
Liu [3]	Some computational challenges in mining social media	Computation of social media mining using big data
Doyle et al. [4]	Mining personal media thresholds for opinion dynamics and social influence	Pattern detection for social media use and behaviour
Rajapaksha et al. [5]	Inspecting interactions: online news media synergies in social media	Predictive model for popularity of news media
Seah et al. [6]	Data mining approach to the detection of suicide in social media: a case study of Singapore	Social sensing on digital traces from Reddit
Shahnawaz et al. [7]	Sentiment analysis: approaches and open issues	Tackling inadequate amount of labelled data
Syarif et al. [8]	Study on mental disorder detection via social media mining	Computational linguistic of depressed people
Vohra et al. [9]	Detection of rumor in social media	Automatic rumour detection on Twitter
Wang et al. [10]	Fine-grained sentiment analysis of social media with emotion sensing	Fine-grained sentiment analysis
Sutar et al. [11]	Intelligent data mining technique of social media for improving health care	Integrated system for medical data
Xu [12]	Joint visual and textual mining on social media	Extract knowledge from visual and textual data
Mohandas et al. [13]	A survey on mining social media data for understanding drug usage	Knowledge on drug usage
Chan et al. [14]	A case study on mining social media data	Decision making research
Savas et al. [15]	Crime intelligence from social media: a case study	Intelligence study on Twitter
Rejeesh et al. [16]	Social media and data mining enabled pre-counseling session: a system to perk up effectiveness of counseling in distance education	Technology-based pre-counselling
Ting et al. [17]	Towards the detection of cyber bullying based on social network mining techniques	Detecting cyber bullying on Twitter, Facebook, ptt and ck101

The tweets are classified into positive and negative based on the words used. If the tweet contains both negative as well as positive words, it will be classified based on the predominant sentiment of the tweet. A dataset from Kaggle was used which had the polarity of both positive and negative tweets. The data from the dataset consist of ID, class and comments. The ID indicates its unique value. The class defines whether

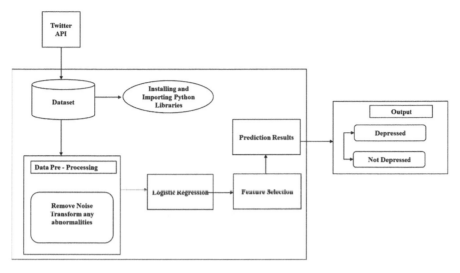

Fig. 1 System design of the proposed solution

the tweet is depressive (1) or non-depressive (0). The comments are the tweets of the users.

The distribution for the dataset is plotted to indicate the positive and negative polarity of the tweets. The process flow diagram is shown in Fig. 2. The data are pre-processed to remove the noise. The CountVectorizer function of the scikit-learn's library is used to perform the pre-processing of data. This function is used to prepare the text for predictive modelling. It parses the text to remove certain words. This is known as tokenization. These words are then encoded as integers, or floating point values which can be easily given as an input to our machine learning algorithm. This process is known as feature extraction or vectorization. The CountVectorizer function converts the text to a set of tokens. It also enables pre-processing of data and converts it to numbers which can be easily understood by the machine and help in text classification.

The model is then fitted using the fit_transform method of scikit-learn library. This enables to build a model from the comments extracted by the CountVectorizer function. A well-fitted machine learning model generates accurate outcomes.

The dataset is split into training data and testing data. 30% of data is taken for training and 70% data is taken for testing. We have chosen logistic regression as the machine learning algorithm. As the data that we are going to predict is binary in nature—that is whether the tweets are depressed or not depressed, logistic regression is used to predict the probability of success or failure. It is easier to implement, interpret and considerably efficient to train the model using logistic regression. This is most appropriate for our project as we want to classify tweets into positive and negative. Our dataset is simple, and hence, this algorithm provides good accuracy.

We have chosen Twitter as the social media platform as it is a better platform where we can get opinions of users very freely compared to websites, blogs and

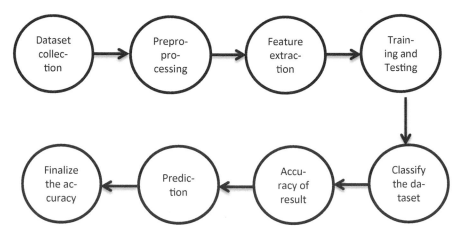

Fig. 2 Process flow of the proposed solution

other social networks. The amount of data is much more organized as it is organized using hashtags which makes it easier to search for relevant content. The response on Twitter is much easier and prompter compared to other platforms as the replies can be given instantly. Sentiment analysis has various applications such as predicting poll results, market analysis, cybercrime and many such fields.

This can be done by understanding the public's opinion on a particular candidate. This would indicate the number of people who are favourable to the person. Market analysis can be done by predicting how a product is received by the people when it is launched. When a lot of people tweet about a particular product it indicates its popularity. Cyber-crime can be detected by understanding the threats sent via tweets. This will help authorities stop crimes and combat terrorism and cyber bullying.

6 Results and Future Enhancements

The dataset that we have chosen has 4439 tweets. We used the logistic regression machine learning algorithm to predict the polarity of the tweets whether they were positive or negative. After training the model using 30% testing data, the accuracy score achieved is 81%.

In future, we would like the system to be integrated with the healthcare services. The healthcare system gets a notification when a social media user continuously posts depressing tweets. This notification helps the healthcare workers to track the depressed person and provide them with the help in case of need. This connectivity from the model to the healthcare system will enhance the future of health care in our country. A mental illness like depression will be detected so that timely help could be provided.

7   Conclusion

In this project, we propose a solution to the detection of depression in social media users through their tweets on Twitter through machine learning using the logistic regression algorithm. We have collected the dataset, and we have cleaned the data. After data cleaning, it will be given to the feature extraction phase in order to extract the essential features. The datasets are collected from various sources, and the essential features are extracted. In the pre-processing phase, null values and other unwanted values are removed from the data and the text is converted to tokens. Data cleaning happens in the data preparation phase and the essential features are extracted in the feature extraction phase. The required features are extracted and the model predicts whether the tweets are depressive or non-depressive. This will help in identifying the mental health of the social media users. Research is underway to use this trained model to be tested for fresh tweets to predict depression. This information is of utmost value as the machine will be able to detect depression before the user takes any extreme step and provide the right guidance to them.

References

1. Kottursamy, K. (2021). A review on finding efficient approach to detect customer emotion analysis using deep learning analysis. *Journal of Trends in Computer Science and Smart Technology, 3*(2), 95–113.
2. Tumasjan, A., Sprenger, T., Sandner, P., Welpe, I. (2010). Predicting elections with twitter: What 140 characters reveal about political sentiment. *Word Journal of the International Linguistic Association, 10*.
3. Liu, H. (2013). Some computational challenges in mining social media. In *2013 IEEE/ACM international conference on advances in social networks analysis and mining (ASONAM 2013)* (pp. xxxvii–xxxvii). https://doi.org/10.1109/ASONAM.2013.6785672
4. Doyle, C., Meandzija, A., Korniss, G., Szymanski, B., Asher, D., & Bowman, E. (2018). Mining personal media thresholds for opinion dynamics and social influence. In *2018 IEEE/ACM international conference on advances in social networks analysis and mining (ASONAM)* (pp. 1258–1265). https://doi.org/10.1109/ASONAM.2018.8508316
5. Rajapaksha, P., Farahbakhsh, R., Crespi, N., & Defude, B. (2018). Inspecting interactions: Online news media synergies in social media. In *2018 IEEE/ACM international conference on advances in social networks analysis and mining (ASONAM)* (pp. 535–539). https://doi.org/10.1109/ASONAM.2018.8508534
6. Seah, J. H. K., & Jin Shim, K (2018). Data mining approach to the detection of suicide in social media: A case study of Singapore. In *2018 IEEE international conference on big data (Big data)*. (pp. 5442–5444). https://doi.org/10.1109/BigData.2018.8622528
7. Shahnawaz, & Astya, P. (2017). Sentiment analysis: Approaches and open issues. In *2017 International conference on computing, communication and automation (ICCCA)* (pp. 154–158). https://doi.org/10.1109/CCAA.2017.8229791
8. Syarif, I., Ningtias, N., & Badriyah, T. (2019). Study on mental disorder detection via social media mining. In *2019 4th International conference on computing, communications and security (ICCCS)* (pp. 1–6). https://doi.org/10.1109/CCCS.2019.8888096
9. Vohra, M., & Kakkar, M. (2018). Detection of rumor in social media. In *2018 8th International conference on cloud computing, data science & engineering (Confluence)* (pp. 485–490). https://doi.org/10.1109/CONFLUENCE.2018.8442442

10. Wang, Z., Chong, C. S., Lan, L., Yang, Y., Beng Ho, S., & Tong, J. C. (2016). Fine-grained senti-ment analysis of social media with emotion sensing. In *2016 Future technologies conference (FTC)* (pp. 1361–1364). https://doi.org/10.1109/FTC.2016.7821783
11. Sutar, S. G. (2017). Intelligent data mining technique of social media for improving health care. In *2017 International conference on intelligent computing and control systems (ICICCS)* (pp. 1356–1360). https://doi.org/10.1109/ICCONS.2017.8250690
12. Xu, J. (2014). Joint visual and textual mining on social media. In *2014 IEEE international conference on data mining workshop (ICDMW)* (pp. 1189–1190), Shenzhen, China.
13. Mohandas, A., Babu, B., Rajan S. D., Suresh, L. P., & Boben, R. (2018). A survey on mining social media data for understanding drug usage. In *2018 Conference on emerging devices and smart systems (ICEDSS)* (pp. 259–261). https://doi.org/10.1109/ICEDSS.2018.8544346
14. Chan, H. K., Lacka, E., Yee, R. W. Y., & Lim, M. K. (2014). A case study on mining social media data. In *2014 IEEE international conference on industrial engineering and engineering management* (pp. 593–596). https://doi.org/10.1109/IEEM.2014.7058707
15. Savaş, S., & Topaloğlu, N. (2017). Crime intelligence from social media: A case study. In *2017 IEEE 14th international scientific conference on informatics* (pp. 313–317). https://doi.org/10.1109/INFORMATICS.2017.8327266
16. Rejeesh, E., & Anupama, M. (2017). Social media and data mining enabled pre-counseling session: A system to perk up effectiveness of counseling in distance education. In *2017 International conference on I-SMAC (IoT in social, mobile, analytics and cloud) (I-SMAC)* (pp. 153–156). https://doi.org/10.1109/I-SMAC.2017.8058328
17. Ting, I., Liou, W. S., Liberona, D., Wang, S., & Tarazona Bermudez, G. M. (2017). Towards the detection of cyberbullying based on social network mining techniques. In *2017 International conference on behavioral, economic, socio-cultural computing (BESC)* (pp. 1–2). https://doi.org/10.1109/BESC.2017.8256403
18. Pandian, A. P. (2021). Performance evaluation and comparison using deep learning techniques in sentiment analysis. *Journal of Soft Computing Paradigm (JSCP), 3*(02), 123–134.
19. Hamilton, M. (1967). Development of a rating scale for primary depressive illness. *British Journal of Social and Clinical Psychology, 6*(4), 278–296.
20. Jordan, M. I., & Mitchell, T. M. (2015). Machine learning: Trends, perspectives, and prospects. *Science, 349*(6245), 255–260.
21. Davcheva, P. (2014). Identifying sports talents by social media mining as a marketing instru-ment. In *2014 Annual SRII global conference* (pp. 223–227). https://doi.org/10.1109/SRII.2014.38
22. Dastanwala, P. B., & Patel, V. (2016). A review on social audience identification on twitter using text mining methods. In *2016 International conference on wireless communications, signal processing and networking (WiSPNET)* (pp. 1917–1920). https://doi.org/10.1109/WiSPNET.2016.7566476
23. Rutledge, R. B., Chekroud, A. M., & Huys, Q. J. (2019). Machine learning and big data in psychiatry: Toward clinical applications. *Current Opinion in Neurobiology, 55*, 152–159.
24. Desai, S., & Patil, S. T. (2015). Efficient regression algorithms for classification of social media data. In *2015 International conference on pervasive computing (ICPC)* (pp. 1–5). https://doi.org/10.1109/PERVASIVE.2015.7087040
25. Chen, X., Vorvoreanu, M., & Madhavan, K. (2014). Mining social media data for understanding students' learning experiences. *IEEE Transactions on Learning Technologies, 7*(3), 246–259. https://doi.org/10.1109/TLT.2013.2296520
26. Executive Board. (2012) *Global burden of mental disorders and the need for a comprehensive, coordinated response from health and social workers at the country level.*
27. Manoharan, J. S. (2021). Study of variants of extreme learning machine (ELM) brands and its performance measure on classification algorithm. *Journal of Soft Computing Paradigm (JSCP), 3*(02), 83–95.
28. Mugunthan, S. R., & Vijayakumar, T. (2021). Design of improved version of sigmoidal function with biases for classification task in ELM domain. *Journal of Soft Computing Paradigm (JSCP), 3*(02), 70–82.

Using Deep Learning Models for Crop and Weed Classification at Early Stage

Akshay Dheeraj and Satish Chand

Abstract Agriculture is essential for human existence, and it plays an important role in the world economy. There is increasing demand for food to feed the ever-increasing world population. Agriculture is affected by climate changes along with weed control. Weeds are unwanted plants that compete with plants for nutrition, and sunlight and adversely affect crop quality and production. Manual weeding is a tedious and labor-intensive task because both crop and weed look the same by visual appearance. Artificial intelligence techniques like deep learning can address this problem of crop and weed classification. In this research work, a deep learning-based classification system has been proposed to classify the weed and crop based on RGB images. We investigated two popular deep learning-based transfer learning models, namely DenseNet169 and MobileNetV2, and assessed their performances for crop and weed recognition. These models perform excellently with an accuracy of 97.14 and 94.92%, respectively. The significant accuracy results make the model an important tool for farmers to identify weeds.

Keywords Plant seedlings · Deep learning · Convolutional neural network · Weed classification · MobileNetV2 · DenseNet169 · Precision agriculture

1 Introduction

Convolutional neural networks (CNNs) have become essential for researchers in computer vision tasks. Large dataset availability and evolution in computing technologies with graphics processing units (GPU) have eased these vision tasks. CNNs are popular because of their applicability to various data and excellent performances

A. Dheeraj (✉)
ICAR-Indian Institute of Soil and Water Conservation, Dehradun, Uttarakhand, India
e-mail: akshay.dheeraj@icar.gov.in

S. Chand
School of Computer and Systems Sciences, Jawaharlal Nehru University, New Delhi, India
e-mail: schand@mail.jnu.ac.in

© The Author(s), under exclusive license to Springer Nature Singapore Pte Ltd. 2023   931
S. Shakya et al. (eds.), *Sentiment Analysis and Deep Learning*,
Advances in Intelligent Systems and Computing 1432,
https://doi.org/10.1007/978-981-19-5443-6_69

[1]. CNN was started with the invention of LeNet-5 which consists of three convolution layers, two sub-sampling layers, and two fully connected layers [2]. CNN achieved further progress with AlexNet in which the number of the kernel was increased and dropout regularization was introduced. After that many networks were introduced that have significant performance [3–5].

Agriculture is the major source of livelihood for world populations. With the evolution in artificial intelligence, the agriculture sector is also being transformed which led to smart farming. Agriculture is being affected by different factors like climate change along with the daunting task of weed control. One of the most important tasks in smart farming is the recognition of weeds from the field and taking necessary steps to control the weeds so that crops get enough water content, solar radiation, and other important nutrient. Manual identification of weeds in the field is a difficult task as they appear to be the same in color, and size as the crop. The objective of this research study is to develop a weed and crop classification system using a deep learning model so that early detection of weeds can be done and crop yield can be increased. Images of weeds and crops have been trained on deep learning models like DenseNet169 and MobileNetV2, and their performances have been compared. A total of 12 classes of weed and crop have been used in the research work with originally 5544 images which have been augmented to increase the size of the dataset. The proposed research study shows the accuracy of 97.14 and 94.92% on test data for DenseNet169 and MobileNetV2 models, respectively. Thus, the significant accuracy of the DenseNet169 model suggests the effectiveness of the proposed system for weed recognition at an early stage in the farmer's field.

2 Literature Survey

A good amount of work has been done in for weed detection using computer vision techniques [6–9]. In [10], the author used SVM on digital images to classify weeds and crops. The authors used a total of 224 test images of six weed species, extracted a total of 14 features that can help to distinguish crops and weeds, and achieved a classification accuracy of 97%. In [11], the authors used fuzzy decision-making and shape descriptors to develop a weed and crop classifier and achieved recognition accuracy of 92.9% on a total of 66 images.

In [12], the authors used a machine learning model specifically supporting vector machine (SVM) for weed detection in the sugar beet field. Four different weed species of sugar beet were considered for research and different shape features like moment invariant and Fourier descriptor were extracted. The research study obtained the classification accuracy of 92.92 and 95% for ANN and SVM respectively.

These days, a convolutional neural network (CNN) is being used in every field [13–18]. In [19], some of the applications of CNN-based deep learning models in agriculture have been surveyed. Research work in [20] focuses on classifying soybean and its weeds using the k-means algorithm for feature extraction and combined with CNN. The authors used 820 images having four classes as soybean and its weeds and achieved an accuracy of 92.89%. In [21], the authors proposed a CNN-based graph convolution neural network (GCN)-based model named GCN-ResNet101 to classify weeds associated with corn, lettuce, and radish crop. Authors used four different datasets having a total of 4200, 560, 280, and 5040 images for corn, lettuce, radish, and mixed weed dataset and achieved 97.80, 99.37, 98.93, and 96.51 percent classification accuracy. In [22], the authors proposed a deep learning-based transfer learning model named ResNet101 to identify nine weed species associated with maize, common wheat, and sugar beet crop and achieved a classification accuracy of 96.04%. The author used augmentation techniques, and a total of 7200 images for 12 classes having 600 images for each class were used.

3 Proposed Methodology

3.1 Data Acquisition

A publicly available dataset has been used in this research work [23]. This dataset has a total of 12 species which includes three crop species maize, common wheat, sugar beet, and their associated nine weed species. Table 1 summarizes the dataset description.

Figure 1 shows some images of the dataset. Figure 2 shows the proposed methodology. The first step of the methodology is data acquisition. Once data is collected, data augmentation techniques are used to enlarge the dataset. The third step is to

Table 1 Detail of dataset used

	Species	Number of original images	Number of augmented images	Total images
1	Black-grass	310	590	900
2	Charlock	452	473	925
3	Cleavers	335	565	900
4	Common Chickweed	713	187	900
5	Common wheat	253	647	900
6	Fat-hen	538	362	900
7	Loose silky-bent	766	134	900
8	Maize	257	643	900
9	Scentless mayweed	607	293	900
10	Shepherd's purse	274	626	900
11	Small-flowered cranesbill	576	324	900
12	Sugar beet	463	437	900
Total		5544	5281	10,825

Fig. 1 Some sample images of the dataset

pre-process the data which includes resizing images according to the input size of the model used. These pre-processed data are the input to the proposed models which classify these crops and weed images.

3.2 Data Augmentation

Originally, this dataset has a total of 5544 images. To enlarge the dataset, data augmentation techniques are used. Table 2 summarizes the data augmentation techniques applied using the Keras library of Python. During data augmentation, images are flipped horizontally, rotated between the angles of 0–45°, width and height are shifted by 20%, zoomed by 20%, and shear by 20%.

Fig. 2 Flowchart of
proposed methodology

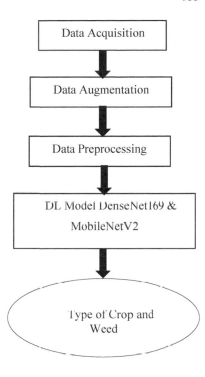

	Technique
1	Horizontal flip = true
2	Rotation = 32
3	Width shift = 0.2
4	Height shift = 0.2
5	Zoom range = 0.2
6	Shear range = 0.2
7	Fill = Nearest

Table 2 Data augmentation
techniques used

3.3 CNN-based Deep Learning Model

Deep learning is being used in every field these days. It is the subfield of artificial
intelligence where feature engineering is done automatically unlike machine learning
where features are extracted manually. Two popular deep learning architectures have
been used in the research.

- DenseNet169 [24]
- MobileNetV2 [25]

DenseNet169 Model

DenseNet consists of a number of dense blocks where within the block, the size of the feature map remains the same but the number of filters keeps on changing. DenseNet169 has four dense blocks. Each dense block defines one 1×1 convolution followed by a 3×3 convolution. The first dense block defines six 1×1 convolutions followed by a 3×3 convolution. The second dense block defines twelve 1×1 convolutions followed by 3×3 convolutions. The third dense block defines thirty-two 1×1 convolutions followed by 3×3 convolution. The fourth dense block defines thirty-two 1×1 convolution followed by 3×3 convolution There are three transition layers which does the convolution and pooling operation and are in place between the dense block. In DenseNet, all layers are connected. Feature maps from previous layers are concatenated.

MobileNetV2 Model

MobileNetV2 model was proposed by Google. This architecture uses depth-wise separable convolution through which efficiency is improved. There are 17 building blocks in MobileNetV2 after that 1×1 convolution is performed followed by the global average pooling and classification layer. One residual connection is used between the input and output layer which learns those features that are already learned and remove irrelevant features. MobileNetV2 model reduces the number of parameters which led to a reduction in computation.

DenseNet169 and MobileNetV2 have total 14.3 M and 3.5 M parameter, respectively, and top-5 accuracy of 93.2 and 90.1% on ILSVRC [26]. These smaller parameter sizes and excellent results have motivated us to choose these models for our research work.

4 Results and Discussions

Two CNN-based deep learning models DenseNet169 and MobileNetV2 have been proposed for crop and weed classification systems. We use the transfer learning approach in which these models that have been already pre-trained on the ImageNet dataset, are used for a new dataset for the research study. The proposed work has been implemented using Google Collaboratory pro which offers GPU and TPU support. These models have been trained on a batch size of 16 with 30 epochs and to optimize the training and validation loss, an Adam optimizer is used. Adam has been used to update the weight of the network iteratively. The dropout value is set to be 0.25 which means that one in four inputs are randomly dropped out during training. The momentum value is set to be 0.9, and rectified linear function (ReLu) has been used as an activation function.

Table 3 summarizes the various hyperparameters used for the proposed work.

For the above model, the same parameters have been used and their performance has been compared. Dataset was divided into 80, 10, and 10% data as training,

Table 3 Hyperparameters used for proposed work

Model	Optimizer	Epoch	Learning rate	Batch size	Dropout	Momentum	Activation function
DenseNet169	Adam	30	0.001	16	0.25	0.99	ReLU
MobileNetV2							

Table 4 Performance of proposed models on a dataset

Model	Avg. training accuracy (%)	Avg. validation accuracy (%)	Training time (sec.)	Avg. testing accuracy (%)
DenseNet169	99.82	97.32	2102	97.14
MobileNetV2	99.30	94.46	2128	94.92

validation, and test data respectively. Table 4 summarizes the performance of these two models on the dataset.

Figures 3 and 4 show the model accuracy and loss curve for DenseNet169 and MobileNetV2 models. From Table 4, it is evident that both models take the almost same time for training but the accuracy of DenseNet169 is higher than that of MobileNetV2. Although the models were trained for 30 epochs, we have used early stopping when results do not improve after three adjustments to the learning rate. Best accuracy results for both DenseNet169 and MobileNetV2 models have been obtained at epochs 22 and 23, respectively. Similarly, minimum loss for both DenseNet169 and MobileNetV2 model occurs at epochs 22 and 21, respectively. It is evident from Figs. 3 and 4 that the accuracy of the DenseNet169 and MobileNetV2 model increases with the increase in the number of epochs.

Initially, both training and validation loss were high but these losses decrease as the number of epochs increase. As shown in Figs. 3 and 4, minimum loss occurs at epochs 22 and 21 for DenseNet169 and MobileNetV2 models, respectively.

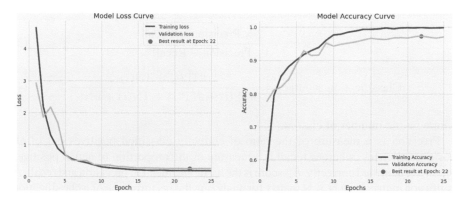

Fig. 3 Model loss and accuracy curve for DenseNet169 model

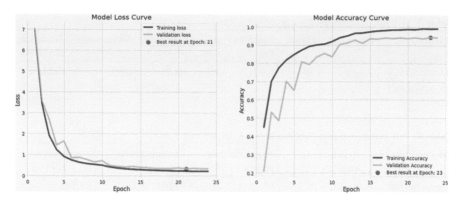

Fig. 4 Model loss and accuracy curve for MobileNetV2 model

For the object classification tasks, various performance metrics such as precision, recall, F_1-score, classification accuracy, and confusion matrix are used and calculated by using Eqs. (1), (2), (3), and (4). There are four terminologies used for calculations of these performance metrics which are true positive (TP), true negative (TN), false positive (FP), and false negative (FN). For this research study, TP means weed image is correctly classified as actual weed class and crop image is classified correctly as actual crop class.

TN means a number of correctly classified images in other classes except for the relevant class. FP means crop image is classified as weed and weed image is classified as crop image. FN means a number of incorrectly classified images in a relevant class.

$$\text{Precision}(i) = \frac{\#\text{TP}(i)}{\#\text{TP}(i) + \#\text{FP}(i)} \tag{1}$$

$$\text{Recall}(i) = \frac{\#\text{TP}(i)}{\#\text{TP}(i) + \#\text{FN}(i)} \tag{2}$$

$$F_1\text{-score} = \frac{2 \times \text{Precision} \times \text{Recall}}{\text{Precision} + \text{Recall}} \tag{3}$$

$$\text{Classification accuracy} = \frac{\#\text{TP}(i) + \text{TN}(i)}{\#\text{TP}(i) + \#\text{FP}(i) + \text{TN}(i) + \text{FN}(i)} \tag{4}$$

where i is the number of classes.

Precision is a measure of the ratio of correctly classified positives out of all positive instances, whereas recall is a measure of the proportion of actual positives that were identified correctly. F_1-score is the harmonic mean of precision and recall. A confusion matrix is an $N \times N$ matrix for analyzing the performance of a classification model where N is several classes. It is a visual representation of actual vs. predicted values.

The proposed model DenseNet169 and MobileNetV2 achieve the classification accuracy of 97.14 and 94.92%, respectively, on test data. Table 5 summarizes the information on precision, recall, and F_1-score of proposed models for 12 species.

The result of the proposed DenseNet169 and MobileNetV2 model for crop and weed classification has been analyzed using a confusing matrix and have been shown in Figs. 5 and 6.

From Table 5, it is evident that avg. precision for both DenseNet169 and MobileNetV2 is 97 and 95%, respectively. DenseNet169 has a precision value of less than 0.90 for two specie namely black-grass and loose silky-bent. Except for these two classes, the model has a precision value of more than 0.97 in the remaining ten classes. Similarly, MobileNetV2 has a precision value of 0.78, 0.87, and 0.89 for three species namely black-grass, loose silky-bent, and scentless mayweed. In the remaining nine classes, the model has a precision value of more than 0.96.

Table 5 Summarized information of precision, recall, and F_1-score of the proposed model for 12 species

	Species	DenseNet169			MobileNetV2		
		Precision	Recall	F_1-score	Precision	Recall	F_1 score
1	Black-grass	0.85	0.86	0.86	0.78	0.86	0.82
2	Charlock	1.00	0.98	0.99	1.00	0.98	0.99
3	Cleavers	0.99	0.99	0.99	0.99	0.99	0.99
4	Common chickweed	1.00	1.00	1.00	0.99	0.95	0.97
5	Common wheat	0.97	0.99	0.98	0.97	0.97	0.97
6	Fat-hen	0.98	0.97	0.97	0.98	0.98	0.98
7	Loose silky-bent	0.88	0.90	0.89	0.87	0.85	0.86
8	Maize	1.00	0.99	1.00	0.98	0.98	0.98
9	Scentless mayweed	0.99	0.99	0.99	0.89	0.95	0.91
10	Shepherd's purse	0.99	0.99	0.99	0.96	0.90	0.93
11	Small-flowered cranesbill	1.00	1.00	1.00	0.99	0.99	0.99
12	Sugar beet	1.00	0.99	0.99	1.00	0.98	0.99
Accuracy		0.97			0.95		
Macro Avg		0.97	0.97	0.97	0.95	0.95	0.95
Weighted Avg		0.97	0.97	0.97	0.95	0.95	0.95

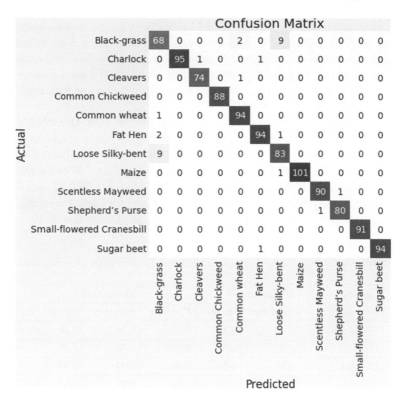

Fig. 5 Confusion matrix for DenseNet169 model

5 Conclusions

This scientific study aimed to develop a classification system for three crops and their associated nine weeds in the field. Transfer learning approaches have been used to train the proposed model. Then, we compared the performances of these two models for classifying weed and crop species into twelve classes. Both DenseNet169 and MobileNetV2 models achieved an accuracy of 97 and 95% with the dataset divided into 80, 10, and 10% as training, validation, and test data, respectively. Some data augmentation techniques were used to augment the data. Both models take almost the same time to train but the performance of DenseNet169 is better than that of MobileNetV2. Good accuracy results make the proposed system applicable to identify weeds at an early stage in the field that can assist the farmers to take action accordingly. In the future, we intend to extend the presented work by incorporating more crop and their associated weed species.

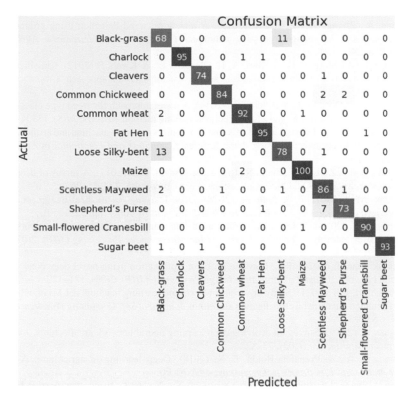

Fig. 6 Confusion matrix for MobileNetV2 model

References

1. Gu, J., Wang, Z., Kuen, J., Ma, L., Shahroudy, A., Shuai, B., Liu, T., Wang, X., Wang, G., Cai, J., & Chen, T. (2018). Recent advances in convolutional neural networks. *Pattern Recognition, 77*, 354–377.
2. LeCun, Y., Bottou, L., Bengio, Y., & Haffner, P. (1998). Gradient-based learning is applied to document recognition. *Proceedings of the IEEE, 86*(11), 2278–2324.
3. Zeiler, M. D., & Fergus, R. (2014). Visualizing and understanding convolutional networks. In *European conference on computer vision* (pp. 818–833). Springer.
4. Simonyan, K., & Zisserman, A. (2014). Very deep convolutional networks for large-scale image recognition. *arXiv preprint*. arXiv:1409.1556.
5. Chollet, F. (2017). Xception: Deep learning with depthwise separable convolutions. In *Proceedings of the IEEE conference on computer vision and pattern recognition* (pp. 1251–1258).
6. Hamuda, E., Glavin, M., & Jones, E. (2016). A survey of image processing techniques for plant extraction and segmentation in the field. *Computers and Electronics in Agriculture, 125*, 184–199.
7. Mennan, H., Jabran, K., Zandstra, B. H., & Pala, F. (2020). Non-chemical weed management in vegetables by using cover crops: A review. *Agronomy, 10*(2), 257.
8. Liakos, K. G., Busato, P., Moshou, D., Pearson, S., & Bochtis, D. (2018). Machine learning in agriculture: A review. *Sensors, 18*(8), 2674.

9. Wang, A., Zhang, W., & Wei, X. (2019). A review on weed detection using ground-based machine vision and image processing techniques. *Computers and Electronics in Agriculture, 158*, 226–240.

10. Ahmed, F., Al-Mamun, H. A., Bari, A. H., Hossain, E., & Kwan, P. (2012). Classification of crops and weeds from digital images: A support vector machine approach. *Crop Protection, 40*, 98–104.

11. Herrera, P. J., Dorado, J., & Ribeiro, Á. (2014). A novel approach for weed type classification based on shape descriptors and a fuzzy decision-making method. *Sensors, 14*(8), 15304–15324.

12. Bakhshipour, A., & Jafari, A. (2018). Evaluation of support vector machine and artificial neural networks in weed detection using shape features. *Computers and Electronics in Agriculture, 145*, 153–160.

13. Liu, W., Wang, Z., Liu, X., Zeng, N., Liu, Y., & Alsaadi, F. E. (2017). A survey of deep neural network architectures and their applications. *Neurocomputing, 234*, 11–26.

14. Razzak, M. I., Naz, S., & Zaib, A. (2018). Deep learning for medical image processing: Overview, challenges and the future. In *Classification in BioApps* (pp. 323–350).

15. Kumar, T. S. (2020). Video based traffic forecasting using convolution neural network model and transfer learning techniques. *Journal of Innovative Image Processing (JIIP), 2*(03), 128–134.

16. Vijayakumar, T. (2020). Posed inverse problem rectification using novel deep convolutional neural network. *Journal of Innovative Image Processing (JIIP), 2*(03), 121–127.

17. Manoharan, J. S. (2021). Study of variants of extreme learning machine (ELM) brands and its performance measure on classification algorithm. *Journal of Soft Computing Paradigm (JSCP), 3*(02), 83–95.

18. Bashar, A. (2019). Survey on evolving deep learning neural network architectures. *Journal of Artificial Intelligence, 1*(02), 73–82.

19. Kamilaris, A., & Prenafeta-Boldú, F. X. (2018). Deep learning in agriculture: A survey. *Computers and Electronics in Agriculture, 147*, 70–90.

20. Tang, J., Wang, D., Zhang, Z., He, L., Xin, J., & Xu, Y. (2017). Weed identification based on K-means feature learning combined with the convolutional neural network. *Computers and Electronics in Agriculture, 135*, 63–70.

21. Jiang, H., Zhang, C., Qiao, Y., Zhang, Z., Zhang, W., & Song, C. (2020). CNN feature-based graph convolutional network for weed and crop recognition in smart farming. *Computers and Electronics in Agriculture, 174*, 105450.

22. Sharma, P. (2019). Crops and weeds classification using convolutional neural networks via optimization of transfer learning parameters. *International Journal of Engineering and Advanced Technology (IJEAT)*, ISSN 2249-8958.

23. Giselsson, T. M., Jørgensen, R. N., Jensen, P. K., Dyrmann, M., & Midtiby, H. S. (2017). A public image database for benchmark plant seedling classification algorithms. *arXiv Preprint*. arXiv:1711.05458.

24. Huang, G., Liu, Z., Van Der Maaten, L., & Weinberger, K. Q. (2017). Densely connected convolutional networks. In *Proceedings of the IEEE conference on computer vision and pattern recognition* (pp. 4700–4708).

25. Sandler, M., Howard, A., Zhu, M., Zhmoginov, A., & Chen, L. C. (2018). Mobilenetv2: Inverted residuals and linear bottlenecks. In *Proceedings of the IEEE conference on computer vision and pattern recognition* (pp. 4510–4520).

26. Russakovsky, O., Deng, J., Su, H., Krause, J., Satheesh, S., Ma, S., Huang, Z., Karpathy, A., Khosla, A., Bernstein, M., Berg, A. C., & Fei-Fei, L. (2015). Imagenet large-scale visual recognition challenge. *International Journal of Computer Vision, 115*(3), 211–252.

A Novel Recommendation System Using the Musical Traits of a User

A. Bharath Kumar Reddy, K. Sai Ruthwik, G. Pavan Kumar,
and N. Damodhar

Abstract In this digital era, the virtual presence of the people has been increasing exponentially. Furthermore, due to the COVID-19 pandemic and lockdowns, almost everything done in day-to-day life has gone virtual. A major portion of screen time is spent on online shopping and streaming content including listening to music and social media. With a steep rise in the number of active users, the content available on the Internet has also been widely increasing. So, finding the right content for the right users is a challenging task which is exactly solved by recommender systems. Therefore, this paper proposes an ingenious approach for recommender systems using musical preference as a metric. The experiments are conducted using unsupervised machine learning (ML) techniques to study if musical preference can act as a psychological parameter for like-mindedness which in turn can be used as a cut above for the existing recommender systems.

Keywords Recommendation system · Collaborative filtering · Content-based filtering · Like-mindedness · Clustering · Silhouette score · Principal component analysis · Correlations

A. Bharath Kumar Reddy · K. Sai Ruthwik (✉) · G. Pavan Kumar · N. Damodhar
Department of Computer Science and Engineering, Amrita Vishwa Vidyapeetham, Amritapuri,
India
e-mail: ksruthwik@am.students.amrita.edu

A. Bharath Kumar Reddy
e-mail: abkumarreddy@am.students.amrita.edu

G. Pavan Kumar
e-mail: gpkumar@am.students.amrita.edu

N. Damodhar
e-mail: ndamodhar@am.students.amrita.edu

1   Introduction

Recommendation systems play a dominant role in enhancing user's experience while using applications beginning from online shopping to news feed in Twitter, where everything is personalized based on the user's activity. Due to the rapid increase in user activity as well as the huge amount of content being generated, it poses a greater challenge to the recommendation systems to understand the new content and to find the target users. Also, recommendation systems can be used for various other types [1, 2].

The existing recommendation systems can be broadly classified into two types such as (1) content-based filtering and (2) collaborative filtering. Content-based filtering uses the data that is generated by the user over a period of time like Websites visited, the past activity of the user, the liking and disliking of the user, ratings that are given by the user for the particular content, etc., to build suggestions to the user. The more the data the better the suggestions are. On the other hand, collaborative filtering works by finding the similarity between the users based on the liking and disliking of each user which can be computed using measures like Pearson correlation, Cosine similarity, etc. The main disadvantage of using these recommendation systems is that they suffer from cold start problems. In order for them to properly recommend the content, they need a lot of data about the user. And the data can only be provided if the user has a significant amount of activity in that particular application which is not the case for new users. For example, if the user has not given any rating to any content, then collaborative filtering does not work effectively. Similarly, content-based filtering will fail if the user is new to the application since there will be no past activity of the user.

Furthermore, the results that are achieved over the collaborative filtering are not consistent. For example, two users from very different locations using a musical application might be matched for similarity, but the same two users might not be suggested as friends to each other in some other social media application even though they have a lot of similarities because here the metric that might be considered is the location of users. Therefore, musical preference can be proposed as a universal metric for grouping people with similar interests, and the results show that musical preference can act as a psychological parameter to group the people who are like-minded and have similar interests. As people are grouped based on their music traits, [3] they can be grouped with the similar idea in social networks using constraint-based clustering algorithm based on global similarity measure.

Psychological studies [4] show that studying a complete musical profile of a user will reveal a lot of characteristic traits of a person, which in turn can be used as a parameter to group like-minded people. Possessing a cluster of like-minded people can extend the capabilities of the recommender systems like cross recommendations can be made possible between different applications. For example, consider two users using some social media platform and they are clustered as like-minded people. Let us say, one of them is new to a video streaming platform but since those two users are like-minded, the already existing user's past activity can be suggested to the new user which eliminates the cold start problem that is faced by the existing recommender systems.

2 Background and Related Work

Paper [5] analyzed the social interaction of a particular user with other users and recommends content based on the people that particular user is connected with. It examined the interaction of a user among his friends and calculated an interaction-based social proximity (IBSP) factor for the user's circle of friends and recommends items, respectively, based on this factor, and the user rates this recommendation too. Zisopoulos et al. [6] used a hybrid approach combining decision trees and Bayesian classifiers to recommend content that is best suited for the user. The decision tree was constructed upon the users' rating on the items, and the tree was divided according to user's favor, similarly, the probability was calculated based on other users' ratings, and then, the recommendation was performed using Bayesian classifiers.

The work mentioned in [7] provided an algorithm to generate recommendations based on the previous data of users, liking, ratings of the users but fails in the case when a user is new to the system where the algorithm fails to recommend the movies, and also when a user has watched multiple movies but rates only a few of them. The model in [8] provided a detailed idea about the existing recommendation systems, but it fails to propose a better recommendation system than the existing ones, which have several problems associated with each recommender system. In order to get a better quality of the clusters, the algorithm proposed in [9] can be considered. This can be highly beneficial for data with many dimensions because it identified a subset of key properties based on the known label for each data instance, and internally uses information gain as a metric for subspace clustering. The below mentioned Table 1 illustrates the research papers and their use cases.

2.1 Agglomerative Clustering [10]

It is an unsupervised learning algorithm and one of the widely used algorithms. It basically uses a bottom-up approach to cluster the data points. Initially, each individual point is considered as a cluster and merged based on some distance metric between the cluster and the data point. To be precise, the closest pair of clusters is merged. This is repeated until all the points are clustered. Also, a dendrogram can be drawn out from the merged clusters to visualize the clusters.

Pseudocode:

Table 1 Research papers and their use case in this paper

Name	Purpose	Method	Limitations
Handling cold start problem in recommender systems by using interaction-based social proximity factor [4]	To handle cold start problems	Recommends content based on social interactions between different users on Facebook	Not effective if the user has no interactions
Content-based recommendation systems [5]	To recommend any type of content to users	Uses decision trees and Bayesian classifier	The method is static and not efficient in dynamic environments
A movie recommender system: MOVREC [6]	Recommend movies to the users	Collaborative filtering, K-means to sort recommendations	Cold start problem (when there are no users), sparsity problem (if users did not rate for some movies they watched)
Recommender systems: an overview of different approaches to recommendations [7]	Currently existing recommendation systems	Collaborative filtering, content-based filtering, hybrid approach	Hybrid approach tries to solve the problems of collaborative and content-based filtering, but in practical use, results are not of much high quality

Pseudocode 1: Agglomerative Clustering

Given a dataset (D1, D2, D3, …, Dn) of size N
Step-1: Calculate the distance matrix
Step-2:
    for i = 1 to N:
        # calculate only lower part of the primary diagonal
        for j = 1 to i:
            dist_mat[i][j] = distance[Di, Dj]
        initially each data point is a single cluster
Step-3: Merge the closest pair of clusters(minimum distance)
Step-4: Update the distance matrix
Step-5: Repeat steps 3 and 4 until a single cluster remains

2.2   K-Means Clustering [11]

As the name suggests, K-means uses the mean of data points in each cluster. It is an unsupervised machine learning algorithm. Initially, k points are chosen as cluster centroids, and distance from each data point to those centroids is found. Based on the computed distance, each point will be assigned to the cluster that it is closer. Once the clusters are assigned for every point, the mean of each point in the cluster is calculated, and it is now chosen as a new cluster centroid, and again, the distances are calculated, and clusters are re-assigned for each data point. This process is continued until there is less number of re-assignment of clusters for data points.

Pseudocode:

Pseudocode 2: K-Means Clustering

Step-1:   Choose the no. of clusters (K) and get the data points
Step-2:   Randomly choose K data points as the centroids
Step-3:   For each datapoint 'i':

        Find the nearest centroid to it by computing the distance
        Assign the point to that cluster

Step-4:   For each cluster 'j':

        New centroid is the mean of all points assigned to that cluster

Step-5:   Repeat steps 3 and 4 until convergence or up to the fixed no. of iterations
Step-6:   Final clusters are formed

2.3   BIRCH Clustering [12, 13]

BIRCH is highly useful in the case of large datasets since the other clustering algorithms are less efficient when compared to BIRCH. In a single scan of data, the large dataset is minimized into summaries, and these summaries are clustered. The algorithm infers groups of CF nodes from the dataset, and these nodes have multiple sub-clusters. A CF tree is a height-balanced tree, which is deduced from the dataset, in which leaf nodes are the sub-clusters.

Pseudocode:

Pseudocode 3: BIRCH Clustering

Step-1: Loading data into memory and fit into the CF trees
Step-2: Resize data
Step-3: Global clustering (clustering using existing clustering algorithms)
Step-4: Refining clusters are formed
Step-5: Final clusters are formed.

3 Proposed Model

3.1 Data

For the purpose of conducting experiments, the data of around 125 people who know each other for a long time are selected and studied. The dataset is categorical, i.e., features are associated with that particular person. The initially collected dataset consists of around 125 rows and 22 features. To finalize the survey questions, the data collection process is iterated multiple times to obtain the best survey questions/features.

In each iteration, multiple survey questions are asked, and the responses are collected and fed to the ML model which gives the clusters of people who are like-minded. Then, the clusters are evaluated manually by the survey participants. In this case, the people are clustered into groups for the proteins as observed in [14], and as observed in [15], the words are clustered into multiple categories, similarly on various basis the data can be clustered.

Here, the evaluation refers to how well the model is able to group like-minded people, and the only way to measure the efficiency of the model is to ask the simple question to surveyors, if the people who they are grouped with are like-minded as them? This question can be answered by surveyors because they already know each other. And this whole process is re-iterated multiple times to come up with the best algorithm that gives maximum accuracy and the best survey questions in order to finalize the features for the dataset. The following are the finalized survey questions/features for the dataset. The underlined ones are the features, and the italicized ones are the options (Table 2).

3.2 Data Visualization and Pre-processing

The basic pre-processing of the data includes converting the text data into numericals, so that the model can be trained on it. Categorical text variables have been

Table 2 Features for the dataset

Parameters for emotion recognition
Name,
Age,
Student *(Yes/No)*,
Education,
Stream,
Gender,
Home State,
Mother Tongue,
My First Language Preference for Streaming Music,
My Second Language Preference,
My Third Language Preference,
My First Favorite Song Language and Album/Artist,
My Second Favorite Song Language and Album/Artist,
My Third Favorite Song Language and Album/Artist,
How likely you enjoy music while traveling *(1 to 5)*,
At what time you listen to music a lot *(Early Morning, Late Morning, AfterNoon, Evenings, Night before sleep)*,
Why do you listen to music *(Just for fun, It makes me relaxed, To manage stress)*,
Do you like listening to old songs *(1 to 5)*,
Do you like listening to devotional songs *(1 to 5)*,
Choose the set that contains your most favorite genres,
I love listening to music *(1 to 5)*

converted using the label encoding technique, and other continuous textual data has been converted using special type of hash functions. These hash functions take the text data and convert them to numerical values. For each text data, it has corresponding numerical value. And the hash functions have been tested for collisions for the entire dataset and found to be 'zero' number of collisions.

Figure 1 gives a glimpse of the dataset distributions for few features present in the dataset.

Since the number of features are high (curse of dimensionality) compared to the number of training samples, correlations have been computed between features using Cramer's rule and correlation coefficient. If two features are highly correlated to each other, then one of them has been dropped from the dataset.

Figures 2 and 3 represent the heatmaps of correlations between the features.

Figures 4 and 5 represent the scatterplot of the dataset; each color in Fig. 4 refers to gender, whereas each color in Fig. 5 refers to a specific age group.

Figure 6 shows an overview of the proposed model.

3.3 Proposed Algorithm

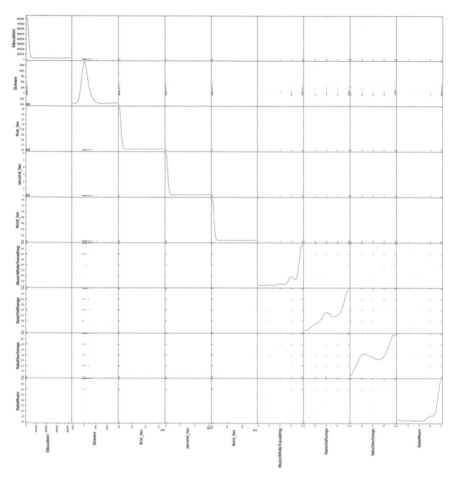

Fig. 1 Dataset distribution

Pseudocode 4: Simulation

Step-1:   Design the survey questions
Step-2:   Pre-process and visualize the data

        Step-3:   Do the following steps till silhouette score is not low
        Step-3a: Choose the number of dimensions
        Step-3b: Do PCA on the dataset
        Step-3c: Calculate silhouette score for

                Agglomerative, K-Means and Birch

Step-4:   Do the following steps till silhouette score is not low
Step-4a: Calculate number of optimal clusters
Step-5:   Do the clustering with Agglomerative, K-Means and Birch

Step-5b: If the surveyor said its poor feedback, restart the model from step-1
Step-6:   Final clusters are formed

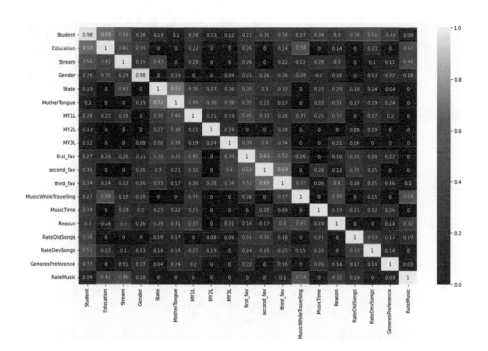

Fig. 2 Heatmap of the correlations-1

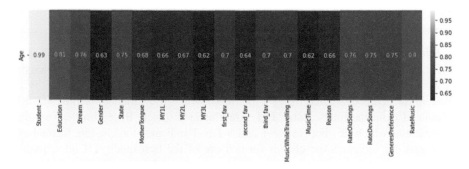

Fig. 3 Heatmap of the correlations-2

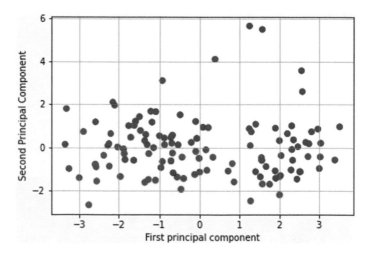

Fig. 4 Scatter plot of gender attribute

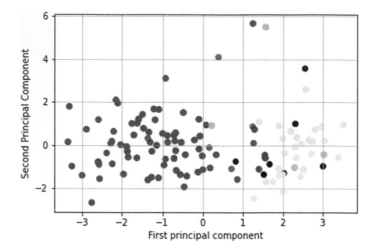

Fig. 5 Scatter plot of age group attribute

Initially, survey responses are taken from the users, and the data are collected. The data are processed and fed to principal component analysis (PCA) to reduce the number of dimensions for effective clustering as discussed in [16]. The quality of clusters formed is evaluated using a metric known as the silhouette score, and it is calculated from the inter-cluster and intra-cluster distance. Based on that, a value is generated which lies between −1 and +1. −1 indicates the clusters formed are poor, and +1 indicates the cluster formed are of the best quality. If poor quality of clusters is observed, then the same PCA is re-iterated with different numbers of

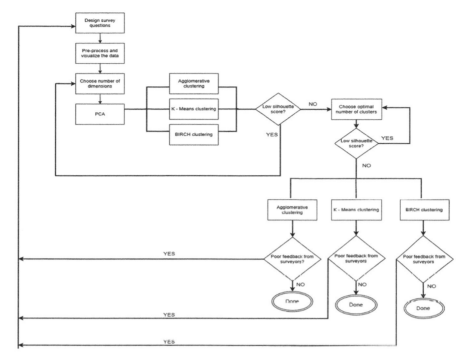

Fig. 6 Flowchart of ML model

dimensions. The same process is repeated until the most optimal dimension which gives the maximum silhouette score is obtained.

Once the optimal number of dimensions for PCA is obtained, this data are then used to form clusters and are fed to the three unsupervised clustering algorithms, namely agglomerative, K-means, and BIRCH. Initially, the number of clusters is taken randomly, and clustering is performed on it. Once the clusters are formed, silhouette score is again calculated, and based on the silhouette score, the same process is repeated multiple times to obtain the optimal number of clusters. Finally, the best possible clusters are formed, and then, these clusters are given to surveyors for evaluating the effectiveness of like-mindedness from the clusters. If the feedback received is poor, then the whole process is repeated starting from redesigning the survey questions.

Figures 7 and 8 correspond to the agglomerative clustering, Figs. 9 and 10 correspond to K-means, and Figs. 11 and 12 correspond to BIRCH clustering (the optimal number of clusters and optimal number of dimensions).

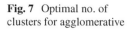

Fig. 7 Optimal no. of clusters for agglomerative

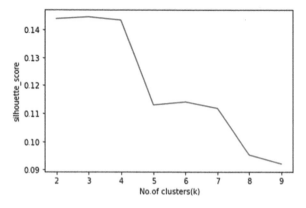

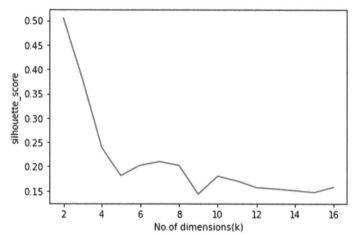

Fig. 8 Optimal no. of dimensions for agglomerative

Fig. 9 Optimal no. of clusters for K-means

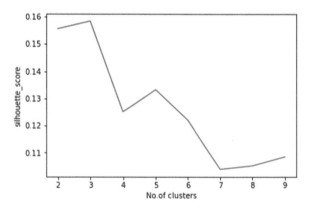

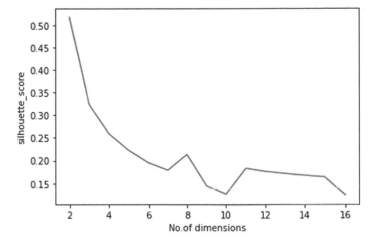

Fig. 10 Optimal no. of dimensions for K-means

Fig. 11 Optimal no. of clusters for BIRCH

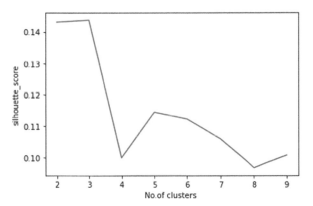

4   Results

Figures 13, 14, and 15 represent the scatterplot of finally formed clusters for agglomerative, K-means, and BIRCH algorithms, respectively. Here, each color represents a particular cluster, and each datapoint is a user from the data. The optimal number of clusters is chosen according to the silhouette score, i.e., optimal number of clusters as 3 gives the maximum silhouette score, which is a performance measure for the cluster quality.

Table 3 shows the optimal number of clusters, optimal number of dimensions, and corresponding silhouette scores for the clustering algorithms chosen. The optimal number of clusters and the optimal number of dimensions are obtained by iteratively checking the silhouette score for each number, and the optimal number of clusters

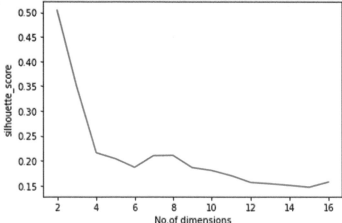

Fig. 12 Optimal no. of dimensions for BIRCH

Fig. 13 Finally formed agglomerative cluster

Fig. 14 Finally formed K-means cluster

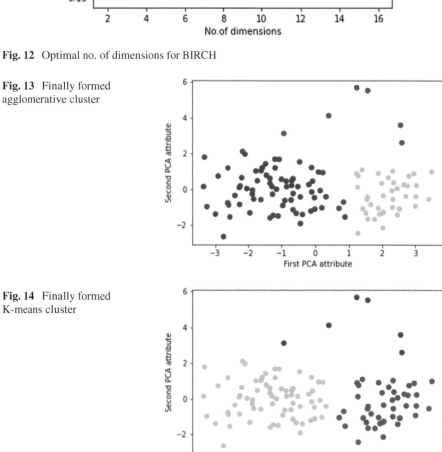

Fig. 15 Finally formed
BIRCH cluster

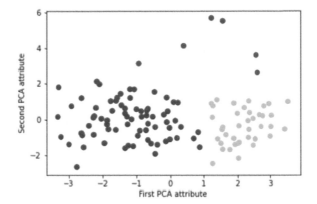

as 3 and the optimal number of dimensions as 2 give the maximum silhouette score
for each of the clustering algorithms.

After the people are clustered, the performance of an algorithm (Table 4) is determined by taking feedback from the same set of surveyors on the effectiveness of the
cluster they are assigned to, i.e., how closely related are the other users in his/her
cluster to that particular person. Since the surveyors know each others beforehand,
it indeed is an efficient way to determine how effective the musical traits determine
the like-mindedness.

It can be clearly observed that K-means has clustered the people better than the
other two clustering algorithms and hence produced the best efficiency than others.
Hence, it is proved that the musical traits of a person can act as a metric in clustering
people based on their like-mindedness.

Table 3 Results obtained

Clustering algorithm	Optimal number of clusters	Optimal number of dimensions	Silhouette score
Agglomerative	3	2	0.50437
K-means	3	2	0.51679
BIRCH	3	2	0.50437

Table 4 Efficiency

Clustering algorithm	Efficiency in determining the like-mindedness (%)
Agglomerative	73
K-means	77
BIRCH	73

5 Conclusion and Future Works

Results have shown that out of the three clustering algorithms, K-means performs better in determining the like-mindedness. All the three clustering algorithms are on par with each other with K-means on a higher side. As a result, it can be certainly said that, musical preference can act as an effective metric for determining the like-mindedness of people, and thus people can be grouped based on their like-mindedness. Hence, recommendations can be given to a particular user 'X' based on the liking of other person in the group/cluster of user 'X'. Similarly, the capabilities of existing recommendation systems can be extended, and also, the same idea could be used for recommendations in various applications like movie recommendations, social media recommendations, etc.

In future, a Web-based GUI application for the users would be developed which is capable of recommending songs as in [17], a common chat option to the entire group and also one-on-one chat between like-minded people that are formed from this algorithm and other features like the one mentioned in [18].

References

1. Kurup, A. R., & Sajeev, G. P. (2017). Task recommendation in reward-based crowdsourcing systems. In *2017 International Conference on Advances in Computing, Communications and Informatics (ICACCI)* (pp. 1511–1518). https://doi.org/10.1109/ICACCI.2017.8126055
2. Devika, P., Jisha, R. C., & Sajeev, G. P. (2016). A novel approach for book recommendation systems. In *2016 IEEE International Conference on Computational Intelligence and Computing Research (ICCIC)* (pp. 1–6).https://doi.org/10.1109/ICCIC.2016.7919606
3. Alsaleh, S., Nayak, R., & Xu, Y. (2012). Grouping people in social networks using a weighted multi-constraints clustering method. In *2012 IEEE International Conference on Fuzzy Systems* (pp. 1–8).https://doi.org/10.1109/FUZZ-IEEE.2012.6250799
4. Greenberg, D., Kosinski, M., Stillwell, D., Monteiro, B., Levitin, D., & Rentfrow, P. (2016). The song is you: Preferences for musical attribute dimensions reflect personality. *Social Psychological and Personality Science, 7*https://doi.org/10.1177/1948550616641473
5. Bedi, P., Sharma, C., Vashisth, P., Goel, D., & Dhanda, M. (2015). Handling cold start problem in recommender systems by using interaction based social proximity factor. In *2015 International Conference on Advances in Computing, Communications and Informatics (ICACCI)* (pp. 1987–1993). https://doi.org/10.1109/ICACCI.2015.7275909
6. Zisopoulos, C., Karagiannidis, S., Demirtsoglou, G., & Antaris, S. (2008). *Content based recommendation systems.*
7. Kumar, M., Yadav, D., Singh, A., & Kr, V. (2015). A movie recommender system: MOVREC. *International Journal of Computer Applications, 124,* 7–11. https://doi.org/10.5120/ijca20159 04111
8. Shah, K., Salunke, A., Dongare, S., & Antala, K. (2017). Recommender systems: An overview of different approaches to recommendations. In *2017 International Conference on Innovations in Information, Embedded and Communication Systems (ICIIECS)* (pp. 1–4). https://doi.org/ 10.1109/ICIIECS.2017.8276172
9. Harikumar, S., & Akhil, A. S. (2018). Semi supervised approach towards subspace clustering. *Journal of Intelligent and Fuzzy Systems, 34,* 1619–1629.

10. Patel, S., Sihmar, S., & Jatain, A. (2015). A study of hierarchical clustering algorithms. In *2015 2nd International Conference on Computing for Sustainable Global Development (INDIACom)* (pp. 537–541).
11. Na, S., Xumin, L., & Yong, G. (2010). Research on k-means clustering algorithm: An improved k-means clustering algorithm. In *2010 Third International Symposium on Intelligent Information Technology and Security Informatics* (pp. 63–67).https://doi.org/10.1109/IITSI.2010.74
12. Jiang, S., & Li, X. (2009). Improved BIRCH clustering algorithm: Improved BIRCH clustering algorithm. *Journal of Computer Applications, 29*, 293–296.https://doi.org/10.3724/SP.J.1087.2009.00293
13. Lorbeer, B., Kosareva, A., Deva, B., Softic, D., Ruppel, P., & Küpper, A. (2017). Variations on the clustering algorithm BIRCH. *Big Data Research, 11*https://doi.org/10.1016/j.bdr.2017.09.002
14. Indulekha, T. S., Aswathy, G. S., & Sudhakaran, P. (2018). *A graph based algorithm for clustering and ranking proteins for identifying disease causing genes* (pp. 1022–1026). https://doi.org/10.1109/ICACCI.2018.8554530
15. Rajesh, R., Gargi, S., & Samili, S. (2016). Clustering of words using dictionary-learnt word representations. In *2016 International Conference on Advances in Computing, Communications and Informatics (ICACCI)*, Jaipur, India.
16. Mishra, S., Sarkar, U., Taraphder, S., Datta, S., Swain, D., Saikhom, R., Panda, S., & Laishram, M. (2017). *Principal component analysis. International Journal of Livestock Research, 1*.https://doi.org/10.5455/ijlr.20170415115235
17. Paul, D., & Kundu, S. (2020). A survey of music recommendation systems with a proposed music recommendation system. In J. Mandal, & D. Bhattacharya (Eds.), *Emerging technology in modelling and graphics. Advances in intelligent systems and computing* (Vol. 937). Springer. https://doi.org/10.1007/978-981-13-7403-6_26
18. Samyuktha, M., & Supriya, M. (2020). Automation of admission enquiry process through Chatbot—A feedback enabled learning system. In V. Bindhu, J. Chen, J. Tavares (Eds.), *International Conference on Communication, Computing and Electronics Systems. Lecture Notes in Electrical Engineering* (Vol. 637). Springer. https://doi.org/10.1007/978-981-15-2612-1_18
19. Shakya, S. (2021). Multi distance face recognition of eye localization with modified Gaussian derivative filter. *Journal of Innovative Image Processing, 3*(3), 240–254.
20. Tesfamikael, H. H., Fray, A., Mengsteab, I., Semere, A., & Amanuel, Z. (2021). Simulation of eye tracking control based electric wheelchair construction by image segmentation algorithm. *Journal of Innovative Image Processing (JIIP), 3*(01), 21–35.

Diabetes Retinopathy Prediction Using Multi-model Hyper Tuned Machine Learning

B. V. Baiju, S. Priyadharshini, S. Haripriya, and A. Aarthi

Abstract Diabetic mellitus is a chronic illness which occurs due to lack of insulin that causes diabetic retinopathy which can incite loss of vision; in case, it is not identified at the very first level. Diabetic retinopathy is also considered as one of the major impermanence diseases in older people. Diabetics can lead to acute complications such as cardiovascular disease and stroke. If preventive measures are not taken, it can lead to further diseases such as nephropathy, diabetic foot, and retinopathy. Data mining plays an important role in diagnosing diabetic retinopathy which can be beneficial for the better health of the society. This model helps to identify the diabetic retinopathy based on classification models and by using machine learning algorithms with hyperparameter tuning aids in predicting the disease with higher accuracy.

Keywords Diabetic · Disease prediction · Machine learning · Retinopathy · Dataset training · Prediction of accuracy

1 Introduction

Diabetic retinopathy (DR) is a severe retinal disorder that occurs because of uncontrollable diabetic mellitus which leads to loss of vision if the disease is not detected at the prompt. Diabetic retinopathy is the majority disorder as per World Health Organization Report [1]. Diabetic retinopathy is anticipated to affect 191 million people by 2030. Retinal disease that is caused due to diabetic retinopathy will subsume glaucoma, cataracts, and retinopathy [2]. There are enormous chances for diabetic infected persons to not be treated for a very long time since in the initial stage, the patient might not be aware of these diseases [3].

B. V. Baiju (✉) · S. Priyadharshini · S. Haripriya · A. Aarthi
Department of Information Technology, Hindustan Institute of Technology and Science, Chennai, India
e-mail: bvbaiju@hindustanuniv.ac.in

© The Author(s), under exclusive license to Springer Nature Singapore Pte Ltd. 2023   961
S. Shakya et al. (eds.), *Sentiment Analysis and Deep Learning*,
Advances in Intelligent Systems and Computing 1432,
https://doi.org/10.1007/978-981-19-5443-6_71

Fig. 1 Anatomy of human eye

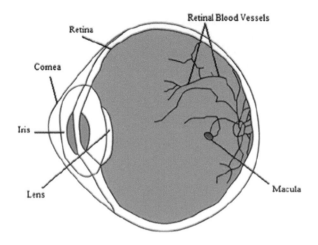

Diabetic retinopathy mainly affects the human red blood cells, and it shrinks the vision of the human eye as shown in Fig. 1. Microvascular disease and macrovascular disease are the most common diabetic complications diseases. Macrovascular disease menaces the tiny blood vessels and arteries. The diabetic retinopathy disease commonly attacks the eye (retinopathy), nerves (neuropathy), and kidney (nephropathy) which are the microvascular diseases. The crucial macrovascular stumbling block includes cardiovascular disease demonstrated as strokes among other serious complications [4]. There are different types of diabetes that can affect a human; however, the most common types of diabetics are type 1 diabetes and type 2 diabetes and gestational diabetes mellitus (GDM). Type 1 diabetic commonly affects children, and type 2 diabetes affects the middle-aged people and infrequently older people. Whereas GDM affects women during their pregnancy [5].

Generally, diabetic retinopathy treatment is conducted by an Ophthalmologist by gathering the retinal images of the patient. Data mining process can be exceedingly helpful for medical practitioners for extracting hidden medical knowledge [15]. Diabetic retinopathy is the most complex disorder that cannot be ignored according to Clinical Research World of Ophthalmology. DR is codified as two different stages non-proliferative DR (NPDR) and proliferative DR (PDR) [6]. Further, DR is classified as mild stage and severe stage.

Light passes through the front of the eye (cornea) to the lens. The cornea and the lens help to focus the light rays onto the back of the eye (retina). The cells in the retina absorb and convert the light to electrochemical impulses which are transferred along the optic nerve and then to the brain. Initial medication of DR can prevent patients from more distant worsening. In today's world, to classify the critical stages of the disease, most experienced doctors perform the identification of DR manually [6]. The foremost aim of this research paper is to construct a working model to detect diabetic retinopathy by using a machine learning algorithm with hyperparameter tuning and to achieve more accuracy in prediction than the existing models. Once the model is trained and tested with the datasets, it can be used to predict diabetic retinopathy at an earlier stage.

The contribution of this research paper is categorized as the enclosed sections. Section 2 discusses the related works. Section 3 contains the existing execution and Sect. 4 briefs the proposed method. Moreover, Sect. 5 describes the implementation of the model, Sect. 6 illustrates its methodology, Sect. 7 recounts the algorithms used, and the algorithms with hyperparameter tuning are explained in Sect. 8. Section 9 analyzes the simulation, and the following section defines the evaluation parameters. Section 11 shows the comparison results between various approaches. Finally, the last section concludes the work.

2 Related Works

Baiju and Aravin [1] proposed multi-attribute interdependency relational clustering of diabetic data with influence measure-based disease prediction. An efficient multi-attribute influence measure-based algorithm was used, and the method has been evaluated for its efficiency using UCI diabetes dataset.

Kamble et al. [2] proposed a retinal pictures dataset by utilizing RBF brain organization. The results showed the delicacy of 71.2, perceptivity 0.83, and particularity 0.043 for DIARETDB0.

Bhatkar [3] proposed a multilayer perceptron neural network (MLPNN) to recognize diabetic retinopathy in retinal pictures. The MLPNN classifier was introduced to group retinal pictures as ordinary and strange. The Train N Times technique was utilized to prepare the MLPNN to track down the best component subset. The preparation and cross approval rates by the MLP NN were 100 percent for location of ordinary and strange retinal pictures.

Fiarni et al. [4] recommended a vaccination model of diabetes entanglement grievance based on information mining style and calculation bunching and characterization. Diabetes clinical information is isolated into four classes by the model: nephropathy, retinopathy, neuropathy, and blended impacts. To develop the ideal rule-based model for vaticination, they measure execution utilizing grouping and order calculations.

Elmogy, Mohammed [5] developed a model to assist ophthalmologists. The ML-CAD framework imagines unmistakable sickness changes and analyzes DR grades. They start by eliminating commotion, working on the quality, and normalizing the retinal picture sizes. Analysts determined the dim level run length network normal in four particular headings to recognize solid and DR people.

Sankaranarayanan and Perumal [15] proposed two significant data mining procedures. FP-Growth and Apriori have been utilized for application to diabetes dataset, and affiliation rules are being produced by both calculations.

Han et al. [6] fostered a PC vision framework for perceiving and computerizing this protest involving a neural organization to give discoveries to an enormous number of cases in a short measure.

Sun and Zhang [7] proposed an assortment of five AI models for distinguishing DR in cases utilizing electronic health rate (EHR) information, as well as a bunch of treatment choices. The last testing discoveries exhibited that irregular backwood is an AI model that accomplishes 92% precision and performs well.

Kalyani et al. [8] suggested that diabetic retinopathy will be distinguished and characterized utilizing the arranged case organization. The convolution and essential case layers were utilized to remove qualities from fundus pictures, while the class case layer and softmax layer were utilized to evaluate the probability that the picture has a place with a particular class. Four execution models were used to check the proposed organization's productivity utilizing the Messidor dataset.

Sharma et al. [9] proposed a PC vision framework via mechanizing. It utilized a neural organization and furthermore identified the grumbling, to give results for an enormous number of cases in a little quantum of time.

Pratt et al. [10] proposed a CNN approach for diagnosing diabetic retinopathy from advanced design pictures and straightforwardly arranging its firmness. Also, he fostered an organization with CNN plan and information expansion that could lay out the complex choices worried, stretches the arrangement task like miniature aneurysms, exudates and hemorrhages on the tissue layer and regularly offer an assignment precisely and keeping in mind that no client finish input. This organization was prepared involving a top-of-the-line GPU for an elevated place arrangement.

Deepa et al. [11] presented a troupe of multi-stage profound CNN models for diabetic retinopathy evaluating upheld picture patches. The pre-handling stage permitted the information pictures to give a great deal of important data than the crude information pictures. Normalization and resizing region units were used because of the pre-handling methods during this work. The arranged multi-stage algorithmic program was executed with three primary stages to execute the order measures of the call organization.

Sreng and Maneerat [12] proposed a strategy to preprocess picture to deflect small commotions and improve the differentiation of the image. Additionally, liquor edge perception was utilized to recognize the splendid sores. Then, at that point, the red sores region unit recognized wishing on formal hat morphological separating ways. Additionally, the remarkable and dull sores region unit were consolidated by exploitation consistent AND administrator. To be left exclusively with neurotic signs, the commotions close vessels are eliminated by exploitation of mass examination. Morphological choices were removed and one more to the SVM classifier.

Suriyaharayananm et al. [13] proposed a most recent limitation for vulnerable side location. At first, sight the vessel and exudate fixes and remove each to actuate point. A few of the choices like vas, exudate, and points are recognized for precise exploitation. The morphological tasks are applied reasonably.

Santos et al. [14] upheld profound neural organization models that perform one-stage object recognition, abuse moderate information increase, and move learning methods to give a model among the assignments of physical design injuries. The model was prepared and upheld the YOLOv5 plan and conjointly the PyTorch structure, accomplishing values for map.

Kavakiotis et al. [15] proposed uses of AI, information mining strategies, and apparatuses as far as expectation and analysis. The utilization of ML and information mining strategies in advanced datasets that incorporate clinical and organic data relied upon to prompt more top to bottom investigation toward analysis and treatment of DM because of the appearance of biotechnology and the tremendous measure of information created.

3 Existing System

The existing work deals with the issue of ailment gage using biomedical diabetic enlightening list that has been particularly analyzed. In like manner, it presents an effective measure-based sickness conjecture estimation. In the first place, the methodology scrutinizes the biomedical enlightening assortment and performs upheaval ejection. Secondly, the features are taken out, and for each data point, a multi-attribute relational similarity measure (MARSM) is evaluated toward different bundles available. Taking into account the MARSM measure evaluated, a lone class has been recognized for the significant data given. The packing estimation is a different evened out one, and it bunches the data centers in an ever-evolving way. For the conjecture of the contamination, the method calculates the effect measure on every trademark perceived from the component. Taking it into account, the multi-attribute influence measure-based (MAIM) system recognizes the possible disease test. The procedure produces higher viability in grouping as well as disease assumption. The procedure diminishes the deceptive request extent and reduces the time unpredictability. The examination can likewise be improved by remembering ecological elements for diabetic subordinate infection expectation. Scientists might additionally add occasions of diabetic retinopathy to the dataset for the forecast of diabetic retinopathy.

4 Proposed System

The proposed work is intended to demonstrate the data mining technique in diabetic disease prediction systems in the medical stream. In order to perform this task, the retinopathy disease-based data are selected for analysis and prediction. In the proposed system, five supervised machine learning algorithms are used to show the comparison and accuracy of the algorithm, and the result shows which supervised algorithm has a high accuracy in detecting the diabetic retinopathy disease. The best model algorithm will predict the diabetic retinopathy based on the user input data. In the model, a Web application with sample data is designed, and so when a user has given any input with regards to the data in the web application, the AI model anticipates the relevant information and concludes whether a person with such data has been impacted with diabetic retinopathy or not. In this system, a hyperparameter tuning method is used to show the accuracy of model algorithm

Pregnancies	Glucose	BloodPressure	SkinThickness	Insulin	BMI	DiabetesPedigreeFunction	Age	Outcome
6	148	72	35	0	33.6	0.627	50	1
1	85	66	29	0	26.6	0.351	31	0
8	183	64	0	0	23.3	0.672	32	1
1	89	66	23	94	28.1	0.167	21	0
0	137	40	35	168	43.1	2.288	33	1

Fig. 2 Dataset image

prediction of diabetic retinopathy, and algorithms with and without hyperparameter tuning have been illustrated. The evaluation parameters are used to identify the best multi-model to detect the patients with diabetic disease.

Figure 2 shows the dataset attributes for training the model and to predict diabetic retinopathy.

5 Implementation

The proposed system is intended to show the data mining techniques in disease prediction for diabetic retinopathy. It is implemented in the following three steps.

- Exploratory data analysis, which is required for spotting hidden patterns, detecting outliers, identifying relevant variables, and identifying any irregularities in data.
- Model building using multiple algorithms, wherein the model is built using several algorithms.
- Prediction of retinopathy, where checkpoints are created for the model which produces best accuracy and prediction.
- Creating a Web application.

5.1 Dataset

The standard dataset of diabetic retinopathy is taken from Kaggle dataset. Table 1 contains the attributes of the dataset.

5.2 Exploratory Data Analysis (EDA)

Exploratory data analysis (EDA) is a vital stage in each examination project. The principal objective of the exploratory examination is to search for appropriation, exceptions, and irregularities in the information to prompt explicit testing. It likewise

Table 1 Attributes of dataset

Attribute 1	Id
Attribute 2	Name
Attribute 3	Pregnancy
Attribute 4	Glucose
Attribute 5	Blood pressure
Attribute 6	High pressure
Attribute 7	Insulin
Attribute 8	BMI
Attribute 9	Diabetes
Attribute 10	Skin thickness
Attribute 11	Age

incorporates instruments for creating speculations by reviewing and understanding information, which is usually done by means of graphical portrayal. EDA is intended to help the expert in recognizing regular examples. At last, highlighted determination approaches are regularly delegated to EDA. EDA is an urgent stage following information assortment and preprocessing in which the information is just shown, plotted, and controlled without making any suspicions to support information quality evaluation and model development. Most presumptions support information quality appraisal and model development.

Most of EDA approaches are graphical in nature as in Fig. 3, with a couple of quantitative strategies tossed in just in case. The reasoning for this broad reliance on designs is on the grounds that the main role of the era is to investigate, and illustrations furnish examiners with the unparalleled ability to do as such while being prepared to get knowledge into the evidence. The information from an investigation is gathered into a rectangular cluster, e.g., accounting page or dataset, most regularly with one line for every trial subject and one section for each subject identifier, result variable, and illustrative variable. Every section contains the numeric qualities for a specific quantitative variable or the levels for an absolute factor (A few more confounded investigations require a more perplexing information design). An examination's information is normally accumulated in a rectangular cluster with one column for each subject recognizable proof, result variable, and logical variable. The numeric qualities for a quantitative variable or the levels for an unmitigated variable are remembered for every section. Most of these methodologies work by covering specific pieces of the information while uncovering others.

There are two strategies to order exploratory information examination. The principal qualification is that each approach is either non-graphical or graphical. Second, every system is univariate or multivariate in nature (normally bivariate). Nongraphical methodologies frequently incorporate the calculation of synopsis measurements, however, graphical techniques plainly sum up information in a diagrammatic manner. Univariate approaches explore connections between one variable (data segment) at a time, and multivariate strategies research relationships between at least

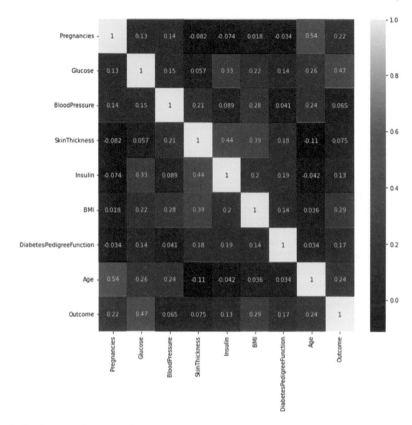

Fig. 3 Exploratory data analysis

two factors simultaneously. Regularly, multivariate EDA will be bivariate, yet at least three factors might be remembered for uncommon events. Prior to finishing a multivariate EDA, it is almost consistently smart to run a univariate EDA on every one of the parts of the multivariate EDA. Every one of the EDA classes has extra divisions relying upon the capacity and sort of the factors being assessed, notwithstanding the four classifications created by the previous cross classifying. The following are EDA's objectives in a nutshell:

- Gain as much insight into the database as possible and understand its structure.
- Create a visual representation of the possible links (direction and magnitude) between exposure and outcome variables.
- Look for outliers and anomalies (numbers that deviate significantly from the rest of the data).
- Build parsimonious models or make a preliminary model selection.
- Generate clinically significant variables by extracting and creating them.

Exploratory facts analysis is a crucial step in studying and analyzing diverse datasets [15], as well as summarizing their key properties. The application of EDA may aid in the discovery of hidden patterns in datasets, and its importance in data science cannot be overstated. Exploratory data analysis is often required for spotting hidden patterns, detecting outliers, identifying relevant variables, and identifying any irregularities in evidence. At this point, the data is cleaned and pre-processed, to get rid of missing and null value records. To achieve an accurate result, preprocessing identifies and eliminates or substitutes missing values in the dataset that only make up a tiny fraction of the total data. After that, the dataset is pre-processed, and the cleaned data are utilized for training. The cleaned data are then examined, and all essential steps such as eliminating noise from the image and setting all resolutions to the same are taken.

Figure 4 shows the implementation of the model for predicting diabetic retinopathy.

There are four stages to data preparation:

- The first phase is data cleaning, which involves fixing or removing duplicated, poorly formatted, or damaged data.
- The second phase is data integration, which involves combining data from several sources into a single perspective.
- The data reduction stage is the third step, in which the data are encoded, scaled, and sorted if necessary.
- The data transformation is the last phase in which the data are turned into the desired format.

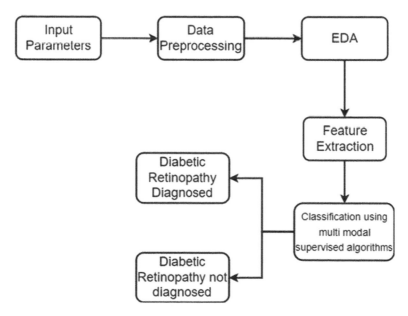

Fig. 4 Implementation of the model

6 Methodology

For building and testing the model, five algorithms which are logistic regression, random forest, decision tree, KNN, and support vector machine are used. These algorithms are used to show the accuracy of individual algorithm by using a hyperparameter tuning. The research paper is intended to show the comparison of the five algorithm without using a hyperparameter tuning and using hyperparameter tuning for identifying the patients with diabetic retinopathy possibilities. The result is designed in such a way that the accuracy of the algorithm using a hyperparameter tuning shows higher accuracy than the algorithm without hyperparameter tuning.

6.1 Hyperparameter Tuning

Figure 5 shows the hyperparameter tuning process to show higher accuracy. In machine attainment, hyperparameter improvement or standardization is the drawback of selecting a group of best hyperparameters for an attainment formula. A hyperparameter could be a parameter whose price is employed to manage the attainment method. For illustration, the terms "model parameter" and "model hyperparameter" are used. Some use cases of model hyperparameters embody the attainment rate for coaching a neural network. The C and alphabetic character hyperparameters are used for support vector machines. TModels will have varied hyperparameters and everchanging fashionable combination of parameters that is treated as a pursuit drawback. 2 fashionable ways for hyperparameter standardization are

- GridSearchCV
- RandomizedSearchCV

In this analysis, hyperparameter standardization is used to point out the high accuracy to identify diabetic retinopathy.

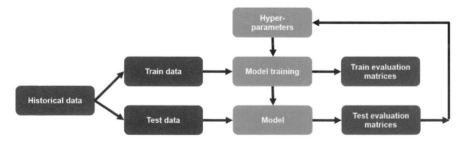

Fig. 5 Hyperparameter tuning

6.2 GridSearchCV

Figure 6 shows GridSearchCV to test and train the dataset. In the GridSearchCV approach, the machine attainment model is calculable for a variety of hyperparameter values. This approach is named GridSearchCV since it searches for a trendy set of hyperparameters from a grid of hyperparameters values.

GridSearchCV could be a performance that comes in Scikit-learn's (or SK-learn) model selection package. So, a vital purpose to notice is that, to own the Scikit-learn library, it has to be put on the PC. This performance helps to circle through predefined hyperparameters and suit the expert (model) on training set.

An online application has been used that facilitates users to notice retinopathy by providing the information. The text box is given wherever the user has the choice to present the values. All the users providing information to the detector could use so as to update the sculpture of model information analysis in future.

7 Algorithms

In this research paper, a supervised ML method is used to detect the early stage of DR in a person. The algorithms used are logistic regression, random forest, decision tree, KNN, and support vector machine.

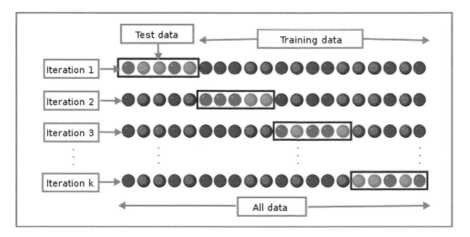

Fig. 6 GridSearchCV

7.1 Logistic Regression Algorithm

Logistic regression is a widely used popular machine learning algorithm, which comes under the supervised learning. It is used for prognosticating the categorical value using a given range of standalone value variables. Logistic retrogression predicts the affair of a categorical dependent variable. The term "logistic" is taken from the logit function that is used in this system of classification. Image result is obtained for logistic retrogression in machine learning.

Logistic retrogression is a supervised machine learning algorithm that can be used to model the class or event. It is used for linear divisible data. That means, logistic retrogression is generally used for double bracket problems. Supervised learning algorithm should have input variables (x) and a target variable (y) when you train the model.

7.2 KNN Algorithm

KNN is otherwise called K-nearest neighbor strategy which is managed by ML. By including all the checkpoints, the basic motive of the class is to predict test data points. As a result, data mining has been used for classification, regression, and missing value estimation. K-means is the another most simple algorithm for segmenting and classifying image into different clusters based on attribute and feature value. There are various way to identify the k-means value but the simplest way to run the algorithm is to choose the one which performs best. Each datapoint has the biggest local neighborhood while developing the model, which encompasses the greatest of data points. The world's biggest data points may be seen as the neighborhood that encompasses all of its data points. To continue the enclosed procedure for data points, the data points are covered by selected representatives.

In any case, the precise k of the technique needs not be shown throughout the creation of the model. The number of data points owned by certain neighborhoods are considered as an ideal k although it varies across representatives. During the creation of the model, the k is created automatically. Furthermore, utilizing the classified model not only decreases the volume of the data to be classified but also enhances the efficiency gradually.

The algorithm can be summarized as:

- A positive is taken along with the new sample.
- New entries which are closer to the new sample are taken.
- The most common classification is found and added as a new sample.

7.3 Decision Tree

Decision tree is the most popular and powerful algorithm for classification and prediction. Decision tree as the name suggests, it is tree like structure which will helps to identify the result based on tree like graph. This research focuses on decision tree algorithm in particular. The numerical weights in the neural network of connections between nodes are substantially more difficult to create conceptual principles. In data mining, decision trees are used as a common classification method. Decision tree offers a wide range of applications due to their ease and analysis of accuracy across many data types. Decision tree is the recursive split of an instance that is used to classify the data. Each test value evaluates a single attribute in the simple and most common situation. The condition refers to the range of cases in numeric properties. Decision tree geometrically understands the collection of hyperplanes. Naturally, decision makers favor less complicated solution as they are more understandable. The reasons to employ decision trees in this model are as follows:

- Decision trees are self-explanatory and simple to follow when condensed.
- Decision trees can deal with datasets which are prone to mistakes.
- Both nominal and numerical input can be handled by a decision tree.
- Decision trees can deal with missing values.
- Decision tree algorithm is a non-parametric approach which implies the assumption about the distribution space and structure of the classifier.
- Training duration is faster related to NN algorithms.
- Any discrete value classifier can be included in the decision tree representatives.

7.4 Random Forest

It is a supervised ML algorithm that helps to solve problems like regression and classification. In the event of a relapse, it assembles decision trees on numerous examples and takes their greater vote in favor of order and normal. The random forest algorithm creates the result by combining the results of many decision trees. It erodes the packing guideline, which creates an alternate preparation subset from test preparation information with substitution, and the final result is mostly dependent on voting.

The tree relies on the benefits of an arbitrary vector evaluated independently and uniformly across the forest. Out to be huge, the speculation mistake for forest meets as far as feasible. The power of the individual trees in the forest, as well as their interrelation, is crucial to the hypothesis blunder of a forest of tree classifiers.

A method for generating random numbers is the random forest algorithm. As the name implies, an unregistered joint distribution value is calculated. The main objective is to identify the prediction function for forecasting value.

7.5 SVM

Support vector machine (SVM) learning formula might be used for each classification similarly as the regression issues. SVM may be a direct model for bracket and retrogression issues. It breaks the direct and non-linear issues and works well for varied sensible issues.

SCIKIT learning is an extensively accessible methodology for imposing cc algorithms. Support vector machine is additionally useful in SCIKIT library appropriate model and validation. SVM works by mapping information to a high-dimensional purpose area so that information points are often distributed, and so once the information is not, it is linearly dissociable. A division between the orders may be a plane, however, conjointly the information square measure regenerates in such a way that the division might be drawn as a hyperplane.

8 Algorithm with Hyperparameter Tuning

8.1 Logistic Regression

The logistic regression category from sklearn forms a model utilizing the training set and utilizes the testing set to accumulate the forecasts for each one in every points of interest within the testing set. The testing set is employed to appraise the exhibition of the model that you just have ready, utilizing the training set. This can be an honest methodology for assessing the model, but it most likely does not give a real plan of the exhibition of the model. As much as you would possibly remember, the data within the testing set may well be discarded and utilizing it to assess the model may provide an authentic result. An instance of the logistic regression category while not passing any initializers is created. The terms used are explained below:

Punishment—Specifies the quality of the penalization.
C—Inverse of regularization strength; lower values confirm a lot of grounded regularization.
Solver—Rule to use within the optimization drawback.
max_iter—Most range of duplications taken for resolution of advanced issues.

Figure 7 shows the accuracy of the logistic regression algorithm with hyperparameter tuning techniques.

8.2 KNN Algorithm

The KNN bracket formula itself is comparatively straightforward and intuitive. Once an information purpose is handed to the formula with a given worth of K, it searches

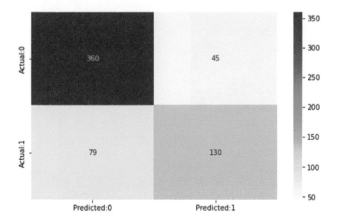

```
[ ] print(classification_report(y_train,knn_predict))

                precision    recall  f1-score   support

           0        0.82      0.89      0.85       405
           1        0.74      0.62      0.68       209

    accuracy                            0.80       614
   macro avg        0.78      0.76      0.77       614
weighted avg        0.79      0.80      0.79       614
```

Fig. 7 Confusion matrix with hyperparameter tuning for logistic regression

for the K-nearest neighbors. The closest neighbors are planted by shrewd space between the given information and therefore the knowledge points within the original dataset.

A KNN classifier case is first developed and conjointly prepare a spread of values of hyperparameter K from one to thirty-one that may be utilized by GridSearchCV to search out the fashionable values. Likewise, cross-validation batch sizes are set as cv = ten, and rating criteria are set as per the demand.

Figure 8 which is the screenshot shows the accuracy of the KNN algorithm with hyperparameter tuning techniques.

8.3 Decision Tree

Decision tree is the most ideal way to arrange classes. They are straightforward or white box classifiers which implies that rationale behind decision tree can be observed. GridSearchCV which is the principal component assists in finding the trendy hyperparameters. The worth of grid search boundary could be a rundown that contains a Python wordbook. The key is the name of the boundary. The worth of the

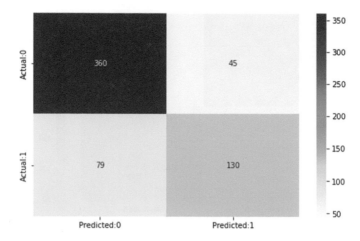

```
]  print(classification_report(y_train,knn_predict))

               precision    recall  f1-score   support

           0       0.82      0.89      0.85       405
           1       0.74      0.62      0.68       209

    accuracy                           0.80       614
   macro avg       0.78      0.76      0.77       614
weighted avg       0.79      0.80      0.79       614
```

Fig. 8 Confusion matrix with hyperparameter tuning for KNN

world book is in the various upsides of the boundary. This creates a table that can be considered as beautiful boundary values. There is an article or model of the decision tree classifier. The grid hunt utilizes brilliant sorts of section execution rules on the scoring styles. For this situation, section blunder and number of groups, the model or item, and the boundary values are used. A portion of the works incorporates various scores for various boundary values. For this situation, section blunder alongside boundary esteems that have the up-to-date score.

Figure 9 shows the accuracy of the decision tree algorithm with hyperparameter tuning techniques.

8.4 Random Forest

Hyperparameters and accordingly the up-to-date way to deal with slender chase gage a wide scope of values for every hyperparameter. Random forest permits to tighten down the reach for each hyperparameter. Since it has become so obvious where to think the chase, each blend of settings can be unequivocally indicated to attempt.

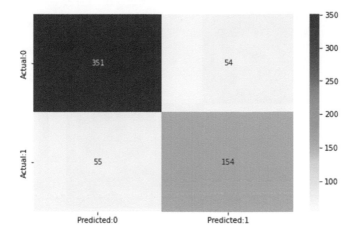

```
[ ]  print(classification_report(y_train,tree_predict))

                 precision    recall  f1-score   support

             0        0.86      0.87      0.87       405
             1        0.74      0.74      0.74       209

      accuracy                            0.82       614
     macro avg        0.80      0.80      0.80       614
  weighted avg        0.82      0.82      0.82       614
```

Fig. 9 Confusion matrix with hyperparameter tuning for decision tree

This is done with GridSearchCV, a framework that, somewhat cut carelessly from an appropriation, assesses all blends defined. Using Scikit-Learn's Randomized-SearchCV framework, a network of hyperparameter ranges can be characterized, and capriciously test from the matrix, performing K-Fold CV with every mix of values.

Figure 10 which is screenshot shows the accuracy of the random forest algorithm with hyperparameter tuning techniques.

8.5 SVM

A machine learning model is characterized as a fine model with varied boundaries that ought to be gained from the data. In any case, there are several boundaries called hyperparameters, and people cannot be foursquare erudite. GridSearchCV takes a reference that depicts the boundaries that might be taken as a stab at a model to

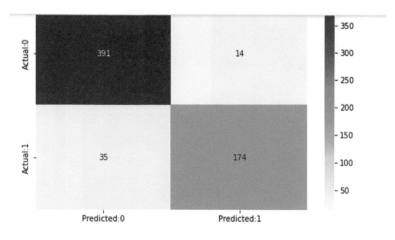

```
print(classification_report(y_train,y_pred))

                 precision    recall  f1-score   support

             0       0.92      0.97      0.94       405
             1       0.93      0.83      0.88       209

      accuracy                           0.92       614
     macro avg       0.92      0.90      0.91       614
  weighted avg       0.92      0.92      0.92       614
```

Fig. 10 Confusion matrix with hyperparameter tuning for random forest

organize it. The framework of boundaries is characterized as a reference, wherever the keys are at the boundaries, and also the qualities are the settings to be tried.

This piece shows how to utilize the GridSearchCV looking through framework to track down ideal hyperactive boundaries and consequently improve the vaccination results. One of the extraordinary impacts about GridSearchCV is that it is a meta-assessor. It takes an assessor like SVC and makes another assessor that acts precisely as something very similar—for this situation, similar to a classifier. Change = True ought to be added and circumlocutory to anything that number needs should be picked. The higher the number's development, the higher the circumlocutory (circumlocutory simply implies the reading material undertaking portraying the interaction).

Figure 11 shows the accuracy of the support vector machine algorithm with hyperparameter tuning techniques.

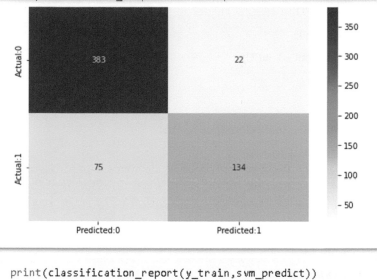

```
print(classification_report(y_train,svm_predict))

              precision    recall  f1-score   support

           0       0.84      0.95      0.89       405
           1       0.86      0.64      0.73       209

    accuracy                           0.84       614
   macro avg       0.85      0.79      0.81       614
```

Fig. 11 Confusion matrix with hyperparameter tuning for SVM

8.6 Pseudocode

Algorithm pseudocode for the machine learning-based load balancing algorithm

Input: task count c, setup pool, of undertaking count c, max assessments nmax, starting example size n

Xout ←sample minna, nmax} distinct configurations from Xp

Yout ← Evaluate parallel (c, Xout)

M ←fit (Xout, Yout)

Xp← Xp Xout

for i←n8+1 to nmax do

Yp← predict (M, Xp)

x_1 ←x Exp with the shortest runtime in p

Yi ← Evaluate (c, zi)

retrain M with (i, y_1)

Xout← Rout Uii Jout - Yout Uyi

Xp ←Xp - Ii

end for

Output: $\in$ out with the shortest runtime in Yout

Algorithm 5 Multi-task tri-training

m train_model(L)

repeat

for i {1..3} do

Li ϕ for x EU do

 if pj(x) = pk (x) (j, ki) then

Li Li U{(x.Pj)(x))}

if i 3 then

mitrain_model (Li)

else

mi +train_model(LUL;)

 until end condition is met

 apply majority vote over ml

9 Simulation and Analysis

An application is able to analyze all of the input data, process the data using distinct techniques, and then accurately predict whether the person is diagnosed with diabetic retinopathy or not has been created. The best model is saved and used for the prediction. The validation results are shown in Figs. 12, 13 and 14. The file given by the user is predicted using the pre-trained model. The accuracy of the prediction model is over 95%.

Retinopathy is finally detected and displayed on to the GUI, respectively, so that the patient can know about the condition of his ailment currently faced by the potential patient.

Checkpoint is created for the model which produces the best accuracy and prediction is done.

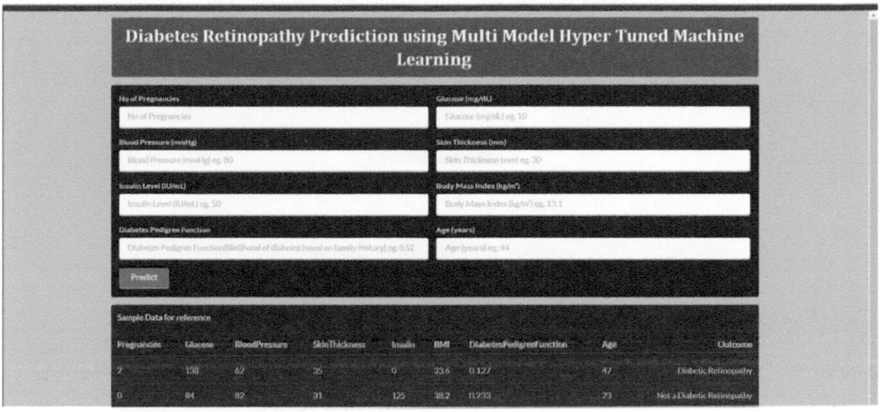

Fig. 12 Diabetic retinopathy prediction

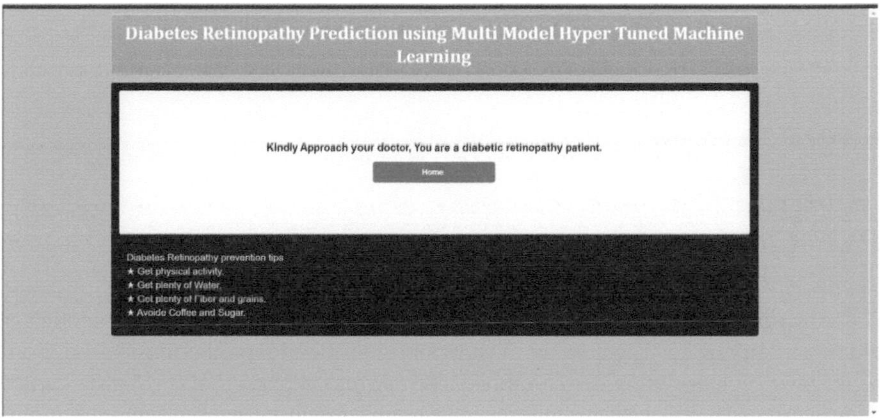

Fig. 13 Diabetic retinopathy affected

The application acts as a virtual assistant in major clinical laboratories, healthcare centers, and medical clinic.

10 Evaluation Parameters

10.1 Precision

Precision is a metric of the machine learning model's performance. It measures the accuracy of a model's positive prediction. Precision is calculated by dividing the total number of positive predictions by the number of genuine positives (i.e., the number of true positives plus the number of false positives).

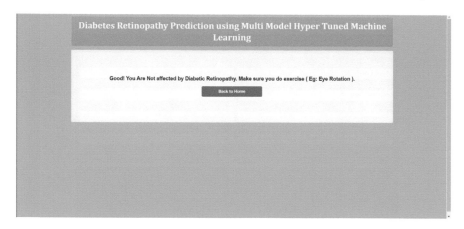

Fig. 14 Diabetic retinopathy unaffected

$$\text{Precision} = \text{TP/TP} + \text{FP}$$

- TP—True positive
- FP—False positive

10.2 Confusion Matrix

A confusion matrix is a summary of classification problem prediction outcomes as mentioned in Fig. 15. The number of right and unsuccessful predictions is totaled and split down by class using count values. The confusion matrix' key is that it depicts various ways in which the categorization model might be misread when it makes estimates. It is perplexed.

- FN—False negative
- TP—True negative

Fig. 15 Confusion matrix formula

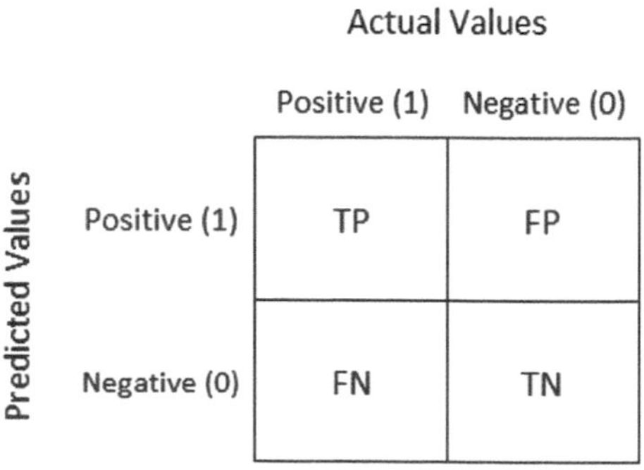

Actual Values

		Positive (1)	Negative (0)
Predicted Values	Positive (1)	TP	FP
	Negative (0)	FN	TN

10.3 Recall

The recall is determined as the proportion between the quantities of positive examples that are accurately delegated positive to the all-out number of positive examples. The recall estimates the model's capacity to identify positive examples. The higher the recall, certain models recognized.

$$Recall = TP/TP + FN$$

10.4 F1-Score

The weighted average of exactitude and recall is that of the F1-score. As a result, this score considers each false positives and false negatives. Though it is not as intuitive as accuracy, F1 is usually additionally helpful than accuracy, particularly if the category distribution is unequal. Once false positives and false negatives have equivalent prices, accuracy works well. It is best to see at each exactitude and recall, if the price of false positives and false negatives is significantly totally different.

$$F1\text{-score} = 2 * (Recall * Precision)/(Recall + Precision)$$

11 Results

The research paper is intended to show the comparison of the five algorithms using a hyperparameter tuning for identifying the patients with diabetic retinopathy possibilities. The result is designed in such a way that the accuracy of the algorithm using a hyperparameter tuning shows higher accuracy.

Figure 16 which is the comparison graph shows the overall accuracy of each algorithm using hyperparameter tuning methods, and it can be concluded that the random forest algorithm has a high accuracy compared to other algorithms.

12 Conclusion

Diabetic retinopathy (DR) is the most well-known disease achieving visual osmic resistance in diabetic patients. DR is the most perceived compromising diabetic eye disorder and causes driving vision hardship and visual weakening. A patient with a diabetic illness must experience rare eye screening. Therefore, this paper proposes a

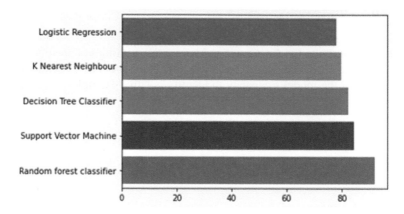

Fig. 16 Comparison graph

model to predict DR at the initial stage. The random forest computation has achieved the most effective execution taking into consideration ten clear factors, compared to other machine learning algorithms with hyperparameter tuning technique.

References

1. Baiju, B. V., & John Aravindhar, D. (2019). Multi attribute inter dependency relational clustering of diabetic data with influence measure based disease prediction. *Journal of Green Engineering (JGE)*, 9(1), 136–148.
2. Kamble, V. V., et al. (2022). Automated diabetic retinopathy detection using radial basis function. In *Published in International Conference on Computational Intelligence and Data Science (ICCIDS 2019)* (Vol. 167, pp. 799–808). https://doi.org/10.1016/j.procs.2020.03.42
3. Kharat, A. P. B. G. U. (2015). Detection of diabetic retinopathy in retinal images using MLP classifier. In *IEEE International Symposium on Nanoelectronic and Information Systems (iNIS)* (pp. 331–335). https://doi.org/10.1109/iNIS.2015.30
4. Fiarni, C., et al. (2019). Analysis and prediction of diabetes complication disease using data mining algorithm. *Procedia Computer Science, 161*, 449–457. https://doi.org/10.1016/j.procs.2019.11.144
5. Khan, F. A., et al. (2021). Detection and prediction of diabetes using data mining: A comprehensive review. *IEEE Access, 9*, 43711–43735. https://doi.org/10.1109/ACCESS.2021.3059343
6. Han, J., et al. (2018). The design of diabetic retinopathy classifier based on parameter optimization SVM. In *International Conference on Intelligent Informatics and Biomedical Sciences (ICIIBMS)* (pp. 52–58). https://doi.org/10.1109/ICIIBMS.2018.8549947
7. Sun, Y., & Zhang, D. (2019). Diagnosis and analysis of diabetic retinopathy based on electronic health records. *IEEE Access, 7*, 2169–3536. https://ieeexplore.ieee.org/stamp/stamp.jsp?tp=&arnumber=8721113
8. Kalyani, et al. (2021). Diabetic retinopathy detection and classification using capsule networks. *Complex and Intelligent Systems*. https://doi.org/10.1007/s40747-021-00318-9
9. Yadav, J., et al. (2017). Diabetic retinopathy detection using feedforward neural network. In *Proceedings of 2017 Tenth International Conference on Contemporary Computing (IC3)* (pp. 1–3). https://doi.org/10.1109/IC3.2017.8284350

10. Pratt, H., et al. (2016). Convolutional neural networks for diabetic retinopathy. *Procedia Computer Science, 90*, 200–205. In *International Conference on Medical Imaging Understanding and Analysis.* https://doi.org/10.1016/j.procs.2016.07.014

11. Deepa, V., et al. (2021). Ensemble of multi-stage deep convolutional neural networks for automated grading of diabetic retinopathy using image patches. *Journal of King Saud University—Computer and Information Sciences,* 1319–1578. https://doi.org/10.1016/j.jksuci.2021.05.009

12. Sreng, S., et al. (2017). Primary screening of diabetic retinopathy based on integrating morphological operation and support vector machine. In *International Conference on Intelligent Informatics and Biomedical Sciences (ICIIBMS)* (pp. 250–254). https://doi.org/10.1109/ICIIBMS.2017.8279750

13. Vijaya Kumari, V., et al. (2010). Feature extraction for early detection of diabetic retinopathy. In *International Conference on Recent Trends in Information, Telecommunication and Computing* (pp. 162–166). https://doi.org/10.1109/

14. Santos, C., et al. (2021). A new method based on deep learning to detect lesions in retinal images using YOLOv5. In *Published in IEEE International Conference on Bioinformatics and Biomedicine (BIBM)* (pp. 3513–3520). https://doi.org/10.1109/BIBM52615.2021.9669581

15. Kavakiotis, I., et al. (2017). Machine learning and data mining methods in diabetes research. *Computational and Structural Biotechnology Journal, 15*, 104–116. https://doi.org/10.1016/j.csbj.2016.12.005

16. Nagpal, D., et al. (2021). A review of diabetic retinopathy: Datasets, approaches, evaluation metrics and future trends. *Journal of King Saud University Computer and Information Sciences.* ISSN 1319-1578.https://doi.org/10.1016/j.jksuci.2021.06.006

17. Anwar, F., et al. (2020). "A comparative analysis on diagnosis of diabetes mellitus using different approaches"—A survey. In *Informatics in Medicine Unlocked* (Vol. 21). ISSN 2352-9148.https://doi.org/10.1016/j.imu.2020.100482

18. Goa, J., et al. (2019). Diabetic retinopathy classification using an efficient convolution neural network. In *IEEE International Conferences on Agents (ICA)* (pp. 80–85). https://doi.org/10.1109/AGENTS.2019.8929191

19. Gadriye, D., & Khandale, G. (2014). Neural network based method for the diagnosis of diabetic retinopathy. In *2014 International Conference on Computational Intelligence and Communication Networks (CICN)*, Bhopal, India (pp. 1073–1077). https://doi.org/10.1109/CICN.2014.2

20. Abdelmaksoud, E., et al. (2021). Automatic diabetic retinopathy grading system based on detecting multiple retinal lesions. *IEEE Access, 9*, 15939–15960. https://ieeexplore.ieee.org/document/93284

Face Mask Detection and Social Distancing Using Machine Learning with Haar Cascade Algorithm

T. Sangeetha, V. Miruthula, C. Kavimalar, and V. Aakash

Abstract COVID-19 is indeed newsworthy as it has reached a global emergency with an impending spread. Wearing a face covering and maintaining physical distance is a smart move recommended by the WHO. People infected with COVID-19 have difficulty breathing with shortness of breath. Environmental elements of those involved can be contaminated by infectious droplets. It is mandatory to wear a cover and follow a physical withdrawal, but many residents ignore the rules. In such situations, regular checks for facial coverings openly places and it are normal to force fines. As article recognition has unfurled to be a receptive biometric process, it has been broadly applied in observation, security, independent driving, and so on. With the fast improvement of profound learning models, object locators are exceptionally reasonable to foster social removing and facial covering indicators to direct the group by means of CCTV and observation cameras. The paper studies different profound learning organizations to foster such locators. In this review, the current article discovery models utilized for reconnaissance and individuals location are examined. The one-stage and two-stage identifiers alongside their applications and execution are framed by exhaustive way.

Keywords Social separating · COVID-19 · Mask prediction

T. Sangeetha (✉) · V. Miruthula · C. Kavimalar · V. Aakash
Department of Information Technology, Sri Krishna College of Technology, Coimbatore, India
e-mail: t.sangeetha@skct.edu.in

V. Miruthula
e-mail: 18tuit054@skct.edu.in

C. Kavimalar
e-mail: 18tuit047@skct.edu.in

V. Aakash
e-mail: 18tuit002@skct.edu.in

© The Author(s), under exclusive license to Springer Nature Singapore Pte Ltd. 2023   987
S. Shakya et al. (eds.), *Sentiment Analysis and Deep Learning*,
Advances in Intelligent Systems and Computing 1432,
https://doi.org/10.1007/978-981-19-5443-6_72

1 Introduction

As per the analysts, this infection can spread while coming in close contact with the individual who is impacted. Consequently, it being similar the wide range of various irresistible respiratory infections, scientists recommend close contact be removed under all possible conditions as the chance of the infection communicating through airborne ways and drops is very prone to happen. Consequently, to beat the spread of COVID, scientists have thought of another accepted practice, social distancing or physical distancing as in Fig. 1. This action has been perceived and recognized to be an extremely helpful measure in controlling the boundless of the infection. It guarantees insignificant human collaboration inside close distances on a public scale. The assessment of the routes in how this social measure has been successful is explored by concentrates by breaking down the affirmed and demise cases in Italy and Spain during the public lockdown and following the upliftment of the lockdown with the assistance of measurements on a time series (Tobías, 2020; Saez, Tobias, Varga and Barceló, 2020). The results portrayed that there were significantly lower announced cases subsequent to executing social separating in the two nations. Coronavirus spreads basically from one individual to another through respiratory beads. Respiratory drops travel high up because of hacking, sniffling, talking, yelling, or singing. These drops can then land in the mouths or noses of individuals or they might inhale these drops in.

In this way, utilization of veils is crucial for check the spread of the infection. Covers are a basic obstruction to assist with keeping the respiratory beads from arriving at others. Concentrates on show that veils lessen the splash of beads when worn over the nose and mouth. Observing the social removing standards and checking facial coverings on individuals physically is not just prohibitive with restricted assets,

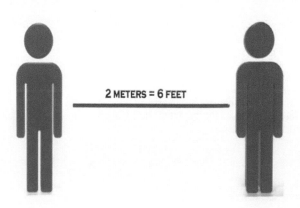

Fig. 1 Social distancing

however, can likewise prompt human mistakes. There is a prompt prerequisite for an answer for manage the infection spread by learning the best friendly separating standards to be trailed by people in general. This incorporates social distance infringement recognition and facial covering grouping to decide the wellbeing of the residents by checking assuming that enough distance is kept up with and assuming facial coverings are utilized. This framework has an extensive variety of pertinence in different public spots with cameras, for example, in supermarkets, petrol bunks, and traffic signals. This gives handling methods to use an observation framework for the majority different applications.

Social separating has been a powerful friendly measure as far as normalizing the bend. It evades any immediate contact among people and helps with lessening such transmission of the drops containing the infection through the human breath. A few not entirely set in stone toward the investigation of directions that communicate the drops by means of human breathing ways like hacking, talking, wheezing, and eating. While a couple of studies considered the drop size to be related with the microbes of the infectious sickness, where they believed enormous drops to be the ones that were to convey the microorganisms from the contaminated individual (Christian et al., 2004; Mangili and Gendreau, 2005), different examinations expected little drops of the particles that were in the core structure to be the ones to remotely spread. Factors like organic, clinical, and designing additionally required to have been considered to address the previously mentioned questions. In this manner, subsequent to thinking about this large number of variables, the proposed distance for a protected social distance was prescribed to be something like 6 feet, or at least, 2 m. Some accepted that this distance was adequately not and required to have been expanded. Albeit, the last distance was viewed as 2 m and was carried out overall making it a fruitful boundary, presently the distance has been diminished to 1 meter3 in a portion of the recuperating nations. Consequently, taking into account this large number of variables, this examination project proposes an AI-based location framework that aides in identifying any sorts of infringement. This examination expects on building a model that can be applied continuously frameworks and consequently help in keeping away from the spread of the infection.

2 Literature Review

Consequently, to forestall fast COVID-19 contamination, numerous arrangements, like constrainment and lockdowns are proposed by most of the world's state run administrations. In any case, this COVID-19 administration inefficacy can be moreover investigated with game-hypothetical situations past the public products game. Specifically, a few scientists have zeroed in on the reluctance of legislatures in sanctioning troublesome, however, essential infection regulation measures (e.g., remain at-home requests and lockdowns), as well as non-cooperation because of reasons other than free riding.

For example, creators in [1] contended that on the grounds that severe stay-at-home measures can significantly affect individuals' livelihoods, the expense of remaining at home (combined with lockdown weakness) can wind up offsetting the gamble of contamination from going out. As individual-level choices straightforwardly affect the general public level adequacy of stay-at-home requests, legislatures might abstain from carrying out them on account of expected low paces of consistence, particularly from financially hindered people who do not have the advantage of remaining at home [2].

A few states might have likewise been confident that crowd insusceptibility from recuperations and immunizations would permit them to try not to force such disagreeable measures through and through [3].

With rising quantities of cases and extended wellbeing offices, as well as the absence of an immunization all through 2020 and troubles related with accomplishing group invulnerability for COVID-19 [4], government inaction turned out to be progressively unviable. Thus, to expand individuals' adherence to severe guidelines, creators in [5] proposed utilizing social projects, for example, crisis alleviation assets and joblessness protection to bring down the expenses of consistence, especially for lower-paid laborers [6].

As immunizations opened up toward the finish of 2020, creators in [7] contended that projects driving inoculation take-up will outperform different perspectives like antibody viability and segregation techniques in significance.

Utilizing EGT, informal community examination, and specialist-based displaying, the creators recommended that people's immunization dynamic will be impacted by "socioeconomics, actual area, the degree of association, the wellbeing of the antibody, pestilence boundaries, and insights about the antibody being presented, and correspondingly, the decision-production of the public authority will be affected by scourge boundaries, the idea of the immunization being presented, coordinated factors, the administration of HR required for the inoculation exertion, and the quantity of antibody portions accessible" [8].

In [9], a creative shrewd strategy in view of a profound convolutional model is utilized to shield individuals from COVID-19. The proposed framework can identify naturally regardless of whether individuals are observing the security rules.

In [10], a definite examination is finished between the different profound gaining ways to deal with screen the illness from clinical imaging. In one more review [11], an IoT framework in view of temperature detecting, veil identification, and social removing is recommended for the security against COVID-19. Arduino Uno is utilized for the infrared detecting of temperature, while Raspberry Pi is utilized for the cover location and social removing utilizing PC vision methods.

In [12], an imaginative veil recognizable proof strategy is proposed by preprocessing of the picture followed by face discovery and picture super goal. The framework is viewed as especially exact, taking everything into account. [13] proposed a deep brain network model to observe individual and social atrophy even under adverse lighting conditions. This strategy is considered superior to many of the past methods in terms of speed and accuracy. [14] uses a deep learning approach based on computer vision to continuously observe the location of veils and social divisions.

This model was created on a Raspberry Pi to demonstrate various walkthroughs. It is designed to preserve opportunities and reduce the spread of COVID-19.

In [15], IoT-based deep learning institutions are suitable for COVID-19 identification. This model is used to detect infectious diseases by applying this model to chest x-rays. Brief recognition of COVID-19 by clinicians has proven to be highly accurate and consequently very convenient. The authors of [16] proposed an identification framework based on cell phone structures to isolate four different types of elements using K-nearest neighbor calculations. This structure is considered excellent in terms of precision, accuracy, and visibility.

3 Existing Method

In an open source procedure, computer vision is proposed to distinguish hidden people. This strategy is considered particularly effective in modern applications. Commitment to IoT and related sensors to track and contain infections. This review provides an in-depth understanding of electronic wellbeing management in light of COVID-19 surveillance sensors and outlines novel IoT networks for the post-epidemic era. This review looks at the innovations that can be used to detect COVID-19. The audit also highlights the challenges ahead in realizing these achievements. Advances reviewed include wearable sensors for screening COVID-19 patients in terms of X-ray-related deep learning, in vitro diagnostics (IVD), and IoT. Identification of ordinary veil wearers is accomplished by considering R-CNN strategies with high glare along with the final goal of district proposals and separation of constraint and characteristic fields. The R-CNN tuning approach is considered very accurate.

4 Proposed Model

To foster as a top priority or truly moderate environmental elements that add to public security from hurt, we recommend a proficient information handling machine situated in close contact with the certifiable in presence time frame made or done by a person to find two dependable public separate themselves and facial coverings really positioned by the model in front of the beginning of the model to screen unique interests or pursuits and find assault through visual hardware [15]. In expansion to introducing a caution to the general population, in this proposed structure as shown in Fig. 2, we have planned veil recognition along which shows individuals to wear their cover appropriately prior to allowing in to the area which they like and the model is made with Haar cascade is utilized for definite step.

Fig. 2 Block diagram of
proposed system used

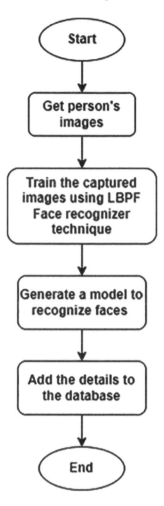

4.1 Face Mask Detector

The method carries out an accuracy of up to 95.77%. We achieve this accuracy
examining by way of MaxPooling. This makes it simple to understand the invariance
of the internal image along with a reduction in the number of edges the model
has to detect. This example of a localized sampling process does not adequately
model news images that contain written descriptions of past events at the cost of
reducing attractiveness. The number of neurons has a progressive value of 100,
which is not overly extreme. As the number of neurons and channels increases, more
false executions can occur. Updated channel principles and pull content will help
filter the body (face) of your account to correctly recognize cover traffic after the
start of retraining. The first sign was rated by 150 people. Our model can handle
acceleration of around 55 ms, giving us typical coverage of 18.1 fps in the US.

Through a continuous experimentation approach, the model shows an accuracy of 100 servings. Assuming a person is covering the hole with their hand, this indicates that the veil is not recognized.

4.2 Face Recognition

Current effect is from COVID vision at the time of appeal. Fit a model accompanying the captured images using the LBPH face tolerance procedure to obtain results between present and past times. After the arrangement, we create a model so that we can see the face. Finally, subscribe the delicacy to our newsbase. While experimenting with the first top-level words to modify a noun, we also shot 150 items to form a face acquisition model. After fitting the model, we rely on the LBPH face recognition system to achieve the effect. The model was able to accurately identify faces. So the model is 90% accurate. The face assertion framework covers three main stages: face assignment, original content, and face assignment. Face acquisition and image constraints are recognized using a Haar overflow classifier. The preprocessing of human images is separated from the elements present in the image. Feature extraction: Erase the base using LBPH on the recognized face.

First register the images of the parallel community model and then plot the histogram. Segmentation factors are controlled by the LBPH model, which tends to assign to previously taken pictures, which apply AI judgments to later view or construct. We train photos taken using the LBPH face authentication process. The LBPH or local binary histogram of the model exhibits inherent absolute honesty. You can affect close focus in your photos and get great results (mostly temperature controlled). Effective for monotonous dark gamma changes. This is probably a unique ability for all OpenCV libraries. LBPH uses unfolding used to assemble a circular dual-environment model and represents a region relative to a focal area. The test suite focuses on building indirectly close doubles. The more instance focus you activate, the higher your compute charges. Container placement with flat titles. The more containers, the better the network, the higher the post-component header coverage. Rich container in the top header. Same container body extension rules as Grid X.

5 Social Distancing

The process flow for the social distancing algorithm is shown in Fig. 3.

Drawing a green box with social distancing as in Fig. 4 is applied. Current findings are derived from COVID vision during the appeal period. All meanings assume a model fitted to the COCO data set. We have showcased our outstanding walkers and are ready to check out green and cardinal ricochet. This model will create flattery in the middle of the two points 2 of the population, even if the distance between the

Fig. 3 Social distancing
block

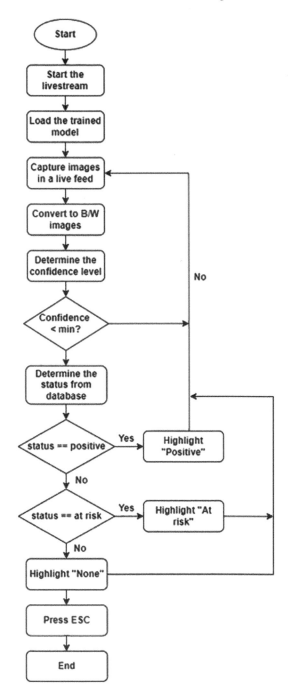

ruling classes is slightly exceeded. Control where unambiguous areas of knowledge are publicly leaked through live broadcasts. The bounds are set to 50 because the data set points increase during program execution. There is no such limitation, and it may change according to the viewpoint of the video camera. When the camcorder is viewed from a distance, the excess due to loss of accuracy should be greatly reduced as the camcorder is installed correctly in the group. To test the detective's accuracy, some scenes were shot live from three CCTV cameras. The first security cameras were launched in schools, the second in taverns, and the second in public areas. Film from each of these three cameras was reliable for two hours. This item was also human explained by three separate commentators to verify the accuracy of the friendly detective.

The mean perception of the three individual analysts was constant. Our fund has an accuracy of 96.49%. The Haar cascade model uses the COCO data set. In a basic step, we query the area model on this data set to remove the bounce field for each person in coverage. The next step is to recognize and systematize the results of Haar cascade. They are usually all in linked boxes, but for the sake of a simple idea, I will outline the focus of the jump box. Measure the Euclidean distance after this step. To degree this distance, we in reality be going to get rid of the factor guidances of every middle. This challenge concede opportunity be accomplished for all edges with inside the composition. Then, we take the whole fee of the end result number. The essential outcomes from those edges are their x and y helps. In the completion, we degree the Euclidean distance taking gain of x and y systematizes. Euclidean distance is calculated out by: Utilizing this specifying, we are able to gauge the distance 'tween stated network with inside the casing groupings. Subsequent to judging the Euclidean distance, we must preserve a base look into conventional for human-to-human distance.

6 Results

The validated result is shown in Figs. 5, 6 and 7.

7 Conclusion

Subsequently, by pursuing it and education about the contamination, this paper can use up two belongings: public segregating can lower the spread of the COVID; masks assist accompanying forestalling the compelling syndrome to please through the air. Subsequently, to help this review, this investigation projected an AI-located continuous methods toward the region of friendly removing and first top. Swarm diameter was examined by gaining the ROI of the television outlines and the count of things ignoring and non-persecuting the conduct was furthermore proved. The

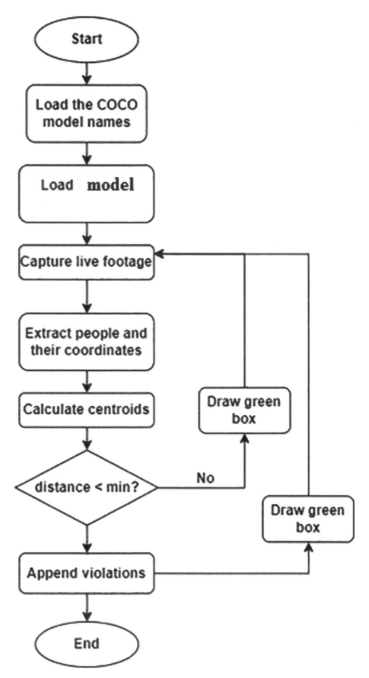

Fig. 4 Social distancing green box drawing

Fig. 5 Without mask

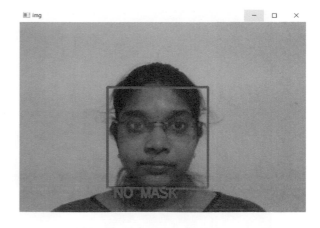

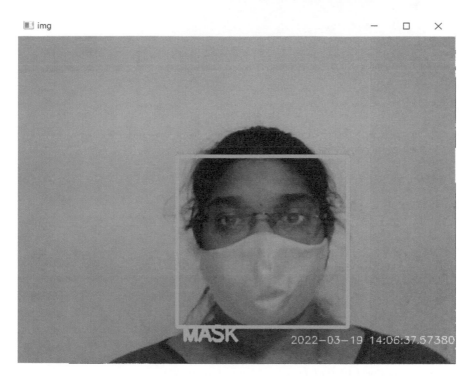

Fig. 6 With mask

consequences received were exact and neverending located. The universal is experiment to extend and is up until now occurrence while in the intervening time collect on this test. Numerous countries with its own government are still in the quarantine stage though any countries with its own government have done the quarantine stage and are held to their public killing measures. Social isolating maybe completed

Fig. 7 Realtime example of social distancing

activity continually in sure countries with its own government and things power need to endure the new common.

References

1. Petrović & Kocić, Đ. (2020). IoT-located scheme for COVID-19 household security listening," Preprint), IcETRAN (pp. 1–6). In press.
2. Qin, B., & Li, D. (2020). Identifying facemask-tiring condition utilizing countenance excellent-determination accompanying categorization network for fear that COVID-19. *Sensors, 20*(18).
3. Rezaei, M., & Azarmi, M. (2020). Deep social: Public outpacing listening and contamination risk appraisal in COVID-19 universal. *Applied Sciences, 10*(21), 7514.
4. Yadav, S. (2020). Deep education located dependable friendly outpacing and face mask discovery honestly districts for COVID-19 security directions devotion. *International Journal for Research in Applied Science and Engineering Technology, 8*(7), 1368–1375.
5. Ahmed, I., Ahmad, A., & Jeon, G. (2020). An IoT located deep education foundation for early estimate of Covid-19. *IEEE Internet of Things Journal, 8*(1).
6. Rao, T. V. N., Sai, G. V., Chenna, M., & Reddy, M. S. (2020). An request for the discovery of COVID-19 face mask utilizing OpenCV. *High Technology Letters, 26*(7), 965–975.
7. Loey, M., Manogaran, G., Taha, M. H. N., & Khalifa, N. E. M. (2021). A composite deep transfer education model accompanying machine intelligence orders for face mask discovery in the stage of the COVID-19 universal. *Measurement, 167*, Article ID 108288.
8. Qin, B., & Li, D. (2020). Identifying facemask-tiring condition utilizing concept excellent-determination accompanying categorization network for fear that COVID-19. *Sensors, 20*(18), 5236.

9. Wang, Z., Wang, P., Louis, P. C., Wheless, L. E., & Huo, Y. (2021) Wearmask: Fast in-portal face mask discovery accompanying serverless edge estimating for COVID-19. https://arxiv.org/antilockbrakingsystem/2101.00784
10. Zhang, X., Saleh, H., Younis, E. M., Sahal, R., & Ali, A. A. (2020). Predicting coronavirus universal in legitimate-period utilizing machine intelligence and grown dossier flooding whole. *Complexity*, 10 pages.
11. Razavi, M., Alikhani, H., Janfaza, V., Sadeghi, B., & Alikhani, E. (2021). An mechanical order to monitor the material distance and face mask tiring of explanation traders in COVID-19 universal. https://arxiv.org/antilockbrakingsystem/2101.01373.
12. Srinivasan, S., Rujula Singh, R., Biradar, R. R., & Revathi, S. (2021). COVID-19 Monitoring system utilizing social distancing and face mask detection on surveillance program datasets. In *2021 International Conference on Emerging Smart Computing and Informatics (ESCI)*, pp. 449–455. https://doi.org/10.1109/ESCI50559.2021.9396783
13. Porumbel, H., & Kuntz. (2011). An effective treasure for calculating the distance middle from two points close partitions. https://doi.org/10.1016/j.dam.2010.09.002
14. McCloskey, B., Zumla, A., Ippolito, G., Blumberg, L., Arbon, P., Cicero, A., Endericks, T., Lim, P. L., & M. Borodina (2020). Mass accumulation occurrences and lowering the further all-encompassing spread of COVID-19: A governmental and community health crisis.
15. Chen, H., Yang, B., Pei, H., & Liu, J. (2019). Next Generation technology for epidemic prevention and control: Data-driven contact tracking. *IEEE Access, 7*, 2633–2642. https://doi.org/10.1109/ACCESS.2018.2882915
16. Chen, K., Wang, J., Pang, J., Cao, Y., Xiong, Y., Li, X., Sun, S., Feng, W., Liu, Z., Xu, J., & Zhang, Z. (2019). MMDetection: Open MMLab detection toolbox and benchmark. arXiv:1906.07155

Efficacy of Autoencoders on Image Dataset

S. Anupama Kumar, Andhe Dharani, and Chandrani Chakravorty

Abstract Autoencoder is an unsupervised learning technique widely used over different types of data that gets the output exactly like the input. The major usage of autoencoders lie in feature extraction, dimensionality reduction, image denoising, compression, etc. This paper concentrates on the implementation of undercomplete, sparse and variational autoencoder over the Modified National Institute of Standards and Technology (MNIST) dataset and analyse the efficiency using loss and activation function. The effect of the number of epochs in building the model is also analysed. For the dataset, sparse autoencoder performs better than the undercomplete when implemented with Adam as an activation function with mean square error as loss function. When compared with the variational autoencoder (VAE), VAE performs better when implemented using Adam function.

Keywords Autoencoders · Sparse encoder · Variational autoencoders (VAE) · Loss · Activation · Epochs

1 Introduction

Autoencoders belongs to unsupervised learning techniques that takes unlabelled data as input and gives the output the same way like inputs. These encoders map the data from higher dimension coordinates to the middle layer where the dimension is reduced and approximates to the higher dimension of output as same as input with minimum loss. Autoencoder is an architecture of neural network which is used to reconstruct data. It consists of two parts: encoder and decoder [1] as shown in Fig. 1.

S. Anupama Kumar (✉) · A. Dharani · C. Chakravorty
Department of MCA, RV College of Engineering, Bengaluru, India
e-mail: anupamakumar@rvce.edu.in

A. Dharani
e-mail: andhedharani@rvce.edu.in

C. Chakravorty
e-mail: chandrani@rvce.edu.in

© The Author(s), under exclusive license to Springer Nature Singapore Pte Ltd. 2023   1001
S. Shakya et al. (eds.), *Sentiment Analysis and Deep Learning*,
Advances in Intelligent Systems and Computing 1432,
https://doi.org/10.1007/978-981-19-5443-6_73

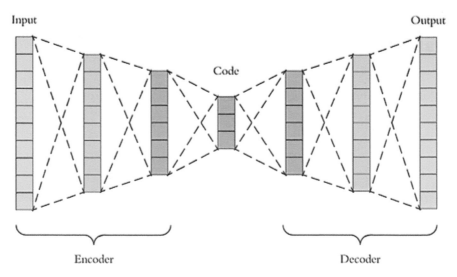

Fig. 1 Architecture of autoencoder

The encoder phase of the autoencoder encodes the data into a lower dimension, i.e. the number of neurons in the encoder phase will be decreasing as they progress towards the bottle neck (code). The decoder phase decodes the compressed data into the original form, i.e. they number of neurons in the decoder will be increasing as they move towards the output. The number of neurons in the bottle neck should be less than the features in the dataset. The major objective of implementing an autoencoder is in understanding the difference between the input and output representations [5].

The layers are trained by the back propagation algorithm [2] to minimize the loss between the original and the re-constructed data. The loss function in a neural network quantifies the difference between the expected outcome and the outcome produced by the machine learning model [3]. The different type of loss functions that can be applied are (i) cross entropy-based loss functions (log loss) that are used for binary classification problems and (ii) mean squared error loss functions that can be used for regression analysis. Activation functions are used in the neural networks to transform the input to output with the help of various hidden layers. Different types of activation functions are especially needed when the neural network has to be trained in a nonlinear convolutional model [4]. Sigmoid activation functions are commonly used in the image datasets to extract the features and get accurate results.

This work focusses on the building autoencoder models using undercomplete autoencoders, variational autoencoders (VAE) and sparse autoencoders on MNIST dataset. The autoencoders were compared based on the choice of hyper parameters, activation functions, loss functions and number of epochs used to build the model. The dataset used in this work comprises of large database of handwritten digits called as MNIST dataset. It's widely used by the researchers for training neural network models and recognize the handwritten digits. The following Fig. 2 gives the sample collection of the dataset.

Fig. 2 MNIST dataset

2 Key Components

This section gives the list of key components and libraries that are imported and used from tensor flow keras models [6] that are implemented in the work.

Sequential Model: A sequential model is applied in this work since the dataset is arranged in the form of stack of images, and each layer has exactly one input tensor and one output tensor.

Flatten: It is used to flatten the input. In this case, the input size is taken as 28×28.

Learning Rate: Learning rate is used to tune how accurately a model converges on a result (classification/prediction, etc.).

Latent Size: This value is a crucial hyperparameter which if set too small will lead to underfitting and if it's too large will lead to overfitting. Overfitting that occurs due to huge latent size can be controlled by using the drop out ratio.

Dense: Dense layer is the regular deeply connected neural network layer. It is most common and frequently used layer. The dense layer is decreased stepwise while building the encoder, the minimum layers being the latent size and increased while building the decoder.

Activation Function: Activation functions are used to map the output functions to 0 to 1 or −1 to 1. The type of activation function to be applied in the network depends on the output of the problem.

For the MNIST dataset, the following activation functions are used:

(i) Sigmoid activation function is used to create the encoder and decoder before building the model. The following Fig. 3 shows the sigmoid function s-curve.
(ii) Rectified Linear Activation Function (ReLu): The rectified linear activation function is a simple calculation that returns the value provided as input directly,

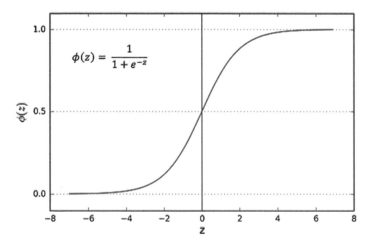

Fig. 3 Sigmoid function

or the value 0.0 if the input is 0.0 or less [7]. The extensions and modifications of ReLu include Leaky ReLU and ELU, where Leaky ReLU modifies the function to allow small negative values when the input is less than zero and the exponential linear unit, or ELU, is a generalization of the ReLU that uses a parameterized exponential function to transition from the positive to small negative values. The following Fig. 4 shows the ReLU curve with the positive and negative values.

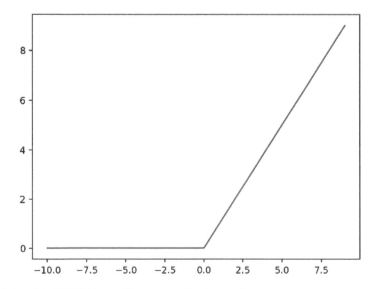

Fig. 4 Line plot of ReLU for positive and negative input values

Table 1 Cross entropy loss function—interpretation

Range of CE values	Interpretation
0.00	No loss of value
<0.02–0.5	Minimum loss
<0.20	Acceptable loss
0.30	Not in range

Loss Functions

Loss functions are used to quantify the closeness of a network towards its training. The loss function or loss determines the minimal error in the model trained. The loss value is calculated by averaging the aggregating the errors over the dataset. They further help to reframe the training model as an optimization function. The loss function plays a role all over the model and down passes a single number that when corrected can build a better model. The following loss functions are used in the proposed work.

(i) Cross Entropy: The cross entropy function is the difference between two probability distribution function and is defined as below:

(ii)

$$\text{Cross Entropy Loss} = -(y_i \log(\hat{y_i}) + (1 - y_i)\log(1 - \hat{y})_i \qquad (1)$$

The range of values for the cross entropy loss function is listed in Table 1.

(iii) Mean Square Error Loss (MSE) (also called L2 regularization)

It is the average of the squared difference between the prediction and actual observations.

$$\text{Mean square error (MSE)} = \frac{\sum_{i=1}^{n} (y_i - (\hat{y_i})^2)}{n} \qquad (2)$$

The authors in [9, 10] have discussed the importance and application mean square as a loss function and its effectiveness in image compression.

Optimization Methods

Optimizers are used to speed up the training speed of the neural network. The popular optimizers are: gradient descent with momentum and Nesterov accelerated gradient, AdaGrad, RMSProp, Adam, and Nadam optimization. In this proposed work, two optimizers namely Adam and Nadam have been used and analysed for its efficiency in training the model.

Adam and Nadam (Adam optimizer with Nesterov technique) are popular optimization models that can handle complex problems and data using sparse gradients. It is capable of handling large number of features with exponential decay average of the past gradients an exponential decay average of past squared gradients (RMS).

$$m \leftarrow \beta_1 m - (1 - \beta_1) \nabla_\theta J(\theta)$$
$$s \leftarrow \beta_2 s - (1 - \beta_2) \nabla_\theta J(\theta)$$
$$\hat{m} \leftarrow \frac{m}{1 - \beta_2^T}; \hat{s} \leftarrow \frac{s}{1 - \beta_2^T} \qquad (3)$$
$$\theta_{\text{next step}} \leftarrow \theta + \frac{\eta \hat{m}}{\sqrt{\hat{s}} + \epsilon}$$

3 Variations of Autoencoders

Autoencoders are trained to learn the lower dimensional representation of the images for higher dimensionality which will result in dimensionality reduction. In case of traditional autoencoders, the number of network layers remains the same in both the input and output layers. The middle layer (bottle neck layer) is responsible for the data compression. This section discusses the architecture and implementation of the undercomplete, sparse and variational autoencoder which are used in this work.

(i) **Under Complete Autoencoders**

In an under complete autoencoder, the output of an image is predicted without using explicit regularization [8]. This can be done by restricting the number nodes in the hidden layer. The encoder and the decoder are built using a sequential model with learning rate as 0.5, Latent size 32 and sigmoid activation function.

(ii) **Sparse Autoencoders**

Sparse autoencoders differ from under complete autoencoders in the way in which they encode the information. The sparse autoencoder learning algorithm, which is one approach to automatically learn features from unlabelled data [11]. They apply LI regularizes to encode the information. The number of nodes at each hidden layer is changed to regulate the sparse encoder [12, 13]. The L1 regularization can be calculated using the function

$$\text{Loss} = \text{Error}(y, \hat{y}) + \lambda \sum_{i=1}^{N} |w_i|, \qquad (4)$$

where,

$$\hat{y} = w_1 x_1 + w_2 x_2 + w_3 x_3 + \cdots + w_N x_N + b \qquad (5)$$

A sparse autoencoder (SAE) is developed by [15] and the model is optimized using the Adam that is capable of adjusting the parameters dynamically where in batch normalization technique is applied to avoid overfitting and improve the performance of the model.

The authors in [16] have conducted two different experiments using SAE, where one of the datasets is MNIST. The author has used L2 regularized and estimated the average activation of hidden units. The presence of redundant hidden units was confirmed which may not be optimal for building the network. For the dataset chosen, the encoder of SPE is trained with ReLu activation function, latent size of 32 and learning rate of 0.5.

(iii) **Variational Autoencoders**

Variational autoencoders generate a realistic image from the given input [17]. The encoder outputs two vectors that represent the mean and standard deviation, a noise is added in the code and the decoder takes this modified image as input. The VAE approach has been intensively applied for synthetic data generation, or representation learning in a broader sense [18]. The application of VAEs in the case of imbalanced datasets is discussed by [19], and the augmentation of MNIST dataset with synthetic samples were generated by a VAE model. The application of VAE on different medical image data augmentation in ECG and EEG records for extracting the features has been discussed in [20–23]. The encoder and decoder were built using ReLu as activation function, learning rate 0.5 and KL divergence as optimizer/regularizer.

The following Sect. 4 discusses the implementation and analysis of the autoencoders implemented over the MNIST dataset.

4 Experimental Setup and Results

Three variations of autoencoders have been implemented in this work, and the results are analysed. The dataset consists of 60,000 images and has been divided into training and test data. The image size is fixed as (28×28) and the shape of the image is dependent on the latent size which is fixed as 32. The efficiency of the autoencoders were measured using the loss functions like binary cross entropy function and mean square error with the activation functions like Adam and Nadam. The number of epochs for undercomplete autoencoder was set to 60 to get a better image where the SPE and VAE could give the output in 50 epochs. The following Table 2 shows the different metrics on which the efficiency of the variants was analysed.

From the analysis, it is understood that application of mean square error as a loss function suits better for the given dataset than the binary cross entropy error. Also the combination of Adam with MSE performs efficiently in case of SPE. In the case of VAE, the combination of Adam as an activation function with the loss functions has performed better than the Nadam function.

Table 2 Analysis of the autoencoders

Type of autoencoder	Type of activation function	Type of loss function	Loss value
Under complete	Nadam	Binary cross entropy	0.1901
	Adam	Mean square error	0.0424
Sparse autoencoder	Nadam	Binary cross entropy	0.0920
	Adam	Mean square error	0.0097
Variational autoencoder	Nadam	KL and (Re construction Loss)	Reconstruction loss: 141.7841 KL_loss: 6.6767
	Adam		Reconstruction loss: 144.1266 KL_loss: 6.5755

5 Conclusion

Autoencoders are highly efficient in reproducing the outputs as the input and enable to extract features without memorizing them. The role of activation function and the loss function along with the number of epochs is analysed to identify the efficiency of the model developed. The following Fig. 5 shows the comparison between the undercomplete and sparse with reference to activation function and loss function. The sparse encoder performs better than the under complete autoencoder with Adam as an activation function and the mean square error as loss function. When compared with the number of epochs used, the undercomplete autoencoder used 60 epochs and the sparse encoder could achieve better performance in 50 epochs itself.

When compared with the VAE as in Fig. 6, VAE performs better than the sparse autoencoder. Even though the type of loss function varies here, the VAE will perform better than the sparse encoder in all the aspects.

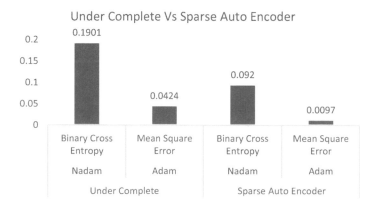

Fig. 5 Under complete autoencoder/sparse autoencoder

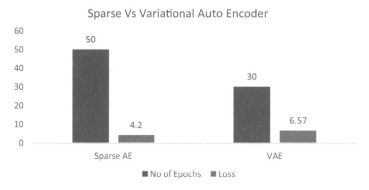

Fig. 6 SPE versus VAE

For the given dataset, the efficiency of the different encoders namely the undercomplete, sparse and VAE have been compared with the help of the activation functions, loss functions and the number of epochs it takes to produce the output and found that the variational autoencoder performs better than the other two variations. The work has been limited to MNIST dataset. Further, the model can be applied over other dataset and implemented for various domains of applications.

References

1. Kovenko, V., & Bogacha, I. (2020). A comprehensive study of autoencoders' applications related to images. In *IT&I-2020 Information Technology and Interactions, 02–03 December 2020, KNU Taras Shevchenko, Kyiv, Ukraine. CEUR Workshop Proceedings (CEUR-WS.org).*
2. le Cun, Y. (1989). A theoretical framework for back-propagation. In *Proceedings of the 1998 Connection Models Summer School*, Carnegie-Mellon University.
3. Seb. (2021). An introduction to neural network loss functions. Posted on 28 September 2021. https://programmathically.com/an-introduction-to-neural-network-loss-functions/
4. Sharma, S., Sharma, S., & Athaiya, A. (2020). Activation functions in neural networks. *International Journal of Engineering Applied Sciences and Technology, 4*(12), 310–316. ISSN 2455-2143, Published Online April 2020 in IJEAST. http://www.ijeast.com
5. Patterson, J., & Gibson, A. (2017). Deep learning: A practitioners approach. O'Reilly Publications. ISBN 978-1-491-91425-0
6. Singh, A., & Ogunfunmi, T. An overview of variational autoencoders for source separation, finance, and bio-signal applications.
7. Article. https://machinelearningmastery.com/rectified-linear-activation-function-for-deep-learning-neural-networks/
8. Jordan, J. (2018). https://www.jeremyjodan.me/autoencoders/
9. Alexandre, D., et al. An autoencoder based learned image compressor: Description of challenge proposal. https://arxiv.org/abs/1902.07385, https://doi.org/10.48550/arXiv.1902.07385
10. Mentzer, F., Agustsson, E., Tschannen, M., Timofte, R., & Van Gool, L. (2018). Conditional probability models for deep image compression. arXiv preprint arXiv:1801.04260
11. Nag, A. Lecture Notes. https://web.stanford.edu/class/cs294a/sparseAutoencoder.pdf
12. Article. https://www.v7labs.com/blog/autoencoders-guide

13. Arpit, D., et al. (2016). Why regularized auto-encoders learn sparse representation. In *Proceedings of the 33rd International Conference on Machine Learning*, New York, NY, USA (Vol. 48). JMLR: W&CP.

14. Mienye, I. D., Sun, Y., & Wang, Z. (2020). Improved sparse autoencoder based artificial neural network approach for prediction of heart disease. *Journal of Informatics in Medicine Unlocked, 18*, 100307. ISSN 2352-9148.https://doi.org/10.1016/j.imu.2020.100307

15. Wan, Z., He, H., & Tang, B. (2018). A generative model for sparse hyperparameter determination. *IEEE Transactions on Big Data, 4*(1), 2–10. https://doi.org/10.1109/TBDATA.2017.2689790

16. Oinar, C. (2021). https://towardsdatascience.com/variational-autoencoder-55b288f2e2e0

17. Elbattah, M., Loughnane, C., Guérin, J.-L., Carette, R., Cilia, F., & Dequen, G. (2021). Variational autoencoder for image-based augmentation of eye-tracking data. *Journal of Imaging, 7*, 83. https://doi.org/10.3390/jimaging7050083

18. Wan, Z., Zhang, Y., & He, H. (2017). Variational autoencoder based synthetic data generation for imbalanced learning. In *Proceedings of the IEEE Symposium Series on Computational Intelligence (SSCI)*, Honolulu, HI, USA, 27 November–1 December 2017 (pp. 1–7).

19. Luo, Y., Zhu, L. Z., Wan, Z. Y., & Lu, B. L. (2020). Data augmentation for enhancing EEG-based emotion recognition with deep generative models. *Journal of Neural Engineering, 17*(5), 056021. https://doi.org/10.1088/1741-2552/abb580 PMID: 33052888.

20. Ozdenizci, O., & Erdogmus, D. (2021). On the use of generative deep neural networks to synthesize artificial multichannel EEG signals. arXiv 2021, arXiv:2102.08061. Available online https://arxiv.org/abs/2102.08061. Accessed on 2 May 2021.

21. Biffi, C., Oktay, O., Tarroni, G., Bai, W., De Marvao, A., Doumou, G., Rajchl, M., Bedair, R., Prasad, S., Cook, S., et al. (2018). Learning interpretable anatomical features through deep generative models: Application to cardiac remodelling. In *Proceedings of the International Conference on Medical Image Computing and Computer-Assisted Intervention (MICCAI)*, Granada, Spain, 16–20 September 2018 (pp. 464–471).

22. Pesteie, M., Abolmaesumi, P., & Rohling, R. N. (2019). Adaptive augmentation of medical data using independently conditional variational auto-encoders. *IEEE Transactions on Medical Imaging, 38*(12), 2807–2820.https://doi.org/10.1109/TMI.2019.2914656. Epub 2019 May 6. PMID: 31059432.

23. Cerrolaza, J. J., Li, Y., Biffi, C., Gomez, A., Sinclair, M., Matthew, J., Knight, C., Kainz, B., & Rueckert, D. (2018). 3D Fetal skull reconstruction from 2DUS via deep conditional generative networks. In J. A. Schnabel, C. Davatzikos, C. Alberola-López, G. Fichtinger, & A. F. Frangi (Eds.), *Medical Image Computing and Computer Assisted Intervention—MICCAI 2018—21st International Conference, 2018, Proceedings* (pp. 383–391). *(Lecture Notes in Computer Science (including subseries Lecture Notes in Artificial Intelligence and Lecture Notes in Bioinformatics)* (Vol. 11070 LNCS). Springer. https://doi.org/10.1007/978-3-030-00928-1_44

Author Index

Printed by Printforce, the Netherlands